Annals of Mathematics Studies

Number 216

# What Determines an Algebraic Variety?

János Kollár
Max Lieblich
Martin Olsson
Will Sawin

PRINCETON UNIVERSITY PRESS

PRINCETON AND OXFORD

2023

Published by Princeton University Press
41 William Street, Princeton, New Jersey 08540
99 Banbury Road, Oxford OX2 6JX
press.princeton.edu

Library of Congress Cataloging-in-Publication Data

Names: Kollár, János, author. | Lieblich, Max, 1978– author. | Olsson, Martin C., author. | Sawin, Will, 1992– author.
Title: What determines an algebraic variety? / János Kollár, Max Lieblich, Martin Olsson, Will Sawin
Description: Princeton: Princeton University Press, 2023. | Series: Annals of mathematics studies; Number 216 | Includes bibliographical references and index.
Identifiers: LCCN 2022059411 (print) | LCCN 2022059412 (ebook) |
ISBN 9780691246802 (hardback) | ISBN 9780691246819 (paperback) |
ISBN 9780691246833 (ebook)
Subjects: LCSH: Algebraic varieties. | BISAC: MATHEMATICS / Geometry / Algebraic
Classification: LCC QA564 .K59 2023 (print) | LCC QA564 (ebook) |
DDC 516.3/53–dc23/eng20230126
LC record available at https://lccn.loc.gov/2022059411
LC ebook record available at https://lccn.loc.gov/2022059412

British Library Cataloging-in-Publication Data is available

Editorial: Diana Gillooly and Kiran Pandey
Production Editorial: Nathan Carr
Jacket/Cover Design: Heather Hansen
Production: Lauren Reese
Publicity: William Pagdatoon
Copyeditor: Bhisham Bherwani

This book has been composed in LaTeX

The publisher would like to acknowledge the authors of this volume for acting as the compositors for this book

10   9   8   7   6   5   4   3   2   1

# Contents

# Preface

Geometry starts with the study of points and lines in the plane. The simplest objects are the points in the plane (as Euclid says, they have no parts) and lines are certain sets of points.

Descartes put coordinates on the plane, so now we usually think of the plane as a 2-dimensional vector space $\mathbb{R}^2$, and the lines as solution sets of linear equations. This is the starting point of algebraic geometry, where one can use algebraic methods to solve geometric problems.

The reworking of the foundations of algebraic geometry, started in the 1930es by van der Waerden, Weil, and Zariski, culminated in Grothendieck's theory of schemes around 1960. This turned 'algebraic *geometry*' into '*algebraic* geometry.' In these treatments the primary object is no longer the $n$-space $\mathbb{R}^n$, but the polynomial ring $\mathbb{C}[x_1, \ldots, x_n]$.

Thus in contemporary algebraic geometry we think of $\mathbb{C}^n$ as consisting of two parts:

- (geometry) the set of points $\mathbb{C}^n$, together with 'closed subsets' given by solution sets of systems of polynomial equations, and
- (algebra) the ring of all polynomial functions $\mathbb{C}^n \to \mathbb{C}$.

Rather sloppily, standard usage does not distinguish between $\mathbb{C}^n$ as a point set, vector space, Zariski topological space, variety or scheme, but for us these distinctions are crucial. We call the geometric object the *Zariski topological space*, and denote it by $|\mathbb{C}^n|$.

By Hilbert's Nullstellensatz, the points of $|\mathbb{C}^n|$ correspond to maximal ideals of $\mathbb{C}[x_1, \ldots, x_n]$, and the closed subsets to other radical ideals. Thus, the ring $\mathbb{C}[x_1, \ldots, x_n]$ uniquely determines the geometry $|\mathbb{C}^n|$.

The main question we aim to study in these notes is a converse: Does geometry determine algebra and function theory?

We compactify $\mathbb{C}^n$ to projective space $\mathbb{CP}^n$ and, for full generality, we work with projective $n$-space $\mathbb{P}^n_K$ over any commutative field $K$.* Any Zariski closed subset $X \subset \mathbb{P}^n_K$ has an underlying geometry, denoted by $|X|$, and a function theory or structure sheaf, denoted by $\mathcal{O}_X$.

Thus the general question that we address is: Given a projective algebraic

---

*If $K$ is not algebraically closed, this is different from projective $n$-space used in discrete geometry or topology; see (1.1.10)–(1.1.12).

set $X \subset \mathbb{P}^n_K$, does the geometry $|X|$ determine the algebra and function theory of $\mathcal{O}_X$?

For algebraic curves the geometry is not rich enough to give any information, and there are a few other, mostly obvious, exceptions. The main theorem—stated in Section 1.2—says that, for projective varieties of dimension $\geq 4$, the geometry does determine the algebra and function theory, at least over fields of characteristic 0. Our answers are less complete in dimensions 2 and 3.

# *Acknowledgments*

During the work on this book, Kollár was partially supported by NSF grant DMS-1901855; Lieblich was partially supported by NSF grants DMS-1600813 and DMS-1901933 and a Simons Foundation Fellowship; Olsson was partially supported by NSF grants DMS-1601940 and DMS-1902251; Sawin served as a Clay Research Fellow and was partially supported by NSF grant DMS-2101491. Part of this work was done while JK, ML, and MO visited the Mathematical Sciences Research Institute in Berkeley, whose support is gratefully acknowledged. We thank Jarod Alper, Giulia Battiston, Daniel Bragg, Charles Godfrey, Johan de Jong, Kristin DeVleming, and Joe Waldron for helpful conversations, Hans Kloss and Stanisław Kolicki for early inspiration, and Brendan Hassett and Yuri Tschinkel for enlightening discussions about earlier versions of this work. Thanks to Noga Alon for pointing out the connection with linearity testing in (3.3.2).

We received help from J.-L. Colliot-Thélène and M. Larsen with algebraic tori, B. Poonen with Bertini and Noether-Lefschetz theorems, J. Silverman with Néron's theorem, K. Smith with discrete valuation rings, and C. Voisin with Noether-Lefschetz theorems; we are grateful to all of them. In preparing his Bourbaki Seminar talk [Čes21] on some of this material, Kęstutis Česnavičius gave us many comments, corrections, and questions. These directly led to numerous clarifications and improvements to the manuscript.

What Determines an
Algebraic Variety?

# Chapter One

## Introduction

### 1.1 FROM LINES AND PLANES TO THE ZARISKI TOPOLOGY OF $\mathbb{P}^N$

Let $K$ be any field. Choosing $K^n$ as our set of points, and solution sets of systems of linear equations as our preferred subsets, we get what is called $n$-dimensional *affine geometry* over $K$, though $n$-dimensional *linear geometry* over $K$ might be a better name. It is frequently denoted by $\mathrm{Aff}_K^n$. Following Kepler (1571–1630) and Desargues (1591–1661) we add points at infinity to get $n$-dimensional *projective geometry* over $K$; the general case appears in the works of von Staudt [vS1857], Fano [Fan1892], and Veblen [Veb1906]. We denote it by $\mathrm{Proj}_K^n$.

Thus the $n$-dimensional affine or projective geometries over $K$ consist of

- point sets

$$\mathrm{Points}(\mathrm{Aff}_K^n) = \{(x_1, \ldots, x_n) \in K^n\}, \text{ or}$$
$$\mathrm{Points}(\mathrm{Proj}_K^n) = \{(x_0 : \cdots : x_n) \in (K^{n+1} \setminus \{\mathbf{0}\})/K^\times\}, \text{ and}$$

- the linear subspaces as distinguished subsets of the point set.

By definition, the algebra of the field $K$ determines the affine and the projective geometries. The Fundamental Theorem of Projective Geometry—which should be called the Fundamental Theorem of *Linear* Geometry—says that, conversely, the geometry of $\mathrm{Aff}_K^n$ or of $\mathrm{Proj}_K^n$ determines the algebra of the field $K$.

The key ideas go back to Menelaus of Alexandria (c. 70–140 AD) and Giovanni Ceva (1647–1734). The first proof is due to von Staudt [vS1857]. A gap was noticed by Klein [Kle1874], and correct versions can be found in Reye's lectures [Rey1866] (starting with the second edition). General forms are given by Russell [Rus1903], Whitehead [Whi1906], and Veblen and Young [VY1908]; see also the books by Baer [Bae52] and Artin [Art57]. We state the two versions separately, although they are really the same.

**Theorem 1.1.1** (Affine form). *Let $K, L$ be fields and $n, m \geq 2$. Let*

$$\Phi\colon \mathrm{Points}(\mathrm{Aff}_K^n) \leftrightarrow \mathrm{Points}(\mathrm{Aff}_L^m)$$

*be a bijection that maps linear subspaces to linear subspaces. Then $n = m$, and* *there is a unique field isomorphism* $\varphi : K \cong L$, *vector* $(c_1, \dots, c_m) \in L^m$ *and matrix* $M \in \mathrm{GL}_m(L)$ *such that*

$$\Phi(x_1, \dots, x_n) = \big(\varphi(x_1) + c_1, \dots, \varphi(x_n) + c_n\big) \cdot M.$$

**Theorem 1.1.2** (Projective form). *Let* $K, L$ *be fields and* $n, m \geq 2$. *Let*

$$\Phi \colon \ \mathrm{Points}\big(\mathrm{Proj}_K^n\big) \leftrightarrow \mathrm{Points}\big(\mathrm{Proj}_L^m\big)$$

*be a bijection that maps linear subspaces to linear subspaces. Then* $n = m$, *and there is a unique field isomorphism* $\varphi : K \cong L$ *and matrix* $M \in \mathrm{PGL}_{m+1}(L)$ *such that*

$$\Phi(x_0{:}\cdots{:}x_n) = \big(\varphi(x_0){:}\cdots{:}\varphi(x_n)\big) \cdot M.$$

**Remark 1.1.3.** The identity is the only automorphism of $\mathbb{R}$, thus for $K = L = \mathbb{R}$ we get that $\Phi(x_0{:}\cdots{:}x_n) = (x_0{:}\cdots{:}x_n) \cdot M$ for some $M \in \mathrm{PGL}_{m+1}(\mathbb{R})$. That is, the coordinatization of $\mathbb{R}\mathbb{P}^n$ is unique, up to linear changes of the coordinates.

By contrast, the automorphism group of $\mathbb{C}$ is huge, of cardinality $2^{|\mathbb{C}|}$.

**Remark 1.1.4.** With some care, one can see that (1.1.1) and (1.1.2) also apply to non-commutative fields, but from now on we consider only commutative fields.

The next natural geometry to consider is *circle geometry,* where we work with lines and circles in the plane. It was discovered by Hipparchus of Nicaea (c. 190-120 BC) that, using stereographic projection, it is better to view this as the geometry whose points are given by the sphere

$$\mathbb{S}^2 := \{(x_1, x_2, x_3) \in \mathbb{R}^3 : x_1^2 + x_2^2 + x_3^2 = 1\},$$

and whose subsets are the circles contained in $\mathbb{S}^2$. These are also the intersections of $\mathbb{S}^2$ with planes.

More generally, let $K$ be any field of characteristic $\neq 2$. Let $\mathrm{Sph}_K^n$ denote *spherical geometry* of dimension $n$ over $K$. That is, its points are

$$\mathbb{S}_K^n := \{(x_0{:}x_1{:}\cdots{:}x_{n+1}) \in \mathrm{Proj}_K^{n+1} : x_1^2 + \cdots + x_{n+1}^2 = x_0^2\},$$

and its distinguished subsets are the intersections of $\mathbb{S}_K^n$ with linear subspaces. (These are spheres if $K$ is a subfield of $\mathbb{R}$. However, if $K = \mathbb{C}$ then the intersection with $(x_3 = \cdots = x_{n-1} = x_n - x_0 = 0)$ is a pair of lines, so the name 'spherical' may be misleading.) The Fundamental Theorem of Spherical Geometry now says the following.

**Theorem 1.1.5.** *Let* $K, L$ *be fields and* $n, m \geq 2$. *Let*

$$\Phi \colon \ \mathrm{Points}\big(\mathrm{Sph}_K^n\big) \leftrightarrow \mathrm{Points}\big(\mathrm{Sph}_L^m\big)$$

*be a bijection that maps linear intersections to linear intersections. Then* $n = m$ *and there is a unique field isomorphism* $\varphi : K \cong L$ *and matrix* $M \in \mathrm{PO}_{n+1,1}(L)$

*such that*

$$\Phi(x_0{:}\cdots{:}x_{n+1}) = \left(\varphi(x_0){:}\cdots{:}\varphi(x_{n+1})\right) \cdot M.$$

Here $\mathrm{PO}_{n+1,1}(L) \subset \mathrm{PGL}_{n+2}(L)$ is the projective othogonal group, that is, the subgroup of those matrices that leave the sphere $\mathbb{S}_K^n$ invariant.

Although this seems like a new result, it easily reduces to the linear geometry case as follows. Fix a point $p \in \mathbb{S}_K^n$. If $K$ is a subfield of $\mathbb{R}$ then stereographic projection shows that $\mathbb{S}_K^n \setminus \{p\}$ (with the spherical subsets containing $p$ as our subsets) is isomorphic to $\mathrm{Aff}_K^n$. For arbitrary $K$, we get the same conclusion for $\mathbb{S}_K^n \setminus \{$all lines through $p$ in $\mathbb{S}_K^n\}$.

The next natural topic could be *conic geometry*. Here we start with sets of points $\mathrm{Points}(\mathrm{Aff}_K^2)$ or $\mathrm{Points}(\mathrm{Proj}_K^2)$, but we work with lines and conics as distinguished subsets.

However, nothing new happens, since we can tell which curves are conics and which are lines. Indeed, in conic geometry, $C$ is a line if and only if $C \cap C'$ consists of at most two points for every other curve $C'$ (which is a conic or a line, not containing $C$). Thus we recover affine geometry.

What if we fix a degree $d$ and consider *degree-d geometry* in the plane? It has the same point set as before, but we use all algebraic curves of degree $\leq d$ as distinguished subsets. That is, solution sets of the form

- $\{(x,y) \in \mathrm{Aff}_K^2 : f(x,y) = 0\}$ where $\deg f \leq d$ (affine case), or
- $\{(x{:}y{:}z) \in \mathrm{Proj}_K^2 : F(x{:}y{:}z) = 0\}$ where $\deg F \leq d$ (projective case).

As before, it is not hard to show that if $|K| \geq d + 1$, then $C$ is a line if and only if it has at least $d + 1$ points and $C \cap C'$ consists of at most $d$ points for every other curve $C'$ (of degree $\leq d$ that does not contain $C$). Thus we get the same fundamental theorems as in the linear case.

While restricting to small values of $d$ may be natural, it is very unlikely that specific large values of $d$ are of much interest. So we should instead let $d$ become infinite and work with all algebraic plane curves and their $K$-points. This is planar algebraic geometry. We focus now on the projective case; see (2.2.15) for some comments on the affine setting. As the natural continuation of (1.1.1)–(1.1.5), the next question to consider is the following.

**Question 1.1.6.** Let $K, L$ be fields and

$$\Phi\colon \mathrm{Points}(\mathrm{Proj}_K^2) \leftrightarrow \mathrm{Points}(\mathrm{Proj}_L^2)$$

a bijection that maps algebraic curves to algebraic curves. Is there a field isomorphism $\varphi : K \cong L$ and a matrix $M \in \mathrm{PGL}_3(L)$ such that

$$\Phi(x_0{:}x_1{:}x_2) = \left(\varphi(x_0){:}\varphi(x_1){:}\varphi(x_2)\right) \cdot M?$$

In a surprising departure from the previous results, the answer is very field-dependent. For illustration, let us see what happens with finite fields, $\mathbb{R}$ and $\mathbb{C}$. For finite fields the answer is negative for trivial reasons (though of course the cardinality of $\mathrm{Points}(\mathrm{Proj}_K^2)$ determines $K$).

**Proposition 1.1.7.** *Let $K$ be a finite field. For every subset $S \subset \mathrm{Points}(\mathrm{Proj}_K^2)$ there is a homogeneous polynomial $F_S$ such that*

$$S = \{(x{:}y{:}z) \in \mathrm{Proj}_K^2 : F_S(x{:}y{:}z) = 0\}.$$

*Thus every subset of $\mathrm{Points}(\mathrm{Proj}_K^2)$ is an algebraic curve, hence every bijection $\mathrm{Points}(\mathrm{Proj}_K^2) \leftrightarrow \mathrm{Points}(\mathrm{Proj}_K^2)$ maps algebraic curves to algebraic curves.*

Clearly, here the problem is that for finite fields $K$, the $K$-points of a high degree curve $C$ tell us very little about $C$. While identifying a line with its $K$-points is harmless over any field, and identifying a (nonempty) conic with its $K$-points works whenever $|K| > 3$, thinking of $C$ as its $K$-points tends to be helpful only if there are infinitely many $K$-points on $C$.

In the real case the answer is again negative, but this is more unexpected. We use $\mathbb{RP}^2$ to denote the real projective plane with its Euclidean topology.

**Theorem 1.1.8** ([KM09]). *Every diffeomorphism $\Psi : \mathbb{RP}^2 \leftrightarrow \mathbb{RP}^2$ can be approximated by diffeomorphisms $\Phi : \mathbb{RP}^2 \leftrightarrow \mathbb{RP}^2$ that map algebraic curves to algebraic curves.*

As an example, the simplest non-linear algebraic diffeomorphisms of $\mathbb{RP}^2$ are given by

$$\begin{aligned}
x &\mapsto x\big((c^6 - 1)y^2 z^2 - c^2(c^2 x^2 + c^4 y^2 + z^2)^2\big), \\
y &\mapsto y\big((c^6 - 1)z^2 x^2 - c^2(c^2 y^2 + c^4 z^2 + x^2)^2\big), \\
z &\mapsto z\big((c^6 - 1)x^2 y^2 - c^2(c^2 z^2 + c^4 x^2 + y^2)^2\big),
\end{aligned}$$

for any $c \in \mathbb{R} \setminus \{\pm 1\}$.

For $\mathbb{C}$, and more generally for algebraically closed fields of characteristic 0, we have a positive answer.

**Theorem 1.1.9.** *Let $K, L$ be algebraically closed fields of characteristic 0, and*

$$\Phi \colon \mathrm{Points}(\mathrm{Proj}_K^2) \leftrightarrow \mathrm{Points}(\mathrm{Proj}_L^2)$$

*a bijection that maps algebraic curves to algebraic curves. Then there is a unique field isomorphism $\varphi : K \cong L$ and a matrix $M \in \mathrm{PGL}_3(L)$ such that*

$$\Phi(x_0{:}x_1{:}x_2) = \big(\varphi(x_0){:}\varphi(x_1){:}\varphi(x_2)\big) \cdot M.$$

It is not unexpected that there could be a difference between the real and complex cases, since we can get only limited information about a real polynomial if we ignore its complex roots. Thus we should not forget about the complex points when dealing with a projective space over $\mathbb{R}$.

Note that if $i$ is a root of a real polynomial, then so is $-i$. In general, working with real polynomials only, we can detect conjugate pairs of complex numbers, but not individual complex numbers. In order to understand how this works for other fields, we need to think about what the basic objects of algebraic geometry are.

**Definition 1.1.10** (Affine $n$-space in algebraic geometry). Let $K$ be a field and $\overline{K} \supset K$ an algebraic closure. We denote the Zariski topological space of affine $n$-space by $|\mathbb{A}_K^n|$. It consists of the following.

(1) A point set, which can be given in two equivalent ways.

    (a) (Geometric form) Points in $\overline{K}^n$ modulo conjugation. That is,

$$|\mathbb{A}_K^n|^{\text{set}} := \{(x_1, \ldots, x_n) \in \overline{K}^n\}/(x_1 \ldots, x_n) \sim (\sigma(x_1), \ldots, \sigma(x_n)),$$

       where $\sigma \in \mathrm{Gal}(\overline{K}/K)$ is any automorphism of $\overline{K}$ that fixes $K$.

    (b) (Algebraic form) The set of maximal ideals of $K[x_1, \ldots, x_n]$.

(2) A topology whose closed sets are the solution sets of systems of equations

$$\{(x_1 \ldots, x_n) \in \overline{K}^n : f_1(x_1, \ldots, x_n) = \cdots = f_r(x_1, \ldots, x_n) = 0\},$$

    where $f_i \in K[x_1, \ldots, x_n]$ are polynomials.

    The advantage of this definition is that the connection between geometry and algebra is now very tight. For example, Hilbert's Nullstelensatz implies that two polynomials $f, g \in K[x_1, \ldots, x_n]$ have the same zero sets in $|\mathbb{A}_K^n|$ if and only if they have the same irreducible factors.

    A disadvantage is that it is no longer clear how to distinguish $K$-points from $\overline{K}$-points. In fact, the arguments of [WK81] show that if $K$ is a finite field, then the group of homeomorphisms is transitive on $|\mathbb{A}_K^2|$; see (10.3.1) for details. While this may be the only such example, we have a good solution of this problem only in the projective case.

**Remark 1.1.11** (Non-closed points). The above is the traditional definition of $\mathbb{A}_K^n$; see [Sha74]. In the modern scheme-theoretic version, the points of $\mathbb{A}_K^n$ correspond to all prime ideals of $K[x_1, \ldots, x_n]$. Thus our $|\mathbb{A}_K^n|$ is the set of closed points of the scheme-theoretic $\mathbb{A}_K^n$; see (2.3.1) for details. For our current purposes, the distinction is not important.

**Definition 1.1.12** (Projective $n$-space in algebraic geometry). Let $K$ be a field and $\overline{K} \supset K$ an algebraic closure. We denote the underlying Zariski topological space of projective $n$-space by $|\mathbb{P}_K^n|$. It consists of

(1) a point set

$$|\mathbb{P}_K^n|^{\text{set}} := \{(x_0 : \cdots : x_n) \in \overline{K}^{n+1} \setminus \{\mathbf{0}\}\}/(x_0 : \cdots : x_n) \sim (c\sigma(x_0) : \cdots : c\sigma(x_n)),$$

    where $c \in \overline{K}^\times$ and $\sigma \in \mathrm{Gal}(\overline{K}/K)$, and

(2) a topology, whose closed sets are

$$\{(x_0{:}\cdots{:}x_n) \in \overline{K}^{n+1} : F_1(x_0{:}\cdots{:}x_n) = \cdots = F_r(x_0{:}\cdots{:}x_n) = 0\},$$

where $F_i \in K[x_0,\ldots,x_n]$ are homogeneous polynomials.

One of our theorems is the following answer to the higher dimensional version of (1.1.6). We prove it in (8.1.4).

**Theorem 1.1.13.** *Let $K, L$ be fields. Assume that* char $K = 0$ *and fix $n, m \geq 2$. Let*

$$\Phi \colon |\mathbb{P}_K^n| \leftrightarrow |\mathbb{P}_L^m|$$

*be a homemorphism (that is, a bijection that maps closed, algebraic subsets to closed, algebraic subsets). Then $n = m$ and there is a unique field isomorphism $\varphi \colon K \cong L$, and a unique matrix $M \in \mathrm{PGL}_{m+1}(L)$ such that*

$$\Phi(x_0{:}\cdots{:}x_n) = \big(\varphi(x_0){:}\cdots{:}\varphi(x_n)\big) \cdot M.$$

**Clarification 1.1.14.** We need to explain what $\varphi(x_i)$ means if $x_i$ is not in $K$. Fix algebraic closures $\overline{K} \supset K$ and $\overline{L} \supset L$. Then $\varphi$ extends (non-uniquely) to a field isomorphism $\overline{\varphi} \colon \overline{K} \cong \overline{L}$. However, if $(x_0{:}\cdots{:}x_n) \in |\mathbb{P}_K^n|$ then

$$\big(\overline{\varphi}(x_0){:}\cdots{:}\overline{\varphi}(x_n)\big) \in |\mathbb{P}_L^m|$$

is independent of the choice of $\overline{\varphi}$ and of the representative of $(x_0{:}\cdots{:}x_n)$.

## 1.2   THE MAIN THEOREM

The book is devoted to extending (1.1.13) from $\mathbb{P}^n$ to other algebraic varieties by proving that in most cases the topological space $|X|$ determines $X$. It is the culmination of several reconstruction results proved in [KLOS20, Kol20]. (The precise definition of a normal, projective, geometrically irreducible variety is given in Section 2.1.)

**Main Theorem 1.2.1.** Let $K$ be a field of characteristic 0, and $X_K, Y_L$ normal, projective, geometrically irreducible varieties over $K$ (resp. over an arbitrary field $L$). Let $\Phi \colon |X_K| \to |Y_L|$ be a homeomorphism. Assume that

(1) dim $X \geq 4$, or

(2) dim $X \geq 3$ and $K$ is a finitely generated field extension of $\mathbb{Q}$, or

(3) dim $X \geq 2$ and $K$ is uncountable.

Then there is a field isomorphism $\varphi \colon K \xrightarrow{\sim} L$ and embeddings $j_K \colon X_K \hookrightarrow \mathbb{P}_K^N$

and $j_L \colon Y_L \hookrightarrow \mathbb{P}_L^N$ for $N = 2\dim X + 1$, such that we get a commutative diagram

$$
\begin{array}{ccc}
|X_K| & \xrightarrow{\ \Phi\ } & |Y_L| \\
{\scriptstyle j_K}\big\downarrow & & \big\downarrow{\scriptstyle j_L} \\
|\mathbb{P}_K^N| & \xrightarrow{\ \Phi'\ } & |\mathbb{P}_L^N|,
\end{array}
$$

where $\Phi'(x_0{:}\cdots{:}x_N) = \big(\varphi(x_0){:}\cdots{:}\varphi(x_N)\big)$.

Examples (2.2.1) to (2.2.6) show that normality and geometric irreducibility are necessary assumptions, but it is possible that $\dim X \geq 4$ can always be weakened to $\dim X \geq 2$.

Projectivity is crucial for our proof, but may not be necessary. The characteristic 0 assumption is also necessary, but there are natural conjectural versions in positive characteristic; we elaborate on these in Section 2.2.

As an illustratrion of some of the methods, we prove a special case of (1.1.13). The key is the following characterization of smooth rational curves in $\mathbb{P}_\mathbb{C}^n$.

**Lemma 1.2.2.** *Let $C \subset \mathbb{P}_K^m$ be an irreducible curve. Consider the properties:*

(1) *$C$ is smooth and rational.*

(2) *For every point $p \in C$ there is a hypersurface $H$ such that $C \cap H = \{p\}$.*

*Then (1) $\Rightarrow$ (2) and if $K = \mathbb{C}$ then (2) $\Rightarrow$ (1).*

*Proof.* Set $c := \deg C$. Fix $d$ such that $H^0(\mathbb{P}^m, \mathcal{O}_{\mathbb{P}^m}(d)) \to H^0\big(C, \mathcal{O}_{\mathbb{P}^m}(d)|_C\big)$ is surjective. If $C$ is smooth and rational then $\mathcal{O}_C(c[p]) \cong \mathcal{O}_{\mathbb{P}^m}(1)|_C$. Thus $\mathcal{O}_{\mathbb{P}^m}(d)|_C$ has a section that vanishes only at $p$ (with multiplicity $cd$), and it lifts to a section of $\mathcal{O}_{\mathbb{P}^m}(d)$ as needed.

Conversely, if (2) holds and $d := \deg H$, then $\mathcal{O}_C(cd[p]) \cong \mathcal{O}_{\mathbb{P}^m}(d)|_C$. Since the smooth points of $C$ generate $\mathrm{Pic}(C)$, this implies that $\mathrm{Pic}(C)/\langle \mathcal{O}_{\mathbb{P}^m}(1)|_C \rangle$ is a torsion group. Equivalently, the connected component $\mathrm{Pic}^\circ(C)$ is a torsion group. Over $\mathbb{C}$ this holds only if $\mathrm{Pic}^\circ(C)$ is trivial, hence $C$ is smooth and rational. $\square$

This already shows why varieties over different fields behave differently. If $K = \overline{\mathbb{F}}_p$ then $\mathrm{Pic}^\circ(C)$ is a torsion group for every curve $C$ over $K$. If $K$ is any field of positive characteristic, then $\mathrm{Pic}^\circ(C)$ is a torsion group whenever $C$ is rational with only cusps. If $K$ is a number field then sometimes $\mathrm{Pic}^\circ(C)$ is a torsion group even if $C$ has large genus.

**Corollary 1.2.3.** *Let $K$ be a field and fix $n, m \geq 2$. Let*

$$
\Phi \colon |\mathbb{P}_K^n| \leftrightarrow |\mathbb{P}_\mathbb{C}^m|
$$

*be a homemorphism. Then*

(1) $\Phi$ *maps smooth rational curves to smooth rational curves, and*

(2) *if $n = 2$ then it maps lines to lines.*

*Proof.* Let $C \subset \mathbb{P}_K^n$ be a smooth rational curve. By (1.2.2) it satisfies (1.2.2(2)). The latter is a purely topological property, so $\Phi(C) \subset \mathbb{P}_{\mathbb{C}}^n$ also satisfies (1.2.2(2)). Thus $\Phi(C)$ is a smooth rational curve by (1.2.2).

If $n = 2$ then lines and conics are the only smooth rational curves; so we are in the case of conic geometry, discussed after (1.1.5). $\qquad\square$

## 1.3   ORGANIZATION OF THE BOOK

Our approach naturally breaks into two, mostly independent, parts.

- Reconstruction of $X$ from $|X|$ together with the additional information of the linear equivalence relation on divisors.
- Reconstruction of linear equivalence of divisors from $|X|$.

We briefly describe each of these two parts.

### 1.3.1   Reconstruction of $X$ from $|X|$ and its divisorial structure

Recall that a (Weil) divisor on a variety is a $\mathbb{Z}$-linear combination of irreducible closed subsets of codimension 1. Since 'irreducible closed subset of codimension 1' is a purely topological notion (the codimension 1 irreducible closed subsets being the maximal proper ones), the group of Weil divisors on $X$ is determined by $|X|$. However, the linear equivalence relation on the group of divisors depends, a priori, on more than just $|X|$.

The *divisorial structure* of $X$ is the topological space $|X|$ together with the linear equivalence relation $\sim$ on the group of Weil divisors of $X$. Our main reconstruction result for varieties together with the divisorial structure is as follows (this is a slightly simplified version—see (4.1.14)).

**Theorem 1.3.1.** *Let $K, L$ be fields and let $X_K, Y_L$ be normal, proper, geometrically integral varieties over $K$ (resp. $L$). Let $\Phi \colon |X_K| \to |Y_L|$ be a homeomorphism such that for $D_1, D_2$ effective divisors on $X$, $\Phi(D_1) \sim \Phi(D_2)$ if and only if $D_1 \sim D_2$. Assume that*

(1) *either $K$ is infinite and $\dim X \geq 2$,*

(2) *or $K$ is a finite field of cardinality $> 2$ and $\dim X \geq 3$,*

(3) *or $K \cong \mathbb{F}_2$, $\dim X \geq 3$, and $X$ is Cohen-Macaulay.*

*Then $\Phi$ is the composite of a field isomorphism $\varphi \colon K \to L$ and an algebraic* *isomorphism of L-varieties $X_L^\varphi \to Y_L$.*

Here, $X_L^\varphi$ refers to the base change of $X_K$ via the isomorphism $\varphi$. See (2.1.6) for a concrete description of this construction.

### 1.3.2 Reconstruction of divisorial structure from $|X|$

Over fields of characteristic 0 one can often recover the linear equivalence relation on divisors from the topological space $|X|$. Our main result in this regard is the following, which is a slightly simplified version of (9.8.18).

**Theorem 1.3.1.** *Let $k$ be a field of characteristic 0 and $X$ a normal, projective, geometrically irreducible $k$-variety. Assume that*

(1) $\dim X \geq 4$, *or*

(2) $\dim X \geq 3$ *and $k$ is a finitely generated field extension of $\mathbb{Q}$, or*

(3) $\dim X \geq 2$ *and $k$ is uncountable.*

*Then $|X|$ determines linear equivalence of divisors.*

**Remark 1.3.2.** Here is a very rough idea why small or very large fields help us. Assume that $f$ is a rational function on a variety $X$, and we know its zero set $Z_0 := (f = 0)$ and its polar set $Z_\infty := (f = \infty)$. Note that if $g^n = c \cdot f^m$ for some $c \in k^\times$ and $m, n \in \mathbb{N}$, then $g$ and $f$ have the same zero and polar sets. If $X$ is normal, projective, and geometrically irreducible, and $Z_0, Z_\infty$ are both irreducible, then the converse also holds. Thus we are in a better situation if there are many rational functions with irreducible zero and polar sets.

If $\dim X \geq 2$ then Bertini's theorem guarantees that almost every rational function is such. If $\dim X = 1$ and $k$ is algebraically closed, there may not be any such functions. However, if $k$ is a finitely generated field extension of $\mathbb{Q}$, then Hilbert's irreducibility theorem guarantees that there are many such functions.

We need to apply such considerations not to the original variety $X$, but in the following setting: $C \subset X$ is a curve, $Y \subset X$ is an irreducible subvariety to which the above considerations apply, and $C \cap Y$ is a single point. Except in rare instances, this can be arranged only if $\dim X > 1 + \dim Y$.

Such considerations lead to the notion of *Bertini-Hilbert dimension* of a field (which is either 1 or 2; see (9.5.5)). Then (1.3.1) holds whenever the dimension of $X$ is greater than $1 + \mathrm{BH}(k)$.

Finally, another problem occurs when the zero and polar sets have 'unexpected' irreducible components. In algebraic geometry it is usually easy to show that 'unexpected' things can happen in only countably many ways. So, over uncountable fields, most functions do not behave in 'unexpected' ways.

In combination with (1.3.1), this yields Main Theorem (1.2.1).

### 1.3.3   Structure of the chapters

The book is broadly organized into two parts, corresponding to Sections 1.3.1 and 1.3.2. In the first part, consisting of Chapters 3 through 6, we prove (1.3.1) by observing that the divisorial structure lets us define linear systems of effective divisors, reconstructing the projective structure on linear systems using variants of the Fundamental Theorem of Projective Geometry, and then reconstructing rings of functions using those linear systems.

In Chapters 8 and 9, we prove (1.3.1) by first reconstructing a weaker equivalence relation for divisors purely from the topology, then using that to reconstruct various types of geometric data, and finally reconstructing the usual linear equivalence relation for divisors. Beforehand, in Chapter 7, we give a simpler argument following a similar strategy for varieties over an uncountable algebraically closed field and also collect various results about pencils that are used in that chapter and subsequent ones.

Chapter 10 includes complements, counterexamples, and conjectures: a topological Gabriel theorem, various types of schemes for which results of the type we describe here fail, and several questions and conjectures about extensions of our results to larger classes of schemes and positive characteristic.

Ancillary results are collected in appendices. These are mostly known but are included as we found it hard to find references for the precise statements that we need. The reader may wish to consult the appendices only as needed while reading the main parts of the book.

The first appendix recalls the definitions and basic properties of locally finite, Mordell-Weil, anti-Mordell-Weil, and Hilbertian fields. This appendix is included at the end of Chapter 8, where these notions are first used. In the second appendix, which appears at the end of Chapter 9, we introduce the notion of weakly Hilbertian fields, (9.9.1). This notion is new and may be of independent interest.

The appendices included in Chapter 11 contain various background material that is used in the book, but follow more standard algebraic geometry terminology. In Sections 11.1 and 11.2 we summarize properties of complete intersections and various Bertini-type theorems. The theory of the Picard group, Picard variety, and Albanese variety is recalled in Section 11.3. The literature is much less complete about the class group and its scheme version, which does not even seem to have a name. Basic results on commutative algebraic groups and the multiplicative groups of Artin algebras are also studied in Section 11.3.

# Chapter Two

## Preliminaries

In this chapter we recall the basic definitions and theorems, leading to the Main Question (2.1.7). After a series of negative examples in Section 2.2, we discuss possible generalizations of Main Theorem (1.2.1), followed by the scheme-theoretic version of Main Theorem (1.2.1) in Section 2.3. We survey related results in Section 2.4 and then fix our terminology in Section 2.5.

## 2.1 ALGEBRAIC VARIETIES

It is quite interesting that the classical algebraic geometry literature—roughly [Sha74] and before—never actually defines what an algebraic variety is. These definitions give a variety $X$ as a subset of $|\mathbb{P}_K^n|$ with no additional structure given. Since the algebraic subvarieties of $X$ were clearly understood, these definitions essentially identify a variety $X$ with the topological space $|X|$.

The morphisms between varieties are then defined by hand. Thus one gets the correct definition of the *category* of algebraic varieties, but not of the individual varieties.

For now this traditional definition is the most natural for us, though soon we switch to the scheme-theoretic version as in [Har77]; see (2.3.1) for further comments and notation.

We need to distinguish a variety $X$ from its underlying set $|X|$, so we are extra careful at the beginning.

**Definition 2.1.1** (Algebraic sets, point set version). Fix a field $K$. Closed subsets of some $|\mathbb{P}_K^n|$ are the *projective algebraic sets*. Thus these are of the form

$$|X| = \{(x_0{:}\cdots{:}x_n) \in |\mathbb{P}_K^n| : F_1(x_0{:}\cdots{:}x_n) = \cdots = F_r(x_0{:}\cdots{:}x_n) = 0\},$$

where the $F_i$ are homogeneous polynomials. We frequently write $|X| = (F_1 = \cdots = F_r = 0)$. The easiest to think of are *hypersurfaces*; these are given by one equation:

$$|X_F| := \{(x_0{:}\cdots{:}x_n) \in |\mathbb{P}_K^n| : F(x_0{:}\cdots{:}x_n) = 0\}.$$

The proofs of our theorems admit only very minor simplifications for hypersurfaces, so nothing is lost if the reader focuses on them.

A projective algebraic set $|X|$ is *irreducible* if it cannot be written as a finite union of projective algebraic sets in a nontrivial way. (That is, if $|X| = \cup_{i \in I}|X_i|$ then $|X| = |X_i|$ for some $i \in I$.) For example, a hypersurface $|X_F|$ is irreducible if and only if $F$ is a power of an irreducible polynomial.

A *quasi-projective algebraic set* is the difference of two projective algebraic sets $|X| = |Y| \setminus |Z|$. $|X|$ is called *irreducible* if $|Y|$ can be chosen irreducible.

Starting with $|\mathbb{A}_K^n|$ instead of $|\mathbb{P}_K^n|$, we get the notion of *affine algebraic sets*. It is quite useful to think of a projective algebraic set $|X| \subset |\mathbb{P}_K^n|$ as covered by the affine algebraic sets $|U_i| := |X| \setminus (x_i = 0)$.

Every quasi-projective algebraic set is a finite union of irreducible ones $|X| = \cup_i |X_i|$. Such irredundant decompositions are unique (up to ordering the $|X_i|$). These $|X_i|$ are the *irreducible components* of $|X|$. For hypersurfaces, the irreducible decomposition corresponds to writing $F = \prod G_i^{m_i}$, where the $G_i$ are irreducible.

**Definition 2.1.2** (Dimension). A point has dimension 0 and an irreducible algebraic set $|X|$ has dimension $d$ if and only if all closed, irreducible algebraic subsets $|Z| \subsetneq |X|$ have dimension $< d$. It is a not obvious that $\dim |\mathbb{P}_K^n| = n$.

**Definition 2.1.3** (Morphisms of algebraic sets, classical version). Fix a field $K$ and let $|X| \subset |\mathbb{P}_K^n|$ be a quasi-projective algebraic set. A *morphism* of $|X|$ to $|\mathbb{P}_K^m|$ is given as

$$\Phi \colon (x_0{:}\cdots{:}x_n) \mapsto \big(F_0(x_0{:}\cdots{:}x_n) : \cdots : F_m(x_0{:}\cdots{:}x_n)\big),$$

where

(1) the $F_i$ are homogeneous of the same degree (so $(x_0{:}\cdots{:}x_n) = (cx_0{:}\cdots{:}cx_n)$ have the same images), and

(2) the $F_i$ have no common zero on $X$ (since $(0{:}\cdots{:}0)$ is not a point of $|\mathbb{P}_K^m|$).

If the image of $\Phi$ lands in a quasi-projective algebraic set $|Y| \subset |\mathbb{P}_K^m|$, then we say that $\Phi \colon |X| \to |Y|$ is a morphism.

Two quasi-projective algebraic sets $|X| \subset |\mathbb{P}_K^n|$ and $|Y| \subset |\mathbb{P}_K^m|$ are *isomorphic* if there are morphisms $\Phi \colon |X| \to |Y|$ and $\Psi \colon |Y| \to |X|$ such that $\Phi \circ \Psi$ and $\Psi \circ \Phi$ are both identities.

**Definition 2.1.4** (Algebraic varieties and sets). Let $|X|$ be a quasi-projective algebraic set over a field $K$. The set of all morphisms $|X| \to |\mathbb{A}_K^1|$ form a $K$-algebra, denoted by $K[X]$ as in [Sha74]. These are the *regular functions* on $X$.

If $|Y| \subset |X|$ then restriction gives a $K$-algebra homomorphism $K[X] \to K[Y]$. Letting $|Y|$ run through all open algebraic subsets of $|X|$, we get the *sheaf of regular functions* on $|X|$, usually denoted by $\mathcal{O}_X$.

From the modern point of view, a *quasi-projective algebraic set* over $K$ should be a pair $X = \big(|X|, \mathcal{O}_X\big)$, where $|X|$ is a quasi-projective algebraic set as in (2.1.1) and $\mathcal{O}_X$ its sheaf of regular functions.

$X = \big(|X|, \mathcal{O}_X\big)$ is called a *variety over $K$* or a *$K$-variety* if $|X|$ is irreducible. The field is frequently omitted if it is clear from the context. Current algebraic geometry considers the pair $X = \big(|X|, \mathcal{O}_X\big)$ as the basic object.

**Warning 2.1.5.** All books we know follow this definition of a variety, but people frequently use 'variety' to refer to a possibly reducible algebraic set, especially if all irreducible components have the same dimension.

**Remark 2.1.6** (The role of the field $K$). One needs to start with a field $K$ in order to define $|\mathbb{A}_K^n|$, $|\mathbb{P}_K^n|$ and their closed algebraic subsets. If $\sigma\colon K \to L$ is a field isomorphism, then

$$(x_0 : \cdots : x_n) \mapsto \big(\sigma(x_0) : \cdots : \sigma(x_n)\big)$$

gives a homeomorphism $|\mathbb{P}_K^n| \sim |\mathbb{P}_L^n|$. On the level of regular functions this is an even more trivial operation: every $K$-algebra $K[U]$ becomes an $L$-algebra. This suggests that for us it would be more natural to view each $K[X]$ simply as ring. (This is essentially what scheme theory does.) Thus to any quasi-projective algebraic set $X$ over $K$ we get a quasi-projective algebraic set $X_L^\sigma$ over $L$ with the 'same' topological space and the 'same' ring of regular functions. To be very concrete, if

$$X = (F_1 = \cdots = F_r = 0) \qquad \text{where} \quad F_i = \sum_{\mathbf{m}} a_{\mathbf{m}}^{(i)} x_0^{m_0} \ldots x_n^{m_n}, \text{ then}$$
$$X_L^\sigma = (F_1^\sigma = \cdots = F_r^\sigma = 0) \quad \text{where} \quad F_i^\sigma = \sum_{\mathbf{m}} \sigma(a_{\mathbf{m}}^{(i)}) x_0^{m_0} \ldots x_n^{m_n}.$$

Ideally we should try to ignore the field as much as possible, but one cannot talk about projectivity without having a field in mind.

Now we can formulate the central problem of our work.

**Main Question 2.1.7.** Let $K, L$ be fields and $X_K, Y_L$ quasi-projective algebraic sets over $K$ (resp. $L$). Let $\Phi\colon |X_K| \to |Y_L|$ be a homeomorphism.
   Is $\Phi$ the composite of a field isomorphism $\varphi\colon K \xrightarrow{\sim} L$ and an algebraic isomorphism of $L$-varieties $X_L^\varphi \xrightarrow{\sim} Y_L$?

It turns out that in this generality the answer is negative. We start by listing the various reasons why we need restrictions on the fields and on the algebraic sets.

## 2.2  EXAMPLES AND SPECULATIONS

We start with a series of negative examples.

**Example 2.2.1** (Dimension 0). An irreducible 0-dimensional $K$-variety consists of a single point. Thus its topology carries no information about the field $K$.

**Example 2.2.2** (Dimension 1). The closed algebraic subsets of a 1-dimensional $K$-variety $C$ are exactly the finite subsets. Thus the only topological information is the cardinality of $|C|$. It is easy to see that this cardinality is $|K|$ if $K$ is infinite and $\omega_0$ if $K$ is finite.

**Example 2.2.3** (Normalization). Let us start with an example. The morphism $t \mapsto (t^2, t^3)$ gives a homeomorphism between the varieties $\mathbb{A}_k^1$ and $(x^3 - y^2 - 0) \subset \mathbb{A}_k^2$. It is, however, not an isomorphism since its inverse is $(x, y) \mapsto y/x$, but $y/x$ is not a polynomial.

The notion of *normalization* was invented to eliminate such examples. It is probably best to think of normal varieties as those where Riemann's extension theorem applies. Thus a $\mathbb{C}$-variety $X$ is normal if and only if the following holds.

- Let $U \subset X$ an open, algebraic subset and $g: U \to \mathbb{C}$ a regular function. Then $g$ extends to a regular function defined at $x \in X \setminus U$ if and only if $g$ is bounded in an open neighborhood (in the Euclidean topology) of $x$.

To be precise, the most relevant algebraic geometry notion here is not normalization, but *seminormalization;* see [Kol96, Sec.I.7.2] for its definition. We comment more about it in (2.2.10).

**Example 2.2.4** (Choosing the wrong field). Let $X$ be the $\mathbb{R}$-variety $(y_{n+1}^2 + 1 = 0) \subset |\mathbb{A}_{\mathbb{R}}^{n+1}|$. Despite appearances, the map $\Phi: |\mathbb{A}_{\mathbb{C}}^n| \to |X|$ given by

$$(x_1, \ldots, x_n) \mapsto (x_1, \ldots, x_n, \sqrt{-1})$$

is a homeomorphism. The explanation becomes clearer algebraically:

$$\mathbb{C}[x_1, \ldots, x_n] \cong \mathbb{R}[y_1, \ldots, y_{n+1}]/(y_{n+1}^2 + 1).$$

Thus they have the same ideals, no matter which field we work over.

The solution is that we should always choose the largest possible field. We discuss this in (2.2.7).

**Example 2.2.5** (Purely inseparable morphisms). Assume that char $K = p$. Then the Frobenius endomorphism

$$(x_0: \cdots : x_n) \mapsto (x_0^p: \cdots : x_n^p)$$

is a homeomorphism of $|\mathbb{P}_K^n|$ to itself, but it is not an isomorphism since $\sqrt[p]{x}$ is not a polynomial. In a similar way, if $L/K$ is a purely inseparable field extension then the identity map gives a homeomorphism $|\mathbb{P}_K^n| \sim |\mathbb{P}_L^m|$.

Thus in positive characteristic the best we can hope for is to get a positive answer up to purely inseparable morphisms and field extensions.

**Example 2.2.6** (Surfaces over finite fields). These unexpected examples are from [WK81]. Let $p, q$ be prime numbers and $K \supset \mathbb{F}_p, L \supset \mathbb{F}_q$ (possibly infinite) algebraic extensions. Then $|\mathbb{P}_K^2|$ and $|\mathbb{P}_L^2|$ are homeomorphic.

More generally, many (though not all) algebraic surfaces over such fields are homeomorphic to projective planes (see (10.3.1)).

**Definition 2.2.7** (Geometric irreducibility). Let $K$ be a field and $X = (F_1 = \cdots = F_r = 0) \subset \mathbb{P}_K^n$ a projective, algebraic set. If $L/K$ is a field extension then the same equations define a projective, algebraic set

$$X_L = (F_1 = \cdots = F_r = 0) \subset \mathbb{P}_L^n.$$

We say that a property of $X$ holds *geometrically* if the property holds for $X_L$ for every algebraic extension $L/K$. In most cases it is enough to check what happens when $L = \overline{K}$, an algebraic closure of $K$.

Consider, for example, a hypersurface $X = (F = 0)$. Then $X$ is irreducible if and only if $F$ is a power of an irreducible polynomial; we may as well assume that $F$ is irreducible.

Then $X$ is *geometrically irreducible* if $X_L$ is irreducible for every algebraic extension $L/K$. (The old literature, for example, [Wei62], uses *absolutely irreducible*.) For example, $X = (x^2 + y^2 = 0) \subset \mathbb{P}_{\mathbb{R}}^2$ is irreducible but

$$X_{\mathbb{C}} = (x + \sqrt{-1}y = 0) \cup (x - \sqrt{-1}y = 0) \subset \mathbb{P}_{\mathbb{C}}^2$$

is reducible.

If char $K = 0$ and $F$ is irreducible over $K$, then $F$ either stays irreducible over $L$ or decomposes as the product of distinct factors. So $X$ is geometrically irreducible if and only if $F$ is irreducible over $\overline{K}$.

However, if char $K = p > 0$, then it can happen that we get factors with multiplicity $p$. For example, start with $K = \mathbb{F}_p(t)$ and the irreducible polynomial $F = x^p + ty^p$. Set $L = \mathbb{F}_p(s)$, where $s^p = t$. Over $L$ we get that $x^p + ty^p = x^p + s^p y^p = (x + sy)^p$.

We say that $X$ is *geometrically integral* if and only if $F$ is irreducible over $\overline{K}$. (This is a rare example where the algebraic terminology diverges from the geometric one.) For non-hypersurfaces; see [Har77, II.Exrc.3.15].

It is quite likely that Main Theorem (1.2.1) is only the first positive result. Below we list natural variants and generalizations. In all cases we try to state the strongest version that is consistent with the known examples. We list them in what we expect to be an increasing order of difficulty.

**Speculation 2.2.8** (Dimension 2). *Let $K, L$ be fields of characteristic 0 and $X_K, Y_L$ normal, proper, geometrically integral varieties of dimension $\geq 2$ over $K$ (resp. $L$). Let $\Phi \colon |X_K| \to |Y_L|$ be a homeomorphism.*

*Then $\Phi$ is the composite of a field isomorphism $\varphi \colon K \xrightarrow{\sim} L$ and an algebraic isomorphism of $L$-varieties $X_L^\varphi \xrightarrow{\sim} Y_L$.*

**Remark 2.2.9.** In [KLOS20] (a preliminary version of this work that was mostly, but not entirely, superseded by it), we used methods similar to (but simpler than) those we use here to prove the above statement when $K$ and $L$ are uncountable algebraically closed fields. The results we describe here use projectivity, but it seems likely that they can be extended to the proper case.

**Speculation 2.2.10** (Reducible varieties). *Let $K, L$ be fields of characteristic 0 and $X_K, Y_L$ projective varieties over $K$ (resp. $L$), all of whose irreducible components have dimension $\geq 2$. Let $\Phi\colon |X_K| \to |Y_L|$ be a homeomorphism. Then $\Phi$ lifts to a homeomorphism of the normalizations $\bar{\Phi}\colon |\bar{X}_K| \to |\bar{Y}_L|$.*

**Remark 2.2.11.** Let $X \subset \mathbb{P}^4_{\mathbb{C}}$ be the union of the 2-planes $X_1 := (x_1 = x_2 = 0)$ and $X_2 := (x_3 = x_4 = 0)$. They meet at the point $(1{:}0{:}0{:}0{:}0)$. Choose $\Phi\colon |X| \to |X|$ to be the identity on $X_1$ and complex conjugation on $X_2$. Then $\Phi$ is not the composite of a field isomorphism and an algebraic isomorphism. The two irreducible components dictate different field isomorphisms. This is the reason of the formulation of (2.2.10). It is, however, possible that such 0-dimensional intersections provide the only counterexamples for seminormal schemes.

**Speculation 2.2.12** (Unequal characteristics). *Let $K$ be a field of characteristic 0, $L$ a field of characteristic $> 0$, and $X_K, Y_L$ normal, projective, geometrically integral varieties of dimension $\geq 2$ over $K$ (resp. $L$). Then $|X_K|$ and $|Y_L|$ are not homeomorphic.*

**Remark 2.2.13.** We prove in (9.6.13) that this holds if $\dim X \geq 4$.

**Speculation 2.2.14** (Quasi-projective varieties). *Let $K, L$ be fields of characteristic 0 and $X_K, Y_L$ normal, geometrically integral varieties of dimension $\geq 2$ over $K$ (resp. $L$). Let $\Phi\colon |X_K| \to |Y_L|$ be a homeomorphism. Then $\Phi$ is the composite of a field isomorphism $\varphi\colon K \xrightarrow{\sim} L$ and an algebraic isomorphism of $L$-varieties $X_L^{\varphi} \xrightarrow{\sim} Y_L$.*

**Remark 2.2.15.** Already the affine version of (1.1.13) seems much harder, even for $n = 2$.

The first thing we notice is that the 'natural' automorphism group is now infinite-dimensional. For any polynomial $g(x) \in K[x]$,

$$(x, y) \mapsto \big(x, y + g(x)\big)$$

is an automorphism of $\mathbb{A}^2_K$ that clearly maps algebraic curves to algebraic curves. The study of such groups of automorphisms is a fascinating subject, with many open problems; see [Kra96].

Our methods give some information about homeomorphisms of $|\mathbb{A}^2_K|$. For example, homeomorphisms of $|\mathbb{A}^2_{\mathbb{C}}|$ map smooth rational curves to smooth rational curves.

A similar statement about $|\mathbb{P}^2_{\mathbb{C}}|$ is very strong, since in $\mathbb{P}^2$ the only smooth rational curves are lines and conics. However, $|\mathbb{A}^2_{\mathbb{C}}|$ has many smooth rational curves, for example, $(y = f(x))$ for any polynomial $f(x)$, so we are still far from getting the desired result.

The situation is quite different for general affine schemes; see (2.4.7).

**Speculation 2.2.16** (Positive characteristic). *Let $K, L$ be fields of characteristic $> 0$ and $X_K, Y_L$ normal, projective, geometrically integral varieties of dimension $\geq 2$ over $K$ (resp. $L$). Let $\Phi \colon |X_K| \to |Y_L|$ be a homeomorphism.*

*Assume that $K, L$ are not algebraic over their prime field.*

*Then $\Phi$ is the composite of a field isomorphism $\varphi \colon K^{\text{ins}} \xrightarrow{\sim} L^{\text{ins}}$ of their purely inseparable closures, and a purely inseparable algebraic morphism of $L^{\text{ins}}$-varieties $X^{\varphi}_{L^{\text{ins}}} \xrightarrow{\sim} Y_{L^{\text{ins}}}$.*

**Remark 2.2.17.** As we noted in (2.2.6), there are many 2-dimensional counterexamples if $K, L$ are algebraic over their prime field. We did not succeed in making 3-dimensional counterexamples, and we are not sure what to expect.

## 2.3   SCHEME-THEORETIC FORMULATION

Much of modern algebraic geometry is written using the language of *schemes;* a standard introduction is [Har77]. For the questions we are considering, the differences between the classical and the scheme-theoretic do not matter, but it useful to know how to switch between the two versions.

**Definition 2.3.1** (Varieties as schemes). A quasi-projective set with its sheaf of regular functions gives a locally ringed space $X = (|X|, \mathcal{O}_X)$. The scheme $X^{\text{sch}} = (|X^{\text{sch}}|, \mathcal{O}_{X^{\text{sch}}})$ associated to it is obtained as follows.

(1) The points of $|X^{\text{sch}}|$ are the closed, irreducible subsets $|Z| \subset |X|$. Let us denote the point corresponding to $Z$ by $\eta_Z$. It is customary to identify a point $p \in |X|$ with $\eta_p$, and view $|X|$ as a subset of $|X^{\text{sch}}|$. We refer to $\eta_Z$ as the *generic point* of $|Z|$.

(2) The closure of $\eta_Z$ is $\bar{\eta}_Z := \{\eta_W : |W| \subset |Z|\}$. This defines a topology on the points of $|X^{\text{sch}}|$. We denote this topological space by $|X^{\text{sch}}|$. The subspace topology on $|X|$ agrees with the previous topology $|X|$, thus $|X|$ and $|X^{\text{sch}}|$ uniquely determine each other.

(3) If $|U^{\text{sch}}| \subset |X^{\text{sch}}|$ is an open set then $|U| := |U^{\text{sch}}| \cap |X| \subset |X|$ is open and we set $\mathcal{O}_{X^{\text{sch}}}(U) = \mathcal{O}_X(U)$. Again, the sheaves $\mathcal{O}_{X^{\text{sch}}}$ and $\mathcal{O}_X$ uniquely determine each other.

Scheme theory also studies much more general objects. The classical quasi-projective sets over $k$ correspond to reduced schemes that are quasi-projective over $k$.

**Convention 2.3.2.** Since $(X, \mathcal{O}_X)$ and $(X^{\text{sch}}, \mathcal{O}_{X^{\text{sch}}})$ determine each other, they are routinely identified in the algebraic geometry literature, and one simply writes $X$ to denote a scheme.

We follow this practice and use $X$ to denote a scheme. Thus, it is a pair $X = (|X|, \mathcal{O}_X)$, where $|X|$ is the underlying Zariski topological space and $\mathcal{O}_X$ its sheaf of rings.

The scheme-theoretic version of Main Theorem (1.2.1) is the following. Its advantage is that the fields do not play a role in its formulation, and so 'geometric irreducibility' is replaced by the simpler 'irreducibility' assumption.

**Theorem 2.3.3.** *Let $X$ and $Y$ be normal, projective, irreducible schemes over fields of characteristic 0. Let $\Phi\colon |X| \xrightarrow{\sim} |Y|$ be a homeomorphism. Assume that*

(1) $\dim X \geq 4$, *or*

(2) $\dim X \geq 3$ *and the fields are finitely generated field extensions of $\mathbb{Q}$, or*

(3) $\dim X \geq 2$ *and the fields are uncountable.*

*Then $\Phi$ extends to an isomorphism $\Phi^{\mathrm{sch}}\colon X \xrightarrow{\sim} Y$ of schemes.*

## 2.4   SURVEY OF RELATED RESULTS

This work has its origins in trying to understand the derived category of coherent sheaves on an algebraic variety and to what extent it determines the variety [LO21]. While this project took quite a different direction, and the work in this book is not directly related to derived categories of coherent sheaves, this idea of categorical invariants determining a variety nonetheless provided significant inspiration. The most classical example of such categorical reconstruction results is the following theorem [Gab62, Ros98].

**Theorem 2.4.1** (Gabriel-Rosenberg). *A quasi-separated scheme is determined by its associated abelian category $\mathrm{QCoh}(X)$ of quasi-coherent sheaves.*

Note that in the case when $X$ is a finite type scheme over a field $k$, the category of quasi-coherent sheaves captures the information of the topological space $|X|$. Indeed, the subcategory of coherent sheaves $\mathrm{Coh}(X)$ can be identified with the finitely presented objects of $\mathrm{QCoh}(X)$, and the skyscraper sheaves of points can be identified in $\mathrm{Coh}(X)$ as the objects $\mathscr{F} \in \mathrm{Coh}(X)$ having the property that any nonzero epimorphism $\mathscr{F} \to \mathscr{F}'$ is an isomorphism. Thus from $\mathrm{QCoh}(X)$ we can recover the set of closed points of $X$. Furthermore, for an object $\mathscr{F} \in \mathrm{Coh}(X)$ we can characterize its support among the closed points as those points for which the associated skyscraper sheaf admits an epimorphism from $\mathscr{F}$. Using this we can therefore recover the Zariski topology on the closed points, and therefore also the entire Zariski topological space $|X|$. The results of this book imply that in many instances there is no loss of information in passing from $\mathrm{QCoh}(X)$ to $|X|$.

Another variant direction one can consider is the problem of reconstructing the function field, or equivalently the birational equivalence class, of a variety from other data. This is a very natural problem to consider in the context of derived categories of coherent sheaves (see, for example, [LO21, §4]. In particular, we mention the program of Bogomolov and Tschinkel [BT11,BT13,BT12,BT09, BRT19]. While this work is in the context of Grothendieck's anabelian geometry, and the work here has a somewhat different emphasis, the ideas presented here are very much extensions of those of the Bogomolov-Tschinkel program. The core idea in this work is to notice that if $K$ is the function field of a variety over a field $k$, then $K^{\times}/k^{\times}$ has the structure of a projective space (of infinite dimension) and a group, and this structure contains a significant amount of information about $K$. In particular, the Bogomolov-Tschinkel program relates this structure to Galois theory. The referenced articles contain many results in this direction. Here we just mention one of them to give the flavor:

**Theorem 2.4.2.** [BT11, Thm.1] *Let $K$ be a function field of transcendence degree at least 2 over the algebraic closure $k$ of a finite field, and let $\ell$ be a prime invertible in $k$. Let $\mathcal{G}_K$ be the pro-$\ell$-completion of the absolute Galois group of $K$, and let $\mathcal{G}_K^c$ be the quotient of $\mathcal{G}_K$ by the second step of the descending central series, so we have an extension*

$$1 \to [\mathcal{G}_K, \mathcal{G}_K]/[\mathcal{G}_K, [\mathcal{G}_K, \mathcal{G}_K]] \to \mathcal{G}_K^c \to \mathcal{G}_K^a \to 1,$$

*where $\mathcal{G}_K^a$ is the abelianization of $\mathcal{G}_K$. Then $\mathcal{G}_K^c$, as a pro-$\ell$-group, determines the pair $(K, k)$.*

**Remark 2.4.3.** Note that the group $\mathcal{G}_K^a$ is closely related to $K^{\times}$ via Kummer theory. In fact, in loc. cit. the formulation is in terms of $\mathcal{G}_K^a$ and certain subgroups, which can be recovered from $\mathcal{G}_K^c$.

The consideration of $K^{\times}/k^{\times}$ as above also naturally leads to studying reconstructions of function fields from Milnor K-theory. This has been done by Bogomolov and Tschinkel [BT09] as well as Cadoret and Pirutka [CP18]. A fundamental result in this direction is the following:

**Theorem 2.4.4.** [BT09, Thm.4] *Let $K$ be a function field of transcendence degree $\geq 2$ over an algebraically closed field $k$. Then $(K, k)$ is determined by the first and second Milnor K-groups of $K$.*

More refined results, including results over non-closed fields are obtained in [CP18]. Instead of Milnor K-theory one might also naturally consider Galois cohomology. This direction was pursued by Topaz in [Top17, Top16]

All of the results mentioned, as well as the work in this book, are focused on dimensions $\geq 2$. In [BKT10] Bogomolov, Korotiaev, and Tschinkel formulated a conjecture for curves over an algebraically closed field. Namely, if $C$ is a smooth projective curve over an algebraically closed field $k$, then one can consider the data $(J(k), P(k), i\colon C(k) \hookrightarrow P(k))$, consisting of:

(1) The $k$-points $J(k)$ of the Jacobian of $C$—an abelian group.

(2) The set $P(k)$ of isomorphism classes of degree 1 line bundles on $C$—a set with a simply transitive action of $J(k)$.

(3) The subset $i\colon C(k) \hookrightarrow P(k)$ given by sending a point to the class of its associated line bundle.

The conjecture is that the data $(J(k), P(k), i\colon C(k) \hookrightarrow P(k))$ determine $(C, k)$. Using deep results in model theory, Zilber addressed this conjecture in [Zil14]. Our own lack of expertise in model theory has rendered us unable to understand the proof. It would be very interesting to understand the situation for curves solely using algebraic geometry.

**Remark 2.4.5.** As we explain in (5.3.2), the assumption that $k$ is algebraically closed is necessary. This is model-theoretically reasonable (in the sense that the model theory of algebraically closed fields is far more tractable than that of other fields), but we believe that an algebro-geometric proof of the curve case could also illuminate what is truly required of the base field for this to be true.

Chow [Cho49] and Hua [Hua49] generalized Theorems 1.1.1–1.1.2 to other homogeneous spaces. In the affine version, Hua studied maps $\varphi$ from matrices to matrices that satisfy $\operatorname{rank}(A - B) = \operatorname{rank}(\varphi(A) - \varphi(B))$ for every $A, B$. Under some natural conditions these are all of the form $\varphi(A) = Q_1 A Q_2 + R$ where $Q_1, Q_2$ are invertible. In the projective version, Chow proves that maps between Grassmannians that send lines to lines come from algebraic isomorphisms. For a detailed treatment see [Wan96], which also discusses other classical homogeneous spaces. See also [dS22] and the references there for the affine case.

Related to this is also the work of Borel and Tits [BT73] showing that in many cases homomorphisms between the $k$-points of algebraic groups are induced by homomorphisms of algebraic groups.

Another direction that has been fruitfully studied concerns reconstruction results for the étale topos, including the following.

**Theorem 2.4.6.** [Voe90, Cor.3.1] *Let $K$ be a field finitely generated over $\mathbb{Q}$ and let $X$ and $Y$ be normal finite type $K$-schemes. If there exists an equivalence $X_{\text{ét}} \simeq Y_{\text{ét}}$ of étale topoi over $(\operatorname{Spec}(K))_{\text{ét}}$ then $X$ and $Y$ are isomorphic (as $K$-schemes).*

In fact, Voevodsky proves stronger results concerning morphisms of schemes, and not just isomorphisms, and also results in low dimensions; for example, in dimension 0 (giving back the Ikeda–Iwasawa–Neukirch–Pop–Uchida theorems [Pop94]) and dimension 1.

The related work of Barwick, Glasman, and Haine on exodromy [BGH18] yields a reconstruction of the étale topos from a category consisting of points together with étale specializations. This builds on work of Lurie [Lur17, Appendix A].

Finally we mention the work of Hochster [Hoc69] characterizing spectra of commutative rings.

**Theorem 2.4.7.** *A topological space $M$ is homeomorphic to $|\operatorname{Spec} A|$ for some commutative ring $A$ if and only if $M$ is $T_0$, quasi-compact, the quasi-compact open subsets are closed under finite intersection and form an open basis, and every nonempty irreducible closed subset has a generic point.*

*Moreover, for every $M$ there are many such rings $A$.*

As a special case one obtains that for every quasi-projective variety $X$, there are many commutative rings $A$ such that $|\operatorname{Spec} A| \sim |X|$. If $X$ is not affine, then these rings are necessarily very far from being finitely generated.

## 2.5   TERMINOLOGY AND NOTATION

At the end of the text we have included indices of terminology and notation. We highlight here a few items of particular importance in the text.

### 2.5.1   Varieties and schemes

As mentioned in Section 2.3, in the writing of this book a choice had to be made in the basic language of algebraic geometry. Since we are primarily interested in quasi-projective varieties over a field, we have chosen to mostly use the classical terminology of varieties. Of course, the reader who wishes can make the translation to the language of schemes. Specifically, we make the convention (which follows other standard treatments such as [Sta22, Tag 020D]) that for a field $K$ a *variety over $K$*, sometimes called a *$K$-variety*, is an integral scheme $X$ over $K$ such that the structure morphism $f\colon X \to \operatorname{Spec}(K)$ is separated and of finite type. Occasionally, we will need to work with possibly nonreduced or reducible schemes, such as when considering zero loci of hyperplane sections of a projective variety, in which case we use the scheme-theoretic language.

The language of varieties, while perhaps making the material accessible to a wider audience, has a drawback when considering morphisms. A morphism $f\colon X \to Y$ of $K$-varieties $X$ and $Y$ is a morphism of schemes over $K$. We will often have occasion to consider morphisms between varieties defined over different fields. If $X$ is a variety over a field $K$ and $Y$ is a variety over a field $L$ and $\varphi\colon L \to K$ is an isomorphism of fields, then a $\varphi$-linear morphism of varieties $\alpha\colon X \to Y$ is a morphism of schemes fitting into a commutative diagram

$$
\begin{array}{ccc}
X & \xrightarrow{\ \alpha\ } & Y \\
\downarrow & & \downarrow \\
\operatorname{Spec}(K) & \xrightarrow{\ \varphi\ } & \operatorname{Spec}(L).
\end{array}
$$

Equivalently, if $Y^{\varphi} := Y \times_{\operatorname{Spec}(L)} \operatorname{Spec}(K)$ then $\alpha$ is a morphism of $K$-varieties $X \to Y^{\varphi}$.

In a few places we will also need to consider morphisms $f\colon X \to Y$ of schemes over $\mathbb{Z}$ (ignoring the ground fields), in which case we say that $f$ is a morphism of schemes. For example, if $X$ and $Y$ are $K$-varieties then a morphism of varieties $X \to Y$ is a $K$-linear morphism of schemes whereas a morphism of schemes $X \to Y$ is a morphism of schemes, without reference to the ground field.

### 2.5.2  Projective spaces

Let $k$ be a field and $V$ a finite-dimensional $k$-vector space.

Following the convention established by Grothendieck, in algebraic geometry the projective space associated to $V$—denoted by $\mathbb{P}(V)$—is a $k$-variety whose $k$-points correspond to 1-dimensional **quotients** of $V$. It has a Zariski topology and a sheaf of functions. We write simply $\mathbb{P}_k^n$ to denote $n$-dimensional *projective space* as a variety over $k$.

In classical projective geometry, the projective $k$-space associated to $V$ is usually the set of 1-dimensional subspaces of $V$, together with its linear subspaces. We use $\mathbf{P}(V)$ to denote the set of 1-dimensional **subspaces** of $V$, together with its set of lines. We call it *discrete projective space* over $k$ if we aim to emphasize the distinction. As sets, $\mathbb{P}(V^{\vee})(k)$ and $\mathbf{P}(V)$ are naturally in bijection. For $V = k^{n+1}$ we write $\mathbf{P}_k^n$. Earlier we also denoted this by $\mathrm{Proj}_k^n$, in order to clarify the exposition.

It is unfortunate that the natural interpretations of projective space, as classifying either quotients or subspaces, differ depending on the point of view. In algebraic geometry the perspective of quotients is more natural, whereas in classical projective geometry or other parts of geometry subspaces are preferable. We opt for the above conventions, which reflect the natural perspective in different contexts, rather than choosing a particular option, which would make certain sections of the book notationally difficult.

At first sight, in going from $\mathbb{P}(V^{\vee})$ to $\mathbf{P}(V)$, we lose a lot of information. The structure sheaf is gone and we keep only the lines from the large collection of subvarieties. Nonetheless, (1.1.13) says that we can recover $\mathbb{P}(V^{\vee})$ from $\mathbf{P}(V)$.

We will also have occasion to consider the set of lines in given projective space $\mathbf{P}(V)$. We will denote this set by $\mathrm{Gr}(1, \mathbf{P}(V))$. This is the $k$-points of a scheme, the Grassmanian of lines in $\mathbb{P}(V^{\vee})$, which we will denote by $\mathbb{G}\mathrm{r}(1, \mathbb{P}(V^{\vee}))$. This scheme represents the functor classifying rank 2 quotients of $V^{\vee}$.

The Zariski topologies on $\mathbb{P}(V^{\vee})$ and $\mathbb{G}\mathrm{r}(1, \mathbf{P}(V^{\vee}))$ induce topologies on the sets $\mathbf{P}(V)$ and $\mathrm{Gr}(1, \mathbf{P}(V))$, which we will occasionally employ. The reader should note, however, that these topologies are not part of the intrinsic structure of these sets.

### 2.5.3  Linear systems

The literature is very inconsistent about linear systems, both conceptually and notationally. Some of the differences are quite important for us, so we set down our conventions.

Let $X$ be a normal, geometrically integral variety over a field $k$, and let $\mathscr{L}$ be a rank 1, reflexive sheaf on $X$. Assume that $X$ is proper, or at least $H^0(X, \mathcal{O}_X) = k$.

**2.5.1** (Complete linear systems). The *complete linear system* associated to $\mathscr{L}$ is usually denoted by $|\mathscr{L}|$. However, in the literature this notation is used for three—closely related but different—objects. We distinguish between them as follows.

(1) $|\mathscr{L}|^{\mathrm{var}}$ is the $k$-variety $\mathbb{P}_k\big(H^0(X, \mathscr{L})^\vee\big)$.

(2) $|\mathscr{L}|$ is the set of $k$-points of $|\mathscr{L}|^{\mathrm{var}}$, viewed as a discrete projective $k$-space in the sense of classical projective geometry. That is, $|\mathscr{L}|$ is a set together with the additional data of the set of lines.

(3) $|\mathscr{L}|^{\mathrm{set}}$ is the set of $k$-points of $|\mathscr{L}|^{\mathrm{var}}$, with no additional structure.

We can also define $|\mathscr{L}|^{\mathrm{set}}$ as

(3') nonzero sections of $\mathscr{L}$ modulo $k^\times$-scalars, or

(3'') the set of effective Weil divisors $D$ on $X$ such that $\mathscr{L} \cong \mathcal{O}_X(D)$.

Note that (3'') is sometimes given as the definition, but in the literature this almost always means (2). That is, the discrete projective $k$-space structure is tacitly understood.

There is usually very little danger of confusion if one considers (1) and (2) the 'same' and authors usually switch between them without mention. We will follow this practice and use $|\mathscr{L}|$ to denote either of these versions, unless the distinction is important.

In many cases the context dictates which variant one means. For example, when one says that the linear system defines a rational map $X \dashrightarrow |\mathscr{L}|^\vee$, then $|\mathscr{L}|^\vee$ must denote the dual of $|\mathscr{L}|^{\mathrm{var}}$ as a $k$-variety.

However, the main theme of our treatment is that if we are given the set $|\mathscr{L}|^{\mathrm{set}}$ for every $\mathscr{L}$, then we can recover their projective $k$-space structures. So using a different notation for the underlying set is very important for us.

If $D$ is a Weil divisor, it is standard to write $|D|$ for $|\mathcal{O}_X(D)|$; we also use $|D|^{\mathrm{set}}$ for $|\mathcal{O}_X(D)|^{\mathrm{set}}$. Unfortunately, the notation $|D|$ is also used for incomplete linear systems; we discuss this issue next.

**2.5.2** (Linear systems). Incomplete linear systems are linear subspaces of complete linear systems. Again the variants (2.5.1 (1–3)) are used usually interchangeably.

Here the classical notation is usually $|D|$; the reader is expected to figure out whether this means the complete linear system or not. The book [Har77, Sec.II.7] uses $\mathfrak{d}$ to denote not necessarily complete linear systems, but this is not in widespread use, and not easily adaptable for linear systems like $|A|$ or $|B|$.

We believe that the following conventions are used in most of the literature.

(1) $|\mathcal{L}|$ denotes the complete linear system for a rank 1, reflexive sheaf $\mathcal{L}$.

(2) *Pencils,* that is, linear systems of dimension 1, are not assumed complete.

(3) If one gives first a divisor $D$, then after that $|D|$ is supposed to be complete. This applies especially to statements like 'Let $D$ be a ... divisor, then $|mD|$ is ... for $m \gg 1$.'

(4) If one starts with a linear system $|D|$ (where $D$ was not previously named), it is allowed to be incomplete.

In questionable cases we will try to clarify whether we use complete or not necessarily complete linear systems, but decades of bad habits are hard to break.

**Warning 2.5.3.** Let $X$ be a proper, normal, irreducible variety over a field $k$ such that $K = H^0(X, \mathcal{O}_X) \neq k$. In this case the sheaf-theoretic and the divisor-theoretic definitions of linear systems are in serious conflict.

For a line bundle $\mathcal{L}$, the natural definition is

$$|\mathcal{L}| := \big( H^0(X, \mathcal{L}) \setminus \{0\} \big) / k^{\times}.$$

However, two sections determine the same divisor $D$ if and only if they differ by multiplication by $K^{\times}$. So the natural thing is to set

$$|D| := \big( H^0(X, \mathcal{O}_X(D)) \setminus \{0\} \big) / K^{\times}.$$

Thus $|D| \neq |\mathcal{O}_X(D)|$. They are over different fields and in fact

$$\dim |\mathcal{O}_X(D)| = \deg(K/k) \cdot \dim |D|.$$

Our main results are about geometrically integral varieties, so we do not need to worry about this discrepancy. However, we do restrict divisors to subvarieties that need not be geometrically integral. Then we naturally end up in $|D|$, not in $|\mathcal{O}_X(D)|$, and care must be taken with regard to this distinction.

# *Chapter Three*

---

## The fundamental theorem of projective geometry

In this chapter, we prove two strengthenings of the classical Fundamental Theorem of Projective Geometry, which roughly states that a discrete projective
space $\mathbf{P}(V)$ is uniquely determined by the incidence relations among points and
lines. In particular, one knows that a bijection of projective spaces that preserves
lines must itself be a semilinear map relative to an isomorphism of base fields.

Our results focus on a weakening of the assumption that we know the full
collinearity relation. In particular, we study linearization of maps between point
sets underlying projective spaces if one only knows that *most* lines are sent to
lines. For infinite fields, 'most' means a Zariski open set of the Grassmannian,
while for finite fields it means a sufficiently large fraction of the lines.

Here is the key example that will be important in Chapter 4. If $X$ is a
projective variety over a field $k$ of dimension at least 2 and $\mathscr{L}$ is an ample
invertible sheaf, then we can characterize certain lines in $|\mathscr{L}|$ purely from the
topological space $|X|$. Indeed, given a subset $Z \subset |X|$, one can consider members
$D \in |\mathscr{L}|$ that contain $Z$. The Bertini theorems show that a general pencil in $|\mathscr{L}|$
is determined in this way if we let $Z$ be its reduced base locus. On the other
hand, it is also easy to show that there are pencils in basepoint-free ample linear
systems that are *not* determined by their base loci. Thus, one only gets 'most'
lines in this way.

The main results of this chapter are (3.1.5) (and its variant (3.2.1)) for infinite
projective spaces, and (3.3.1) for finite projective spaces. The key to the proof of
the classical fundamental theorem of projective geometry is the reconstruction
of the base field of the projective space by describing the arithmetic operations
on a line with three distinguished points (assigned to play the roles of 0, 1,
and $\infty$) using the geometry of various configurations of lines. Our proof here
is similar, but the fact that we only know a general line means that we must
combine the classical approach with various general position arguments.

## 3.1   THE FUNDAMENTAL THEOREM OF DEFINABLE
## PROJECTIVE GEOMETRY

Here we discuss a variant of the Fundamental Theorem of Projective Geometry, in which one only knows distinguished subsets of 'definable' lines in the
projective structures and one still wishes to produce a semilinear isomorphism

between the underlying vector spaces that induces the isomorphism on a dense open subset. In Section 4.3 and Chapter 5 we explain how to use this theory to reconstruct varieties.

Let $\mathbf{P}(V)$ be the abstract projective space arising from a vector space $V$ and let $\mathrm{Gr}(1, \mathbf{P}(V))$ be the set of lines as in Section 2.5.2. Both $\mathbf{P}(V)$ and $\mathrm{Gr}(1, \mathbf{P}(V))$ may be endowed with Zariski topologies, in the classical sense of the Zariski topology on the $k$-points of a variety.

**Definition 3.1.1.** A *definable projective space* is a triple $(k, V, U)$ consisting of an infinite field $k$, a $k$-vector space $V$, and a subset $U \subset \mathrm{Gr}(1, \mathbf{P}(V))$ that contains a dense Zariski open subset of the space $\mathrm{Gr}(1, \mathbf{P}(V))$ of lines in the projective space $\mathbf{P}(V)$. The *dimension* of $(k, V, U)$ is defined to be

$$\dim(k, V, U) := \dim_k V - 1.$$

In other words, a definable projective space is a projective space together with a collection of lines that are declared 'definable' subject to some conditions.

**Definition 3.1.2.** Let $k$ be a field and $V$ a $k$-vector space. The *sweep* of a subset $U \subset \mathrm{Gr}(1, \mathbf{P}(V))$, denoted by $S_U(\mathbf{P}(V))$, is the set of $p \in \mathbf{P}(V)$ that lie on some line parametrized by $U$.

**3.1.3.** Let $(k, V, U)$ be a definable projective space. Then there exists a maximal subset $U^\circ \subset U$ that is a Zariski open subset of $\mathrm{Gr}(1, \mathbf{P}(V))$. Furthermore, $(k, V, U^\circ)$ is again a definable projective space. This is immediate from the definition.

**Example 3.1.4.** Fix a projective $k$-variety $(X, \mathcal{O}_X(1))$ of dimension $d$ at least 2. Given a closed subset $Z \subset X$, we can associate the subspace $V(Z) \subset |\mathcal{O}(1)|$ of divisors that contain $Z$. The lines of the form $V(Z)$ give a subset of $\mathrm{Gr}(1, |\mathcal{O}(1)|)$ (see Section 4.3). These are the definable lines we will consider.

The main goal of this section is to prove the following result.

**Theorem 3.1.5.** *Suppose $(k_1, V_1, U_1)$ and $(k_2, V_2, U_2)$ are finite-dimensional definable projective spaces of dimension at least 2. Given an injection*

$$\varphi \colon \mathbf{P}(V_1) \to \mathbf{P}(V_2)$$

*that induces an inclusion*

$$\lambda \colon U_1 \to U_2,$$

*there is an isomorphism $\sigma \colon k_1 \to k_2$ and a $\sigma$-linear injective map of vector spaces $\psi \colon V_1 \to V_2$ such that $\mathbf{P}(\psi)$ agrees with $\varphi$ on a Zariski-dense open subset of $\mathbf{P}(V_1)$ containing the sweep of $(k_1, V_1, U_1^\circ)$.*

**Remark 3.1.6.** In (3.1.5) we can without loss of generality assume that

$$U_2 = \mathrm{Gr}(1, \mathbf{P}(V_2)).$$

However, we prefer to formulate it as above to make it a statement about definable projective spaces.

**Remark 3.1.7.** If either the dimensions of $V_1$ and $V_2$ are equal or we assume that $\lambda(U_1^\circ) \subset \mathrm{Gr}(1, \mathbf{P}(V_2))$ is dense, then the map $\psi$ is an isomorphism. In the case when the dimensions are equal this is immediate, and in the second case observe that if $V_1 \otimes_{k_1,\sigma} k_2 \subsetneq V_2$ is a proper subspace then there exists a dense open subset $W \subset \mathrm{Gr}(1, \mathbf{P}(V_2))$ of lines that are not in the image of $\mathbf{P}(\psi)$, contradicting our assumption that $\mathbf{P}(\psi)(U_1^\circ) = \lambda(U_1^\circ)$ is dense.

**Remark 3.1.8.** Observe that two lines in a projective space are coplanar if and only if they intersect in a unique point. This enables us to describe the map $\mathbf{P}(\psi)$ as follows. Let $U' \subset U_1$ be any dense Zariski open subset of $\mathrm{Gr}(1, \mathbf{P}(V_1))$, and let $P \in \mathbf{P}(V_1)$ be a point. Choose any line $\ell \subset \mathbf{P}(V_1)$ corresponding to a point of $U'$ and not containing $P$ (this is possible since $U'$ is an open subset of $\mathrm{Gr}(1, \mathbf{P}(V_1))$), and let $Q, R \in \ell$ be two distinct points. Let $L_{P,Q}$ (resp. $L_{P,R}$) be the line through $P$ and $Q$ (resp. $P$ and $R$), and choose points $S \in L_{P,Q} - \{P, Q\}$ and $T \in L_{P,R} - \{P, R\}$ such that the line $L_{S,T}$ through $S$ and $T$ is also given by a point of $U'$ (it is possible to choose such $S$ and $T$ since $U'$ is an open set). The lines $L_{S,T}$ and $L_{Q,R} = \ell$ are then coplanar and therefore intersect in a unique point $E$. It follows that $\varphi(L_{S,T})$ and $\varphi(L_{Q,R})$, which are lines since $L_{S,T}$ and $L_{Q,R}$ are definable, are coplanar since they intersect in $\varphi(E)$. It follows that the lines in $\mathbf{P}(V_2)$ given by $L_{\varphi(Q),\varphi(T)}$ and $L_{\varphi(S),\varphi(R)}$ are coplanar and consequently intersect in a unique point, which is $\mathbf{P}(\psi)(P)$. This argument is summarized in the following diagram:

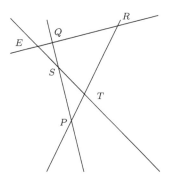

Figure 3.1

This description will play an important role in Section 3.3 below.

*Proof of (3.1.5).* This proof is very similar to the classical proof, as described (for example) in [Jac85, Section 8.4].

We may without loss of generality assume that $U_1 = U_1^\circ$.

Let us begin by showing the existence of the isomorphism of fields $\sigma\colon k_1 \to k_2$. The construction will be in several steps.

First we set up some basic notation. Let $V$ be a vector space over a field $k$. For a nonzero element $v \in V$ let $[v] \in \mathbf{P}(V)$ denote the point given by the line spanned by $v$. For $P \in \mathbf{P}(V)$ write $\ell_P \subset V$ for the line corresponding to $P$, and for two distinct points $P, Q \in \mathbf{P}(V)$ write $L_{P,Q} \subset \mathbf{P}(V)$ for the projective line connecting $P$ and $Q$. If $P = [v]$ and $Q = [w]$ then $L_{P,Q}$ corresponds to the 2-dimensional subspace of $V$ given by

$$\mathrm{Span}(v, w) := \{av + bw \,|\, a, b \in k\}.$$

If $L \subset \mathbf{P}(V)$ is a line and $P, Q, R \in L$ are three pairwise distinct points then there is a unique $k$-linear isomorphism $L \xrightarrow{\sim} \mathbb{P}^1(k)$ sending $P$ to 0, $Q$ to 1, and $R$ to $\infty$. For a collection of data $(L, \{P, Q, R\})$ we therefore have a canonical identification

$$\epsilon^{P,Q,R}\colon k \xrightarrow{\sim} L - \{R\}.$$

In the case when $L = L_{[v],[w]}$ for two non-collinear vectors $v, w \in V - \{0\}$, we take $P = [v]$, $Q = [v+w]$, and $R = [w]$. Then the identification of $k$ with $L - \{R\}$ is given by

$$a \mapsto v + aw.$$

Suppose we have $(L, \{P, Q, R\})$ as above, and fix a basis vector $v_P \in \ell_P$. Then one sees that there exists a unique basis vector $v_R \in \ell_R$ such that $[v_P + v_R] = Q$. This observation enables us to relate the maps $\epsilon^{P,Q,R}$ for different lines as follows.

Consider a second line $L'$ passing through $P$ and equipped with two additional points $\{S, T\}$, and let $a, b \in k - \{0\}$ be two scalars. We can then consider the two lines

$$L_{T,R}, \quad L_{\epsilon^{P,Q,R}(a),\epsilon^{P,S,T}(b)},$$

which will intersect in some point

$$\{O\} = L_{T,R} \cap L_{\epsilon^{P,Q,R}(a),\epsilon^{P,S,T}(b)}.$$

The situation is summarized in the following picture, where to ease notation we write simply $a$ (resp. $b$) for $\epsilon^{P,Q,R}(a)$ (resp. $\epsilon^{P,S,T}(b)$):

If we fix a basis element $v_P \in \ell_P$, we get by the above observation a basis vector $v_Q$ (resp. $v_R$, $v_S$, $v_T$) for $\ell_Q$ (resp. $\ell_R$, $\ell_S$, $\ell_T$), which in turn gives an identification

$$\epsilon^{[v_T],[v_T+v_R],[v_R]}\colon k \xrightarrow{\sim} L_{T,R} - \{R\}.$$

An elementary calculation then shows that

$$O = \epsilon^{[v_T],[v_T+v_R],[v_R]}(-a/b).$$

In particular, if $a = b$ then the point $O$ is independent of the choice of $a$, and furthermore it follows from the construction that $O$ is also independent of the

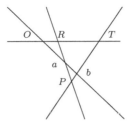

Figure 3.2

choice of the basis element $v_P$.

Consider now a definable projective space $(k, V, U)$, and let $L_0 \subset \mathbf{P}(V)$ be a definable line with three points $P, Q, R \in L_0$.

Let $M_P$ denote the variety classifying data $(L, \{S, T\})$, where $L$ is a line through $P$ and $\{S, T\}$ is a set of two additional points on $L$ such that $P$, $S$, and $T$ are all distinct. The variety $M_P$ has the following description. The point $P$ corresponds to a line $\ell_P \subset V$ and the set of lines passing through $P$ is given by $\mathbb{P}((V/\ell_P)^\vee)$ (recall our conventions about projective spaces Section 2.5.2). If $\mathscr{L} \to \mathbb{P}((V/\ell_P)^\vee)$ denotes the universal line in $\mathbb{P}(V^\vee)$ passing through $P$, then there is an open immersion

$$M_P \subset \mathscr{L} \times_{\mathbb{P}((V/\ell_P)^\vee)} \mathscr{L},$$

whence $M_P$ is smooth, geometrically connected, and rational. Since $k$ is infinite it follows that the $k$-points of $M_P$ are dense.

**Lemma 3.1.9.** *Fix $(k, V, U)$ and $(L_0, P, Q, R)$ as above and let $a \in k$ be an element. Then there exists a nonempty open subset $U_{P,a} \subset M_P$ such that if $(L, \{S, T\})$ is a twice-pointed line through $P$ corresponding to a point of $U_{P,a}$, then the lines*

$$L = L_{P,T}, \quad L_{T,R}, \quad L_{\epsilon^{P,Q,R}(a), \epsilon^{P,S,T}(a)} \tag{3.1.9.1}$$

*are all definable.*

*Proof.* We may without loss of generality assume that $U = U^\circ$.

Let $Q_0 \in M_P$ denote the point corresponding to $(L_0, \{Q, R\})$. The procedure of assigning one of the lines in (3.1.9.1) to a pointed line $(L, \{S, T\})$ is a map

$$q \colon M_P \to \mathbb{G}\mathrm{r}(1, \mathbf{P}(V)).$$

Note that the image of this map contains the point corresponding to the line $L_0$, and therefore the inverse image $q^{-1}(U)$ is nonempty. Since $M_P$ is integral it follows that the intersection of the preimages of $U$ under the three maps defined by (3.1.9.1) is nonempty.                                                                  □

A variant of the above lemma is the following, which we will use below.

**Lemma 3.1.10.** *With notation as in (3.1.9), let* $P, Q \in \mathbf{P}(V)$ *be two points in the sweep of* $U^\circ$. *Then there exists a definable line* $L_P$ *through* $P$ *and a definable line* $L_Q$ *through* $Q$ *such that* $L_P$ *and* $L_Q$ *intersect at a point* $R$.

*Proof.* Let $N_P \subset \mathbf{G}_1(1, \mathbf{P}(V))$ denote the set of lines through $P$, so $N_P \simeq \mathbf{P}(V/\ell_P)$ for the line $\ell_P \subset V$ corresponding to $P$. Let $\mathscr{L} \to N_P$ denote the universal line through $P$, and let $s\colon N_P \to \mathscr{L}$ denote the tautological section. Then the natural map

$$\mathscr{L} - \{s(N_P)\} \to \mathbf{P}(V) - \{P\}$$

is an isomorphism, since any two distinct points lie on a unique line. The set of points of $\mathbf{P}(V) - \{P\}$ that can be connected to $P$ by a line given by a point of $U^\circ$ is under this isomorphism identified with the preimage of $U^\circ \cap N_P$. In particular, this set is nonempty and open. It follows that the set of points of $\mathbf{P}(V)$ that can be connected to both $P$ and $Q$ by lines given by points of $U^\circ$ is the intersection of two dense open subsets, and therefore is nonempty.                □

With these preparations we can now proceed with the proof of (3.1.5). With the notation of that theorem, let us first define the map $\sigma\colon k_1 \to k_2$. Choose a definable line $L_0 \subset \mathbf{P}(V_1)$ together with three points $P, Q, R \in L_0$ such that $\varphi(L_0) \subset \mathbf{P}(V_2)$ is also a definable line. We then get a map

$$k_1 \xrightarrow{\epsilon^{P,Q,R}} L_0 - \{R\} \xrightarrow{\varphi} \varphi(L_0) - \{\varphi(R)\} \xrightarrow{\left(\epsilon^{\varphi(P),\varphi(Q),\varphi(R)}\right)^{-1}} k_2,$$

which we temporarily denote by $\sigma^{(L_0,\{P,Q,R\})}$.

**Claim 3.1.11.** *The map* $\sigma^{(L_0,\{P,Q,R\})}$ *is independent of* $(L_0, \{P, Q, R\})$.

*Proof.* Let $(L_0', \{P', Q', R'\})$ be a second definable line with three points. Given $a \in k_1$, we will show that

$$\sigma^{(L_0,\{P,Q,R\})}(a) = \sigma^{(L_0',\{P',Q',R'\})}(a).$$

From the definition, we see that this holds for $a = 0$ and $a = 1$, so we assume that $a \neq 0$ in what follows. First consider the case when $P = P'$. By (3.1.9) we can find a line $L$ with two points $\{S, T\}$ such that the lines (3.1.9.1) are all definable, as well as the lines (3.1.9.1) obtained by replacing $(L_0, \{P, Q, R\})$ with $(L_0', \{P, Q', R'\})$

The picture in Figure 3.2 is taken by $\varphi$ to the corresponding picture in $\mathbf{P}(V_2)$. Looking at the intersection point it follows that

$$\sigma^{(L_0,\{P,Q,R\})}(a) = \sigma^{(L,\{P,S,T\})}(a) = \sigma^{(L_0',\{P,Q',R'\})}(a).$$

It follows, in particular, that the map $\sigma^{(L_0,\{P,Q,R\})}$ is independent of the points $Q$ and $R$. Since $\sigma^{(L_0,\{R,Q,P\})}$ is given by the formula

$$\iota_{k_2} \circ \sigma^{(L_0,\{P,Q,R\})} \circ \iota_{k_1},$$

where $\iota_{k_j}$ denotes the involution of $k_j^\times$ given by $u \mapsto u^{-1}$, it follows that the map $\sigma^{(L_0,\{P,Q,R\})}$ is independent of the triple $\{P,Q,R\}$, so we get a well-defined map $\sigma^{L_0}\colon k_1 \to k_2$. Now, for a second definable line $L_0'$ which has nonempty intersection with $L_0$, the point $P:=L_0 \cap L_0'$ is on both lines, so we can apply the preceding discussion with the two lines $L_0$ and $L_0'$ and $Q, R$ and $Q', R'$ chosen arbitrarily to deduce the independence of the choice of $(L_0, \{P,Q,R\})$. Finally, for an arbitrary definable line, we can by (3.1.10) find a chain (in fact of length two) of definable lines that connect the two, which concludes the proof.    □

Let us write the map of (3.1.11) as $\sigma\colon k_1 \to k_2$.

**Claim 3.1.12.** *The map $\sigma$ is an isomorphism of fields.*

*Proof.* First, note that by construction the map $\sigma$ sends 1 to 1 and is compatible with the inversion map $a \mapsto a^{-1}$. Indeed, the statement that $\sigma(1) = 1$ is immediate from the construction and the compatibility with the inversion map can be seen as follows. Let $\iota_j\colon k_j^\times \to k_j^\times$ ($k = 1,2$) denote the map $a \mapsto a^{-1}$, and let $(L, \{P,Q,R\})$ be a definable line with three marked points. Write $L^\times$ (resp. $\varphi(L)^\times$) for $L - \{P,R\}$ (resp. $\varphi(L) - \{\varphi(P),\varphi(R)\}$). Then by the independence of the choice of marked line in the definition of $\sigma$, we have that the diagram

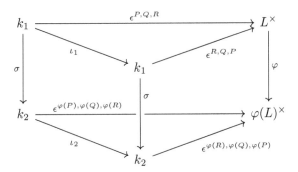

commutes. The compatibility with the multiplicative structure again follows from contemplating Figure 3.2, and the observation that by construction the map $\sigma$ takes 1 to 1. Indeed, given $a, b \in k_1^\times$ such that all the lines in Figure 3.2 are definable, we must have

$$\sigma(-a/b) = -\sigma(a)/\sigma(b) \qquad\qquad (3.1.12.1)$$

since this fraction is given by the point $O$. Since the condition of being definable is open (by our initial reduction to $U_1 = U_1^\circ$), the fact that for any definable $(L, \{P,Q,R\})$ the line through $\epsilon^{P,Q,R}(a)$ and $\epsilon^{P,Q,R}(b)$ is definable implies that the same is true after deforming $(L, \{P, Q, R\})$. Thus we get the formula (3.1.12.1) for all $a$ and $b$. In particular, taking $b = 1$ we get that $\sigma(-a) = -\sigma(a)$ for all $a$, and since $\sigma$ is compatible with the inversion maps, we get that

$$\sigma(ab) = \sigma(a)\sigma(b)$$

for all $a, b \subset k^\times$. Since 0 is also taken to 0 by $\sigma$ we in fact get this formula for all $a, b \in k$.

For the verification of the compatibility with additive structure, consider a marked line $(L, \{P, Q, R\})$. Let $S$ be a point not on the line and let $T$ be a third point on $L_{S,R}$. The lines $L_{P,T}$ and $L_{Q,S}$ intersect at a point we call $V$, and then the line $L_{V,R}$ intersects $L_{P,S}$ at a point we call $W$. This is summarized in the following picture, where we write simply $a$ (resp. $b$) for $\epsilon^{P,Q,R}(a)$ (resp. $\epsilon^{S,T,R}(b)$).

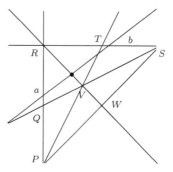

Figure 3.3

A straightforward calculation done by choosing a basis $v_R \in \ell_R$ then shows that the point of intersection marked with the bullet is the point

$$\epsilon^{W,V,R}(a + b).$$

To prove that $\sigma$ is compatible with the additive structure it suffices to show the following lemma, which concludes the proof. □

**Lemma 3.1.13.** *For any $a, b \in k$ there exists a pointed line $(L, \{P, Q, R\})$ and points $S$ and $T$ such that all the lines in Figure 3.3 are definable.*

*Proof.* The collections of data

$$(L, \{P, Q, R\}, \{S, T\}) \tag{3.1.13.1}$$

defining a diagram as in Figure 3.3 are classified by an irreducible scheme $M$; each line in the diagram gives a morphism

$$t \colon M \to \mathbb{Gr}(1, \mathbf{P}(V_1)).$$

It therefore suffices to show that for any particular choice of line in Figure 3.3, there exists a choice of (3.1.13.1) for which that line is definable. Indeed, then the set of choices of data (3.1.13.1) for which that line is definable is nonempty and open in $M$. Since the $M$ is irreducible the intersection of nonempty open sets is nonempty and we conclude that there exists a point for which all the lines in Figure 3.3 are definable.

For the line through $R$, $V$, and $W$ this follows from noting that the data of the collinear points $S$ and $T$ is equivalent to the data of the points $\{V, W\}$. Indeed given these two collinear points, the lines $\overline{QV}$ and $\overline{PQ}$ are coplanar and therefore intersect in a unique point, which defines $S$, and the intersection of $\overline{SR}$ and $\overline{PV}$ then defines $T$. Therefore the map $t$ is smooth and dominant in this case, so the preimage of $U_1$ is nonempty.

For the other lines in Figure 3.3, note that we can extend the map $t$ to the bigger (but still irreducible) scheme $\overline{M}$ classifying collections of data of the form

$$(L, \{P, Q, R\}, \{S, T\}),$$

where as before $L$ is a point, $\{P, Q, R\}$ are three points on $L$, and $\{S, T\}$ are two additional points that are collinear with $R$, but where we no longer insist that the line through $T$ and $S$ be distinct from $L$, but only that the points $\{P, Q, R, S, T, a, b\}$ be distinct. Now it is clear that the preimage in $\overline{M}$ of $U_1$ is nonempty since we can take all the points to lie on the same definable line $L$.                                                                                  $\square$

Now that we have constructed the isomorphism $\sigma$, it remains to construct the map $\psi \colon V_1 \to V_2$.

First note that we can choose a basis $e_1, \ldots e_n$ for $V_1$ with the property that the span of $e_i$ and $e_j$ is a definable line for any $i \neq j$. Define $e'_1, \ldots, e'_n \in V_2$ as follows. For $e'_1$ we take any basis element in $\ell_{\varphi([e_1])}$. Now for each $e_i$, $i \geq 2$, the line in $\mathbf{P}(V_1)$ associated to the plane $\mathrm{Span}(e_1, e_i)$ is definable, and therefore the image under $\varphi$ is a definable line and contains the points $\varphi([e_1])$, $\varphi([e_i])$, and $\varphi([e_1 + e_i])$. The choice of the representative $e'_1$ for $\varphi([e_1])$ defines a representative $e'_i$ for $\varphi([e_i])$ such that $\varphi([e_1 + e_i]) = e'_1 + e'_i$. Consider the map

$$\gamma \colon V_1 \to V_2$$

defined by

$$\gamma(a_1 e_1 + \cdots + a_n e_n) := \sigma(a_1) e'_1 + \cdots + \sigma(a_n) e'_n.$$

**Claim 3.1.14.** *For general $(a_1, \ldots, a_n)$ we have*

$$\varphi([a_1 e_1 + \cdots + a_n e_n]) = [\gamma(a_1 e_1 + \cdots + a_n e_n)].$$

*Proof.* By the construction of $\sigma$, if for each $2 \leq i \leq n$ the vectors

$$a_1 e_1 + \cdots + a_{i-1} e_{i-1}, \quad a_i e_i \qquad\qquad (3.1.14.1)$$

span a definable line, then we get by induction on $i$ that

$$\varphi([a_1 e_1 + \cdots + a_i e_i]) = [\gamma(a_1 e_1 + \cdots + a_i e_i)].$$

Let $\mathbb{A} \subset \mathbb{A}_{k_1}^n$ be the open subscheme classifying tuples $(a_1, \ldots, a_n)$ for which all the pairs (3.1.14.1) span a line, and let $A$ denote the $k_1$-points of $\mathbb{A}$. For each $i$ the map sending a vector $(a_1, \ldots, a_n)$ to the span of the elements (3.1.14.1) defines a morphism of schemes $\mathbb{A} \to \mathbb{Gr}(1, \mathbf{P}(V_1))$ whose image meets $U_1$. Taking the common intersection of the preimages of $U_1$ under these maps, we get a nonempty open subset $A^\circ \subset A$ of tuples $(a_1, \ldots, a_n) \in A(k_1)$ for which the vectors (3.1.14.1) span a definable line.    □

As a consequence, the map $\gamma$ defined above is uniquely associated to $\varphi$, up to scalar, and is thus independent of the general choice of basis $e_1, \ldots, e_n$.

To complete the proof of (3.1.5) it suffices to show that $\mathbf{P}(\gamma)$ agrees with $\varphi$ on the entire sweep of $(k_1, V_1, U_1)$. By the above remark, to show this for a particular point $p$, it suffices to work with any general basis. To prove this we show that given a point $p \in S_{U_1}(\mathbf{P}_{k_1}(V_1))$ there exists a basis $e_1, \ldots, e_n$ for $V_1$ as above for which $p$ lies in the resulting subset $A^\circ$. Reviewing the above construction, one sees that it suffices to show that we can find a basis $e_1, \ldots, e_n$ for $V_1$ such that the following hold:

(1) $p$ is the point corresponding to the line spanned by $e_1$.

(2) Any two elements $e_i$ and $e_j$, with $i \neq j$, span a definable line.

(3) For any $2 \leq i \leq n$ the vectors $e_1 + \cdots + e_{i-1}, \quad e_i$ span a definable line.

For this, start by choosing $e_1$ so that (1) holds. Since $p$ lies in the sweep, we can then find $e_2$ such that $e_1$ and $e_2$ span a definable line. Now observe that given $2 \leq r \leq n$ and a basis $e_1, \ldots, e_r$ satisfying (2) and (3) with $i, j \leq r$, we can find $e_{r+1}$ such that (2) and (3) hold with $i, j \leq r + 1$. Indeed a general choice of vector in $V_1$ will do for $e_{r+1}$ since for given fixed vector $v_0$ lying in the sweep there is a nonempty Zariski open subset of vectors $w$ such that $w$ and $v_0$ span a definable line.

This completes the proof of the Theorem.    □

## 3.2    A VARIANT FUNDAMENTAL THEOREM

Suppose $(k_1, V_1, U_2)$ and $(k_2, V_2, U_2)$ are finite-dimensional definable projective spaces. Write $P_i = \mathbf{P}_{k_i}(V_i)$ for the associated projective space for $i = 1, 2$.

In this section we prove the following result, weakening the assumptions of (3.1.5). This is included primarily for technical reasons related to Section 4.3; a reader interested in working only over algebraically closed fields can ignore this section on a first reading.

**Theorem 3.2.1.** *Assume $P_1$ and $P_2$ have dimension at least 2. Suppose $\sigma: P_1 \to P_2$ is a bijection such that each line in $U_1$ is sent under $\sigma$ to a linear subspace of $P_2$ and each line in $U_2$ is sent under $\sigma^{-1}$ to a linear subspace of $P_1$. Then $\sigma$ sends elements of $U_1$ to lines and it agrees with a linear isomorphism $P_1 \to P_2$ on the sweep of $U_1$.*

Without loss of generality, we assume that $U_i$ is the $k_i$-points of an open subset of the appropriate Grassmannian, and we make this assumption for the remainder of the proof.

The proof of (3.2.1) appears at the very end of this section, after several requisite precursors about data $(P_1, P_2, \sigma)$, as in the theorem, are developed (assuming $U_1$ and $U_2$ are Zariski open subsets).

**Remark 3.2.2.** Note that in the statement of (3.2.1), we only assume that lines are sent to *linear subspaces*, not to lines. This comes up naturally when one seeks to define lines in linear systems using incidence relations: given a subset $Z$ of a variety $X$, the members of a linear system that contain $Z$ is a linear subspace. Detecting the dimension of this linear subspace is quite subtle when the field of constants of $X$ is not algebraically closed. In particular, while it may be obvious that such a subset is a line on one side of an isomorphism, it is not generally clear that it remains a line on the other side. A trivial (and not particularly informative) example comes from the existence of an abstract bijection between a line and a projective space of arbitrary positive dimension. As we will see in the proof, one needs control over a large set of lines to avoid this situation.

**Definition 3.2.3.** A pair $(D_0, D_1) \in P_1^2$ is *good* if it lies in the inverse image of $U_1$ under the natural span map $P_1^2 \setminus \Delta \to \mathrm{Gr}(1, P_1)$.

**Definition 3.2.4.** A collection of elements

$$\mathbf{D} := (D_0, \ldots, D_s)$$

of $P_1$ is *admissible* if for any two $0 \le i, j \le s$ the pair $(D_i, D_j)$ is good and if the $D_i$ span a linear subspace of $P_1$ of dimension $s$.

We fix an admissible collection $\mathbf{D}$ in what follows.

**Definition 3.2.5.** A point $Q \in P_1$ is $\mathbf{D}$-*good* if $(D_i, Q)$ is good for all $i$.

For a pair $i, j$ let $\ell_{ij}^1 \subset P_1$ be the line spanned by $D_i$ and $D_j$ and let $\ell_{ij}^2 \subset P_2$ denote the line spanned by $\sigma(D_i)$ and $\sigma(D_j)$. Note that since $\ell_{ij}^1$ is definable, which implies that $\sigma(\ell_{ij}^1)$ is a linear subspace $T_{ij}^2 \subset P_2$ containing $\sigma(D_i)$ and

$\sigma(D_j)$, we have

$$\ell_{ij}^2 \subset T_{ij}^2.$$

**3.2.6.** For $t \leq s$ define $W_t^2 \subset P_2$ inductively as follows. For $t = 0$ we define $W_0^2 := \sigma(D_0)$. Then inductively define $W_{t+1}^2$ to be the linear span of $W_t^2$ and $\sigma(\ell_{t,t+1})$. When we want to be unambiguous, we will write $W_t^2(\mathbf{D})$ to denote the dependence upon $\mathbf{D}$. Note that it is a priori possible for $\mathbf{D}$ and $\mathbf{D}'$ to have the same span in $P_1$ while $W_s^?(\mathbf{D}) \neq W_s^?(\mathbf{D}')$.

Let $Q \in P_1$ be a $\mathbf{D}$-good point.

**Theorem 3.2.7.** *If $\sigma(Q) \in W_s^2(\mathbf{D})$ then $Q$ is in the linear span of the $D_i$.*

**3.2.8.** We first identify mutations of $\mathbf{D}$ that leave (3.2.7) invariant. In each of the following two cases, we have that if the assumptions hold for $\mathbf{D}$, then they hold after replacing $\mathbf{D}$ by $\mathbf{D}'$, and if the conclusions hold for $\mathbf{D}'$, then they hold for $\mathbf{D}$, and therefore in the proof we may replace $\mathbf{D}$ by $\mathbf{D}'$.

- Suppose $\mathbf{D}'$ is an admissible tuple gotten by replacing $D_s$ by a point $D_s' \in \ell_{s-1,s}$ such that $Q$ is $\mathbf{D}'$-good and $\sigma(D_s')$ lies in the linear span

$$\langle W_{s-1}^2(\mathbf{D}), \sigma(Q) \rangle.$$

  Then we have that $\sigma(Q) \in W_s^2(\mathbf{D}')$ and $Q$ is $\mathbf{D}'$-good. Moreover, we have that $\langle \mathbf{D} \rangle = \langle \mathbf{D}' \rangle$.
- Suppose $\mathbf{D}'$ is an admissible tuple gotten by replacing $D_{s-1}$ by a point $D_{s-1}' \in \ell_{s-1,s}$ such that $Q$ is $\mathbf{D}'$-good and $\sigma(D_{s-1}) \in \langle \sigma(D_s), \sigma(Q) \rangle$.

These mutations will arise as follows: the set of choices $D_s'$ or $D_{s-1}'$ will range through a line contained in the definable subspace $\sigma(\ell_{s-1,s})$. Since the base field is infinite, such a line is infinite, so its preimage in the line $\ell_{s-1,s}$ hits every open subset. This is main way in which we use the fact that the definable subspaces $\ell_{i-1,i}$ are lines.

**3.2.9.** We assume that $\sigma(Q) \in W_s^2$ and show that $Q$ is in the linear span of the $D_i$.

The basic idea is to work inductively by projection from $D_s$ to the lower dimensional subspace. To get things into appropriately general position, however, we will do this along with modifying our original configuration $(D_0, \ldots, D_s)$ so as to obtain a contradiction.

First of all, by our assumptions the line $\ell$ through $Q$ and $D_s$ is definable, so $\sigma(\ell) \cap W_s^2$ is a linear subspace of positive dimension.

Furthermore, proceeding by induction we may assume that the theorem holds for collections of elements $(D_0, \ldots, D_t)$ with $t < s$. Note here that the statement for $s = 0$ is trivial.

**Lemma 3.2.10.** *The following hold.*

(1) $\sigma(\ell_{s-1,s}) \cap W_{s-1}^2 = \{\sigma(D_{s-1})\}$.

(2) *The intersection of $\sigma(\ell_{s-1,s})$ with the linear span of $W_{s-1}^2$ and $\sigma(Q)$ is a positive dimensional linear subspace of $\sigma(\ell_{s-1,s})$.*

*Proof.* Note that for all but finitely many points $U \in \ell_{s-1,s}$ the collection

$$(D_0, \ldots, D_{s-1}, U)$$

is admissible, and $U$ does not lie in the linear span of $(D_0, \ldots, D_{s-1})$ (recall that $U_1$ is assumed open). By the induction hypothesis it follows that the intersection

$$\sigma(\ell_{s-1,s}) \cap W_{s-1}^2$$

is finite. Since this is also a linear space, it follows that it consists of exactly one point, namely $\sigma(D_{s-1})$. This proves (1).

For (2), let $\delta$ be the dimension of the linear space $\sigma(\ell_{s-1,s})$. Then using (1) we have

$$\dim(W_{s-1}^2) + \delta = \dim(W_s^2).$$

On the other hand, the dimension of the linear span of $\sigma(Q)$ and $W_{s-1}^2$ is equal to

$$\dim(W_{s-1}^2) + 1.$$

Therefore the intersection in question in (2) is the intersection of a space of dimension $\delta$ with a space of dimension $\dim(W_{s-1}^2)+1$ inside a space of dimension $\dim(W_{s-1}^2) + 1$. From this (2) follows. $\qquad \square$

**3.2.11.** For all but finitely many points $D_s' \in \ell_{s-1,s}$ the collection of elements

$$\mathbf{D}' := (D_0, \ldots, D_{s-1}, D_s')$$

is admissible and each of these elements is pairwise good with $Q$. We can therefore find $D_s' \in \ell_{s-1,s}$ such that $\mathbf{D}'$ is admissible, $Q$ is $\mathbf{D}'$-good, and $\sigma(D_s')$ lies in the linear span of $W_{s-1}^2(\mathbf{D}')$ and $\sigma(Q)$. Replacing $\mathbf{D}$ by such a $\mathbf{D}'$ is an allowable mutation of type I, as in (3.2.8).

**3.2.12.** Consider the projection

$$q_s : \langle \sigma(D_s), W_{s-1}^2 \rangle \dashrightarrow W_{s-1}^2$$

from the linear span of $\sigma(D_s)$ and $W_{s-1}^2$, sending an element $R$ to the intersection of the line through $\sigma(D_s)$ and $R$ with $W_{s-1}^2$. This is defined in a neighborhood of the line through $\sigma(D_{s-1})$ and $\sigma(Q)$. In particular, it is defined at $\sigma(Q)$.

Let $Q_{s-1} \in W_{s-1}^2$ denote $\sigma^{-1}(q_s(\sigma(Q)))$. (We will write $Q_{s-1}(\mathbf{D})$ when we want to remember the dependence on $\mathbf{D}$.)

**3.2.13.** If $\sigma(Q_{s-1}) \in W_{s-2}^2$ then we are done by applying our induction hypothesis to

$$(D_0, \ldots, D_{s-2}, D_s)$$

and $Q$.

So assume $\sigma(Q_{s-1})$ is not in $W^2_{s-2}$.

**3.2.14.** The space $W^2_{s-1}$ is the linear span of $W^2_{s-2}$ and $\sigma(\ell_{s-1,s-2})$. Since the space spanned by $W^2_s$ and $\sigma(Q_{s-1})$ is assumed to be strictly bigger than $W^2_{s-2}$, this space meets a line $T \subset \sigma(\ell_{s-1,s-2})$. For all but finitely many elements $D'_{s-1} \in \ell_{s-1,s-2}$ the collection

$$\mathbf{D'} := (D_0, \ldots, D_{s-2}, D'_{s-1}, D_s)$$

is again admissible. Replacing $D_{s-1}$ by a suitable element of $\sigma^{-1}(T)$, we may therefore further assume that the line through $\sigma(D_{s-1})$ and $\sigma(Q_{s-1})$ meets $W^2_{s-2}$. Call this point $R_0 \in W^2_{s-2}$. (We will write $R_0(\mathbf{D})$ when we want to remember the dependence on $\mathbf{D}$.)

Note that the construction ensures that

$$\sigma(Q_{s-1}(\mathbf{D})) = \langle \sigma(D_{s-1}), R_0(\mathbf{D}) \rangle \cap \langle \sigma(D_s), \sigma(Q) \rangle$$

for any $\mathbf{D}$ satisfying the assumptions.

**3.2.15.** We claim that for a suitably chosen element $D'_{s-1} \in \ell_{s-1,s}$ we can arrange that $Q_{s-1}$ is $\mathbf{D'}$-good.

The linear span of $\sigma(\ell_{s-1,s})$ and $R_0$ contains the line connecting $\sigma(D_s)$ and $\sigma(Q_{s-1})$. It follows that the plane spanned by $R_0$, $\sigma(Q)$, and $\sigma(Q_{s-1})$ meets $\sigma(\ell_{s-1,s})$ in a line $M$.

Observe that $R_0$ does not lie in $M$. To see this, note that if that were not the case then $R_0 \in \sigma(\ell_{s-1,s})$, which would imply $\sigma(Q_{s-1}) \in \sigma(\ell_{s-1,s})$, which would imply $\sigma(Q) \in \sigma(\ell_{s-1,s})$, so $Q$ is in $\ell_{s-1,s}$, which we are assuming is not the case.

We then have an infinitude of elements $D'_{s-1} \in \sigma^{-1}(M)$ such that the collection

$$\mathbf{D'} := (D_0, \ldots, D'_{s-1}, D_s)$$

is admissible.

Keeping $R_0$ fixed, we see that the set of points of the form

$$\sigma^{-1}\left(\langle \sigma(D'_{s-1}), R_0 \rangle \cap \langle \sigma(D_s), \sigma(Q) \rangle\right)$$

is an infinite subset of the definable line $\langle D_s, Q \rangle$. Since $\mathbf{D'}$ is admissible, any such infinite subset contains infinitely many points that are $\mathbf{D'}$-good, as desired.

Replacing $\mathbf{D}$ with $\mathbf{D'}$ is an allowable mutation of type II, as in (3.2.8). Therefore for suitable chosen $D'_{s-1} \in \sigma^{-1}(M)$, we get that $Q_{s-1}$ is $\mathbf{D''}$-good.

**3.2.16.** Applying our induction hypothesis we conclude that $Q_{s-1}$ is in the linear span of $D_0, \ldots, D_{s-1}$. Since $\sigma(Q)$ lies in the line $\langle \sigma(D_s), \sigma(Q_{s-1}) \rangle$, we have that $\sigma(Q) \in \sigma(\langle D_s, Q_{s-1} \rangle)$, and thus $Q$ lies in the definable line $\langle D_s, Q_{s-1} \rangle$. This completes the proof of (3.2.7). $\qquad \square$

**Corollary 3.2.17.** *Let* $(D_0, \ldots, D_s)$ *be an admissible collection of elements of* $P_1$. *Then we have*

$$s \leq \dim W_s^2 = \sum_{i=1}^{s} \dim \sigma(\ell_{i-1,i}),$$

*with equality if and only each space* $\sigma(\ell_{i-1,i})$ *is a line.*

*Proof.* This follows immediately from (3.2.10)(1). Note that this lemma uses the induction hypothesis of the proof of (3.2.7), but with the proof of that theorem complete we can apply the lemma unconditionally. $\qquad \square$

**Corollary 3.2.18.** *The dimensions of* $P_1$ *and* $P_2$ *are equal and for a good pair of points* $(E, F)$ *in* $P_1$ *the line through* $E$ *and* $F$ *is sent under* $\sigma$ *to a line in* $P_2$.

*Proof.* By the preceding corollary we see that the dimension of $P_2$ is at least that of the dimension of $P_1$, and by consideration of $\sigma^{-1}$ we see that they must be equal.

With the equality of dimensions established, note that if $(E, F)$ is a good pair that spans a line $\ell$ such that the dimension of $\sigma(\ell)$ is $> 1$, then we can extend $(E, F)$ to an admissible collection

$$\mathbf{D} = (D_0, \ldots, D_{\dim(P_1)})$$

with

$$(D_0, D_1) = (E, F),$$

and (3.2.17) gives

$$\dim W_{\dim(P_1)}^2 > \dim(P_1),$$

contradicting the equality of dimensions. $\qquad \square$

*Proof of (3.2.1).* This follows from (3.2.18) and (3.1.5). $\qquad \square$

## 3.3   THE PROBABILISTIC FUNDAMENTAL THEOREM OF PROJECTIVE GEOMETRY

In this section, we prove that knowing most lines also determines linearity of a map of *finite* projective spaces.

To state the main result consider the following functions of four variables (whose origin will be explained in the proof):

$$A(q,N,G,\epsilon):=\tag{3.3.0.1}$$
$$\frac{3(2(q-1)^2(\epsilon N(N-1)+G)+G(q-1)^3)}{(q-1)G}+q^3(q+1)$$

$$B(q,N,G,\epsilon):=\tag{3.3.0.2}$$
$$2(q-1)\frac{(q-1)\epsilon G(G-1)}{(q-1)(G-1)-A(q,N,G,\epsilon)-q(q+1)}+A(q,G,N,\epsilon)$$

The main result of this section is the following:

**Theorem 3.3.1.** Let **F** be a finite field with $q$ elements, and let $P_1$ and $P_2$ be projective spaces over **F** of dimension $n > 3$. Let $\{V_i\}_{i \in B}$ be a transverse collection of proper linear subspaces of $P_1$. Let $N$ be the number of points in $P_1$ and let $G$ be the number of points in $P_q \setminus \bigcup_{i \in B} V_i$.*

Let $f: P_1 \to P_2$ be an injection of sets. Assume we have $\epsilon > 0$ such that the number of lines $L \subset P_1, L \not\subset \bigcup_{i \in B} V_i$ for which $f(L) \subset P_2$ is not a line is at most $\epsilon G(G-1)/q(q+1)$, and assume that

$$2A(q,N,G,\epsilon)+q(q+1) < (q-1)(G-1)$$

and

$$\frac{9B(q,G,N,\epsilon)}{q^{n+1}\prod_{i \in B}(1-2q^{-\,\mathrm{codim}\,V_i})}+\frac{q^{n+4}}{q^{2n+2}\prod_{i \in B}(1-2q^{-\,\mathrm{codim}\,V_i})^2} < 1.$$

Then there is an injection $f': P_1 \to P_2$ that takes lines to lines, and such that the proportion of elements of $P_1$ on which $f$ and $f'$ agree is at least

$$1-\frac{(q-1)\epsilon G(G-1)}{N((q-1)(G-1)-A(q,N,G,\epsilon)-q(q+1))}.$$

**Remark 3.3.2.** Theorem (3.3.1) is similar in spirit to the Blum-Luby-Rubinfeld linearity test [BLR93, Lemmas 9–12], part of the theory of property testing in computer science. The arguments of that paper show that given a function $f: G \to G'$ for groups $G, G'$, if the proportion of $x, y \in G$ such that $f(x)f(y) = f(xy)$ is close enough to 1, then there exists a group homomorphism $f': G \to G'$ such that $f(x) = f'(x)$ for a proportion of $x$ close to 1. Their methods are computational and give an approach to finding $f'$. Theorem (3.3.1) solves the analogous problem where, instead of a group homomorphism, we have an injective map of projective spaces of large enough dimension. The strategy of [BLR93] relies on choosing $f'(x)$ so that $f'(x) = f(xy^{-1})f(y)$ for a proportion of $y$ close to 1, and our strategy uses a similar, but more complex, formula adapted

---

*$B$ stands for the 'bad' set and $G$ stands for the number of 'good' points.

to the case of projective spaces. Further adjustments must be made to handle the bad set $B$ of linear subspaces, which might contain a high proportion of all points and lines, where nothing is assumed—essentially, we must keep track of the condition that certain points do not lie in any of these subspaces.

Before proving (3.3.1), we prove the following lemma, which guarantees that the assumptions hold as long as $n \to \infty$ and $\epsilon \to 0$ with $q, \#B$ fixed.

**Lemma 3.3.3.** *Fix $q$ a prime power, $\#B$ a natural number, and $\delta > 0$. Assume that either $q > 2$ or $\#B = 0$.*

*Then there exists a natural number $n_0$ and $\epsilon_0 > 0$ such that, for any projective space $P_1$ of dimension $n$ over $\mathbb{F}_q$ and set $B$ of proper linear subspaces of $P_1$ with cardinality $\#B$, as long as $n \geq n_0$ and $\epsilon \leq \epsilon_0$ we have*

$$2A(q, N, G, \epsilon) + q(q+1) < (q-1)(G-1),$$

$$\frac{9B(q, G, N, \epsilon)}{q^{n+1} \prod_{i \in B}(1 - 2q^{-\operatorname{codim} V_i})} + \frac{q^{n+4}}{q^{2n+2} \prod_{i \in B}(1 - 2q^{-\operatorname{codim} V_i})^2} < 1,$$

*and*

$$1 - \frac{(q-1)\epsilon G(G-1)}{N((q-1)(G-1) - A(q, N, G, \epsilon) - q(q+1))} > 1 - \delta.$$

*Proof.* We have

$$G = \frac{1}{q-1}q^{n+1} \prod_{i \in B}(1 - q^{-\operatorname{codim} V_i})$$

if $B$ is nonempty and

$$G = \frac{1}{q-1}(q^{n+1} - 1)$$

if $N$ is empty. In either case $\frac{q^n}{G}$ is bounded by a constant depending only on $q, \#B$. The same is true for the ratios $\frac{q^n}{N}, \frac{G}{q^n}, \frac{N}{q^n}$—in fact, in these cases we can take the constant to be 2.

Thus in the expression

$$A(q, N, G, \epsilon) = \frac{3(q-1)G(2(q-1)^2(\epsilon N(N-1) + G) + G(q-1)^3)}{(q-1)^2 G^2} + q^3(q+1)$$

the denominator is at least a nonzero constant times $q^{2n}$ and the numerator is at most a constant times $\epsilon q^{3n} + q^{2n}$, so $A(q, N, G, \epsilon)$ is at most a constant times $\epsilon q^n + 1$. Thus

$$\frac{2A(q, N, G\epsilon) + q(q+1)}{(q-1)(G-1)}$$

is at most a constant times $\epsilon + q^{-n}$, thus at most a constant times $\epsilon_0 + q^{-n_0}$, so by choosing $\epsilon_0$ sufficiently small and $n_0$ sufficiently large we can ensure it is at most 1, verifying the first inequality.

In fact, we choose $\epsilon_0$ and $n_0$ slightly larger, to ensure

$$2A(q, N, G, \epsilon) + 2q(q+1) < (q-1)(G-1)$$

and so

$$B(q, N, G, \epsilon) = 2(q-1)\frac{(q-1)\epsilon G(G-1)}{(q-1)(G-1) - A(q, N, G, \epsilon) - q(q+1)} + A(q, G, N, \epsilon)$$

$$\geq 2(q-1)\frac{(q-1)\epsilon G(G-1)}{(q-1)(G-1)/2} + A(q, G, N, \epsilon),$$

so the denominator in the fraction is at least a positive constant times $q^n$ and thus the fraction is at most a constant times $\epsilon q^n$ and hence $B(q, G, N, \epsilon)$ is at most a constant times $\epsilon q^n + 1$.

Now, because either $q > 2$ or $B$ is empty, we can lower bound $\prod_{i \in B}(1 - 2q^{-\text{codim } V_i})$ by the positive constant $(1 - 2/q)^{\#B}$, and so

$$\frac{9B(q, G, N, \epsilon)}{q^{n+1} \prod_{i \in B}(1 - 2q^{-\text{codim } V_i})}$$

is at most a constant times $\epsilon + q^{-n}$ while the second term

$$\frac{q^{n+4}}{q^{2n+2} \prod_{i \in B}(1 - 2q^{-\text{codim } V_i})^2}$$

is at most a constant times $q^{-n}$, thus choosing $n_0$ sufficiently large and $\epsilon_0$ sufficiently small, we can guarantee the second inequality.

For the third inequality, we have already forced

$$(q-1)(G-1) - A(q, N, G, \epsilon) - q(q+1) \geq (q-1)(G-1)/2,$$

so the denominator

$$N((q-1)(G-1) - A(q, N, G, \epsilon) - q(q+1)) \geq N(q-1)(G-1)/2$$

is at least a positive constant times $q^{2n}$ and thus the ratio

$$\frac{(q-1)\epsilon G(G-1)}{N((q-1)(G-1) - A(q, N, G, \epsilon) - q(q+1))}$$

is at most a constant times $\epsilon$, so choosing $\epsilon_0$ sufficiently small, we can ensure it is at most $\delta$. $\qquad\square$

After building up suitable technical material (including the definition of the map $f'$), we will record the proof of (3.3.1) in (3.3.26) below.

**3.3.4.** Let $k$ be a finite field with $q$ elements. Let $P_1$ and $P_2$ be projective spaces over $k$ of dimension $n > 3$ and let

$$f : P_1 \to P_2$$

be an injection of sets. Let $\mathscr{L}_{P_i}$ be the set of lines in $P_i$. Let $\mathscr{L}_{P_1, B}$ be the set of lines in $P_1$ that are not contained in $\bigcup_{i \in B} V_i$. As in the statement of (3.3.1), let $G$ be the cardinality of $P_1 \setminus \bigcup_{i \in B} V_i$ and and let $N$ be the cardinality of $P_1$. Since each point in $P_1 \setminus \bigcup_{i \in B} V_i$ is contained in $(N-1)/q$ lines, the cardinality of $\mathscr{L}_{P_1, B}$ is at most $G(N-1)/q$ and at least $G(N-1)/(q^2 + q)$.

Let $\mathscr{L}^f_{P_1, B} \subset \mathscr{L}_{P_1, B}$ be the subset of lines $L \subset P_1$ for which $f(L)$ is a line in $P_2$. We make the following assumption.

**Assumption 3.3.5.** For a given $\epsilon > 0$, we have

$$\#\mathscr{L}_{P_1, B} - \#L^f_{P_1, B} \le \epsilon \frac{N(N-1)}{q(q+1)}.$$

The relevance of the quantity $\frac{N(N-1)}{q(q+1)}$ is that it is the total number of lines, not necessarily in $\mathscr{L}_{P_1, B}$.

Under the conditions of (3.3.5), we will explain how to construct a new map

$$f' : P_1 \to P_2$$

that agrees with $f$ on a large proportion of points. This construction will yield a linear map agreeing with $f$ at most points by applying the usual fundamental theorem of projective geometry to $f'$, giving us the desired approximate linearization.

**3.3.6.** The construction of $f'$ follows the recipe described in (3.1.8): Starting with $x \in P_1$ choose two points $y_1, y_3$ in $P_1 \setminus \bigcup_{i \in B} V_i$, not equal to $x$, at random. Let $L_1$ be the line through $x$ and $y_1$ and let $L_2$ be the line thorugh $x$ and $y_3$. Let $y_2$ be a random point on $L_1$ other than $x$ and $y_1$ and let $y_4$ be a random point on $L_2$ other than $x$ and $y_3$. Let $M_1$ (resp. $M_2$) be the line in $P_2$ through $f(y_1)$ and $f(y_2)$ (resp. $f(y_3)$ and $f(y_4)$). Then we will argue that, with high probability, $M_1$ and $M_2$ intersect in a unique point $z$, and define $f'(x) := z$.

To make this precise, let us begin with some calculations. For two points $y_1, y_2 \in P_1$ we can consider the linear span $Sp(y_1, y_2) \subset P_1$, which is either a line (if the points are distinct) or a point. Let $P_1^{2,f} \subset (P_1 \setminus \bigcup_{i \in B} V_i) \times P_1$ be the subset of pairs of distinct points $y_1, y_2$ for which $Sp(y_1, y_2) \in \mathscr{L}^f_{P_1, B}$.

**Lemma 3.3.7.** *We have*

$$\#((P_1 \setminus \bigcup_{i \in B} V_i) \times P_1) - \#P_1^{(2,f)} \le \epsilon N(N-1) + G.$$

*Proof.* We have a map

$$((P_1 \setminus \bigcup_{i \in B} V_i) \times P_1) \setminus \Delta \to \mathscr{L}_{P_1, B}, \quad (y_1, y_2) \mapsto Sp(y_1, y_2),$$

which has fibers of cardinality at most $q(q + 1)$. Here $\Delta \subset (P_1 \setminus \bigcup_{i \in B} V_i) \times P_1$ denotes the diagonal $P_1 \setminus \bigcup_{i \in B} V_i$. Therefore the number of pairs $(y_1, y_2) \in ((P_1 \setminus \bigcup_{i \in B} V_i) \times P_1) \setminus \Delta$ for which $f(Sp(y_1, y_2))$ is not a line is at most

$$q(q + 1) \frac{\epsilon N(N - 1)}{q(q + 1)} = \epsilon N(N - 1).$$

Furthermore the cardinality of the diagonal is at most $G$. □

Fix a point $x \in P_1$ and let $\mathscr{L}_x$ denote the set of lines through $x$. For

$$(i, j) \in \{(1, 3), (1, 4), (3, 2)\}$$

let

$$\pi_{ij} \colon (\mathscr{L}_x^{(2)})^2 \to (P_1 \setminus \bigcup_{i \in B} V_i) \times P_1$$

be the map given by

$$((L_1, y_1, y_2), (L_2, y_3, y_4)) \mapsto (y_i, y_j).$$

Let

$$(\mathscr{L}_x^{(2)})^{2, (i,j) - \text{good}} \subset (\mathscr{L}_x^{(2)})^2$$

denote the subset of data $((L_1, y_1, y_2), (L_2, y_3, y_4))$ for which $y_i$ and $y_j$ are distinct and span a line in $\mathscr{L}_{P_1}^f$.

**Lemma 3.3.8.** *For $(i, j) \in \{(1, 4), (3, 2)\}$,*

$$\#(\mathscr{L}_x^{(2)})^2 - \#(\mathscr{L}_x^{(2)})^{2, (i,j) - \text{good}} \le (q - 1)^2 (\epsilon N(N - 1) + G)$$

*Proof.* Indeed, this follows from (3.3.7) and the observation that the map $\pi_{ij}$ has fibers of cardinality at most $(q - 1)^2$. □

Let $\mathcal{S} \subset (\mathscr{L}_x^{(2)})^2$ denote the subset of data $((L_1, y_1, y_2), (L_2, y_3, y_4))$ such that

$$Sp(y_1, y_3), Sp(y_2, y_4) \in \mathscr{L}_x^f$$

and $Sp(f(y_1), f(y_2))$ and $Sp(f(y_3), f(y_4))$ have a unique intersection point.

**Lemma 3.3.9.**

$$\#(\mathscr{L}_x^{(2)})^2 - \#\mathcal{S} \le 2(q - 1)^2 (\epsilon N(N - 1) + G) + G(q - 1)^3$$

*Proof.* Two lines in $P_i$ are coplanar if and only if they intersect in exactly one point. From this it follows that for data

$$((L_1, y_1, y_2), (L_2, y_3, y_4)) \in (\mathcal{L}_x^{(2)})^{2,(1,4)-\text{good}} \cap (\mathcal{L}_x^{(2)})^{2,(3,2)-\text{good}} \qquad (3.3.9.1)$$

the points $(f(y_1), f(y_2), f(y_3), f(y_4))$ are coplanar. Indeed, because the points $(y_1, y_2, y_3, y_4)$ are coplanar, the lines $Sp(y_1, y_3)$ and $Sp(y_2, y_4)$ intersect in a unique point from which it follows that the lines

$$Sp(f(y_1), f(y_3)) = f(Sp(y_1, y_3)), \quad Sp(f(y_2), f(y_4)) = f(Sp(y_2, y_4))$$

are coplanar (since they intersect in a unique point).

Let $\mathcal{S}^c \subset (\mathcal{L}_x^{(2)})^{2,(1,3)-\text{good}} \cap (\mathcal{L}_x^{(2)})^{2,(2,4)-\text{good}}$ be the subset of the collections of data (3.3.9.1) for which

$$Sp(f(y_1), f(y_2)) = Sp(f(y_3), f(y_4)).$$

From this discussion we then have

$$\#(\mathcal{L}_x^{(2)})^2 - \#\mathcal{S} \le 2(q-1)^2(\epsilon N(N-1) + G) + \#\mathcal{S}^c.$$

It therefore suffices to show that

$$\mathcal{S}^c \le G(q-1)^3. \qquad (3.3.9.2)$$

The set $\mathcal{S}^c$ is contained in the set of collections of data (3.3.9.1) for which $f(y_3)$ and $f(y_4)$ are each points of the line $Sp(f(y_1), f(y_2))$. Since $f$ is an injection the cardinality of this set is less than or equal to

$$(\#\mathcal{L}_x^{(2)}) \cdot (q-1)^2,$$

and $\#\mathcal{L}_x^{(2)} \le G(q-1)$, so we obtain the inequality (3.3.9.2). $\qquad \square$

**3.3.10.** We now introduce a third line, and we are interested in the probability that it contains $z$.

For $z \in P_2$ let

$$\mathcal{S}_z \subset \mathcal{L}_x^{(2)}$$

be the subset of $(L, y_1, y_2) \in \mathcal{L}_x^{(2)}$ such that $z \in Sp(f(y_1), f(y_2))$.

**Lemma 3.3.11.** *There exists $z \in P_2$ such that*

$$\#(\mathcal{L}_x^{(2)})^2 - \#\mathcal{S}_z \le A(q, N, G, \epsilon).$$

*Proof.* Let $\mathcal{T}$ be the collection of triples $((L_1, y_1, y_2), (L_2, y_3, y_4), (L_3, y_5, y_6)) \in (\mathcal{L}_x^{(2)})^3$ such that either $((L, y_1, y_2), (L, y_3, y_4)) \notin \mathcal{S}$ or the unique intersection point $z$ of $Sp(f(y_1), f(y_2))$ and $Sp(f(y_3), f(y_4))$ does not lie in $Sp(f(y_5), f(y_6))$.

We will show

$$\#\mathscr{T} \le 3(q-1)G(2(q-1)^2(\epsilon N(N-1)+G)+G(q-1)^3)+q^3(q+1)(q-1)^2G^2.$$

To do this, note that by Lemma 3.3.9 and the fact that $\#\mathscr{L}_x^{(2)} \le G(q-1)$, the number of triples such that $((L, y_1, y_2), ((L, y_3, y_4)) \notin \mathcal{S}$ is at most

$$(q-1)G(2(q-1)^2(\epsilon N(N-1)+G)+G(q-1)^3).$$

By symmetry, the same holds for every other pair of two of the three lines

$$(L_1, y_1, y_2), (L_2, y_3, y_4), (L_3, y_5, y_6).$$

So the number of triples such that at least one of these three pairs fails to be in $\mathcal{S}$ is at most

$$3(q-1)G(2(q-1)^2(\epsilon N(N-1)+G).$$

Thus, it suffices to show that the number of triples with all three pairs in $\mathcal{S}$ but where $z \notin Sp(f(y_5), f(y_6))$ is at most

$$q^3(q+1)(q-1)^2G^2.$$

So it suffices to show that for each of the $(q-1)^2G^2$ choices of lines

$$((L, y_1, y_2), (L, y_3, y_4)),$$

there are at most $q^3(q+1)$ choices of $(L_3, y_5, y_6)$ such that $Sp(f(y_5), f(y_6))$ has a unique intersection with $Sp(f(y_1), f(y_2))$ and a unique intersection with $Sp(f(y_3), f(y_4))$ but does not contain their intersection point $z$.

For each $a_1 \in Sp(f(y_1), f(y_2))$ and $a_2 \in Sp(f(y_3), f(y_4))$ there exists a unique line $L_{a_1, a_2} \subset P_2$ through $a_1$ and $a_2$, and there are $(q+1)q$ pairs of ordered points $(w_5, w_6)$ on this line. Now if $(L_3, y_5, y_6) \in \mathscr{L}_x^{(2)}$ is such that the intersections

$$Sp(f(y_5), f(y_6)) \cap Sp((f(y_1), f(y_2))$$

and

$$Sp(f(y_5), f(y_6)) \cap Sp((f(y_3), f(y_4))$$

consist of single points not equal to $z$, then we must have

$$(y_5, y_6) = (f^{-1}(w_5), f^{-1}(w_6))$$

for some such pair $(w_5, w_6)$ on the line $L_{a_1, a_2}$, where $a_1$ and $a_2$ are the two respective intersections. Since $L$ is determined by $(y_5, y_6)$, this shows that the number of such triples $(L_3, y_5, y_6)$ is bounded by $q(q+1)$ for a given $(a_1, a_2)$. Since there are $q$ points $a_1$ in $Sp(f(y_1), f(y_2))$ other than $z$, and $q$ points $a_2$ in $Sp(f(y_3), f(y_4))$ other than $z$, the total number of possibilities is $(q+1)q^3$, as desired.

Now $\mathcal{T}$ maps to $(\mathcal{L}_x^{(2)})^2$ by projecting onto the first two factors. Since the image has size at least $(q-1)^2(G-1)^2$, the fiber over some point

$$((L_1, y_1, y_2), (L_2, y_3, y_4))$$

must contain at most

$$\frac{3(q-1)G(2(q-1)^2(\epsilon N(N-1)+G)+G(q-1)^3)+q^3(q+1)(q-1)^2G^2}{(q-1)^2G^2}$$
$$= \frac{3(2(q-1)^2(\epsilon N(N-1)+G)+G(q-1)^3)}{(q-1)G} + q^3(q+1) = A(q, G, N, \epsilon)$$

elements. If $((L_1, y_1, y_2), (L_2, y_3, y_4)) \in \mathcal{S}$, let $z$ be the unique intersection point of $Sp(f(y_1), f(y_2))$ and $Sp(f(y_3), f(y_4))$, and otherwise take $z$ arbitrary. In either case, if

$$z \notin Sp(f(y_5), f(y_6))$$

then

$$((L_1, y_1, y_2), (L_2, y_3, y_4), (L_3, y_5, y_6)) \in \mathcal{T},$$

so there are at most $A(q, G, N, \epsilon)$ lines with that property.  $\square$

**Corollary 3.3.12.** *If*

$$2A(q, N, G\epsilon) + q(q+1) < (q-1)(G-1) \qquad (3.3.12.1)$$

*then there exists a unique point $z \in P_2$ such that*

$$\#\mathcal{L}_x^{(2)} - \#\mathcal{S}_z \le A(q, N, G, \epsilon)$$

*and for every $z' \ne z$ we have*

$$\mathcal{S}_{z'} \le A(q, N, G, \epsilon) + q(q+1).$$

*Proof.* That there exists $z$ with $\#\mathcal{L}_x^{(2)} - \#\mathcal{S}_z \le A(q, N, G\epsilon)$ follows from (3.3.11). Take such a $z$. We will show the bound on $z' \ne z$ and use it to deduce uniqueness.

To show that

$$\mathcal{S}_{z'} \le A(q, N, G, \epsilon) + q(q+1)$$

it suffices to show that

$$\mathcal{S}_{z'} \cap \mathcal{S}_z \le q(q+1)$$

since we have

$$\mathcal{S}_z^c \le A(q, N, G, \epsilon).$$

To do this, note that $z \in Sp(f(y_1), f(y_2))$ and $z' \in Sp(f(y_1), f(y_2))$ together with $z \ne z'$ implies that $f(y_1), f(y_2) \in Sp(z, z')$, since $Sp(z_1, z_2)$ is the unique line between $z$ and $z'$. Thus there are at most $q(q+1)$ possibilities for $f(y_1), f(y_2)$ and hence at most $q(q+1)$ possibilities for $y_1, y_2$ since $f$ is injective. Since

$L = Sp(y_1, y_2)$, there are at most $q(q+1)$ possibilities for $(L, y_1, y_2)$, as desired. For uniqueness, if $z' \neq z$ also satisfied

$$\# \mathscr{L}_x^{(2)} - \# \mathcal{S}_{z'} \leq A(q, N, G\epsilon),$$

then we would have

$$(q-1)(G-1) \leq \# \mathscr{L}_x^{(2)}$$
$$= (\# \mathscr{L}_x^{(2)} - \# \mathcal{S}_{z'}) + \# \mathcal{S}_{z'}$$
$$\leq A(q, N, G, \epsilon) + A(q, N, G, \epsilon) + q(q+1),$$

contradicting our assumption (3.3.12.1). □

**Assumption 3.3.13.** Assume for the rest of the discussion that the inequality (3.3.12.1) holds.

**3.3.14.** We define a map

$$f' \colon P_1 \to P_2$$

by sending $x \in P_1$ to the point $z \in P_2$ given by (3.3.12).

Let $P_1^{f=f'} \subset P_1$ be the set of points $x$ for which $f(x) = f'(x)$.

**Lemma 3.3.15.**

$$(\# P_1 - \#(P_1^{f=f'})) \leq \frac{(q-1)\epsilon G(G-1)}{(q-1)(G-1) - A(q, N, G, \epsilon) - q(q+1)}$$

*Proof.* If $f(x) \neq f'(x)$ then by (3.3.12), the number of $(L, y_1, y_2) \in L_{(x)}^2$ such that $f(x) \in Sp(f(y_1), f(y_2))$ is at most $A(q, N, G, \epsilon) - q(q+1)$ and thus the number with $f(x) \notin Sp((f(y_1), f(y_2))$ is at least

$$(q-1)(G-1) - A(q, N, G, \epsilon) - q(q+1).$$

So there are at least

$$(\# P_1 - \#(P_1^{f=f'}))((q-1)(G-1) - A(q, N, G, \epsilon) - q(q+1))$$

triples $(L, x, y_1, y_2)$, with $x, y_1, y_2$ three distinct points on $L$, and $y_1 \notin \bigcup_{i \in B} V_i$, such that $f$ does not take $L$ to a line.

On the other hand, there are at most $\frac{\epsilon G(G-1)}{(q+1)q}$ lines which contain a point $\notin \bigcup_{i \in B} V_i$ and which $f$ does not take to a line, and each such line contains at most $(q+1)q(q-1)$ triples of points, so we obtain

$$(\# P_1 - \#(P_1^{f=f'}))((q-1)(G-1) - A(q, N, G, \epsilon) - q(q+1)) \leq (q-1)\epsilon G(G-1)$$

and thus

$$(\#P_1 - \#(P_1^{f=f'})) \leq \frac{(q-1)\epsilon G(G-1)}{(q-1)(G-1) - A(q,N,G,\epsilon) - q(q+1)}. \qquad \square$$

**3.3.16.** Fix a point $x \in P_1$. Let us calculate a lower bound for the number of elements $(L, y_1, y_2) \in \mathcal{L}_x^{(2)}$ for which the following conditions hold:

(1) $f(y_1) = f'(y_1)$ and $f(y_2) = f'(y_2)$.

(2) $(L, y_1, y_2) \in \mathcal{S}_{f'(x)}$.

By (3.3.15) we have $f(y) \neq f'(y)$ for most

$$\frac{(q-1)\epsilon G(G-1)}{(q-1)(G-1) - A(q,N,G,\epsilon) - q(q+1)}$$

values of $y_1$. Because we have $y_1 = y$ for at most $q-1$ elements of $L_x^{(2)}$, and $y_2 = y$ for at most $q-1$ elements of $L_x^{(2)}$, the number of elements of $L_x^{(2)}$ at which condition (1) fails is

$$2(q-1)\frac{(q-1)\epsilon G(G-1)}{(q-1)(G-1) - A(q,N,G,\epsilon) - q(q+1)}.$$

By (3.3.12) the number of elements for which condition (2) fails is at most $A(q,G,N,\epsilon)$ and thus the number of elements at which conditions (1) or (2) fail is at least

$$2(q-1)\frac{(q-1)\epsilon G(G-1)}{(q-1)(G-1) - A(q,N,G,\epsilon) - q(q+1)} - A(q,G,N,\epsilon) = B(q,N,G,\epsilon).$$

**Lemma 3.3.17.** *We have*

$$\#L_x^{(2)} - \#\{(L, y_1, y_2) \in \mathcal{L}_x^{(2)} \mid f'(x), f'(y_1), f'(y_2) \text{ are collinear}\} \leq$$
$$B(q,N,G,\epsilon).$$

*Proof.* This follows from the preceding discussion. $\qquad \square$

**3.3.18.** We will use (3.3.17) to show that $f'$ takes lines to lines and that $f'$ is injective. For this we will use Desargues's theorem, which is a consequence of Pappus's axiom, and the notion of Desargues configurations.

Recall that a *Desargues configuration* is a collection of ten points and ten lines such that any line contains exactly three of the points and exactly three lines pass through each point.

Desargues's theorem can be stated as follows. Consider two collections of three points $\{A, B, C\}$ and $\{D, E, F\}$, usually thought of as the vertices of two triangles, and consider the nine lines

$$\{AB, AC, BC, DE, DF, EF, AD, BE, CF\}.$$

**Theorem 3.3.19** (Desargues). *If the three lines AD, BE, and CF meet at a common point G then the three intersection points*

$$H := AB \cap DE, \quad I := AC \cap DF, \quad J := BC \cap EF$$

*are collinear, and conversely if these three points are collinear then the lines AD, BE, and CF meet at a common point.*

In other words, the ten points and ten lines obtained in this way form a Desargues configuration.

**3.3.20.** To show that $f'$ takes lines to lines, it therefore suffices to show that for any three collinear points $(t_1, t_2, t_3)$ there exists a Desargues configuration as above with $(H, I, J) = (t_1, t_2, t_3)$ such that $f'$ takes all the lines other than $Sp(t_1, t_2)$ to lines in $P_2$. For then, by Desargues's theorem, it follows that $(f'(t_1), f'(t_2), f'(t_3))$ are collinear. We will produce such a Desargues configuration using basic linear algebra. We fix the collinear points $\{x, y, t\}$ in what follows.

**Notation 3.3.21.** Let $V_1$ be an **F**-vector space with $\mathbb{P}V_1 = P_1$, and choose vectors $a, b \in V_1$ such that $(t_1, t_2, t_3)$ is given by the three elements $(a, b, a-b) \in V_1$.

**Construction 3.3.22.** For $c, d \in V_1$, consider the ordered set of five elements $\{0, a, b, c, d\}$. Let $\mathscr{P}(c, d)$ denote the set of points of $P_1$ given by the differences of two elements

$$\mathscr{P}(c, d) := \{[a], [b], [c], [d], [b-a], [c-a], [d-a], [c-b], [d-b], [c-d]\},$$

and let $\mathscr{M}(c, d)$ denote the set of lines obtained by taking for each subset of three elements $T \subset \{0, a, b, c, d\}$ the linear span $L_T$ of differences of elements of $T$.

**Lemma 3.3.23.** *As long as the set of four elements $\{a, b, c, d\}$ are linearly independent, the ten points and ten lines $(\mathscr{P}(c, d), \mathscr{M}(c, d))$ form a Desargues configuration.*

*Proof.* The proof is routine linear algebra.                                    $\square$

Figure 3.4 shows a typical configuration generated by (3.3.22) (on a true set of randomly generated data). The bold line shows the collinear points $t_1, t_2,$ and $t_3$, together with the auxiliary points given by the choices of $c$ and $d$. Some of the lines naturally come in pairs, corresponding to the construction of the map $f'$ in (3.3.14). (For example, the dotted line connecting $t_1$ to $d$ and $d-a$ and the dotted line connecting $x$ to $c$ and $c-a$ serve to define $f'(x)$, under the

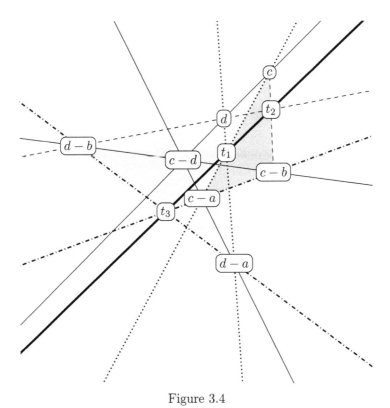

Figure 3.4

assumption that those two lines are mapped to lines under $f$.) The remaining solid lines complete the Desargues configuration. The two perspective triangles are shaded in gray. The center of perspectivity lies at $c - d$, and the axis of perspectivity is the line spanned by $t_1$, $t_2$, and $t_3$.

**Notation 3.3.24.** Let $W \subset V_1^{\times 2}$ be the subset of pairs $(c, d)$ such that the following conditions hold.

(1) $[c], [c - a], [d], [d - b] \notin \bigcup_{i \in B} V_i$.

(2) The set of ten lines and ten points $(\mathcal{P}(c, d), \mathcal{M}(c, d))$ of (3.3.22) is a Desargues configuration.

(3) The map $f'$ takes every line in $\mathcal{M}(c, d) \setminus \{Sp(x, y)\}$ to a line in $P_2$.

We can show $W$ is nonempty as follows. Recall the function $B$ from (3.3.0.2).

**Proposition 3.3.25.** *If*

$$\frac{9B(q, G, N, \epsilon)}{q^{n+1} \prod_{i \in B}(1 - 2q^{-\operatorname{codim} V_i})} + \frac{q^{n+4}}{q^{2n+2} \prod_{i \in B}(1 - 2q^{-\operatorname{codim} V_i})^2} < 1$$

*then $W \neq \emptyset$.*

*Proof.* Let $N_c$ be the number of elements $c \in V$ such that $[c], [c-a] \notin \bigcup_{i \in B} V_i$. Then $N_c \geq q^{n+1} \prod_{i \in B}(1 - 2q^{-\operatorname{codim} V_i})$ since we must avoid the two conditions $c = 0 \mod V_i$ or $c = a \mod V_i$ and these conditions are independent for different $i$ since the $V_i$ are transverse. The same logic holds for $d$, and so the number of pairs $c, d$ satisfying (1) is at least

$$N_c N_d \geq q^{2n+2} \prod_{i \in B}(1 - 2q^{-\operatorname{codim} V_i})^2.$$

Let $\mathcal{F}$ be the set of pairs $c, d$ with $a, b, c, d$ linearly independent, and $[c], [c-a] \notin \bigcup_{i \in B} V_i$, i.e., satisfying conditions (1) and (2).

The number of pairs for which $a, b, c, d$ are linearly dependent is at most $q^{n+4}$. So

$$\#\mathcal{F} \geq N_c N_d - q^{n+4}.$$

Now let $\mathcal{F}^{0ac}$ be the set of $([c], [d]) \in \mathcal{F}$ such that $f'([a]), f'([c])$, and $f'([c-a])$ are not collinear. Similarly let $\mathcal{F}^{acd}$ be the set of $([c], [d]) \in \mathcal{F}$ such that $f'([c-a]), f'([d-c]), f'([d-a])$ are not collinear, and similarly for all nine sets of three symbols from $\{0, a, b, c, d\}$ other than $0ab$.

Let us show

$$\#\mathcal{F}^{0ac} \leq N_d B(q, N, G, \epsilon).$$

To do this, note that $\mathcal{F}$ projects to $\mathcal{L}^2_{[a]}$ by sending $([c], [d])$ to $(Sp([a], [c]), [c], [c-a])$. The key fact here is that $[c] \notin \bigcup_{i \in B} V_i$ by the definition of $\mathcal{F}$.

If $(c, d) \in \mathcal{F}^{0ac}$, then the image of $(c, d)$ under this projection fails the condition of (3.3.17). So the image of $\mathcal{F}^{0ac}$ under this projection has cardinality at most $B(q, N, G, \epsilon)$. From the three points $[a], [c], [c-a]$ and the vector $a$ we can reconstruct $c$, so the fiber of each point under this projection is the number of possible choices for $d$, which is $N_d$. So indeed

$$\#\mathcal{F}^{0ac} \leq N_d B(q, N, G, \epsilon).$$

Similarly we can show

$$\#\mathcal{F}^{0ad} \leq N_c B(q, N, G, \epsilon)$$

using the projection $(c, d) \mapsto (Sp([a], [d]), [d], [d-a])$ that sends $\mathcal{F}$ to $\mathcal{L}^{(2)}_x$. The key point is again that $[d] \notin \bigcup_{i \in B} V_i$ and thus we can always take $y_1 = d$.

The projections

$$x = [b],, y_1 = [c], y_2 = [c-b], L = Sp([b], [c])$$

$$x = [b],, y_1 = [d], y_2 = [d-b], L = Sp([b], [d])$$

$$x = [b-a],, y_1 = [d-b], y_2 = [d-a], L = Sp([b-a], [d-b])$$

$$x = [b - a], , y_1 = [c - a], y_2[c - b], L = Sp([b - a], [c - a])$$

give

$$\#\mathcal{F}^{0bc} \le N_d B(q, N, G, \epsilon),$$
$$\#\mathcal{F}^{0bd} \le N_c B(q, N, G, \epsilon),$$
$$\#\mathcal{F}^{abd} \le N_c B(q, N, G, \epsilon),$$
$$\#\mathcal{F}^{abc} \le N_d B(q, N, G, \epsilon).$$

To bound $\#\mathcal{F}^{0cd}, \#\mathcal{F}^{acd}, \#\mathcal{F}^{bcd}$ we need a slightly different argument. Let us consider $\mathcal{F}^{0cd}$ first. There are $N_c$ possible values of $c$. For each value of $c$, every value of $d$ with $(c, d) \in \mathcal{F}$ defines a triple $(Sp([c], [d]), [d], [d - c]) \in L_{[c]}^{(2)}$, and if $(c, d) \in \mathcal{F}^{0cd}$ then $(Sp([c], [d]), [d], [d - c])$ fails the conditions of (3.3.17). Hence for each vector $c$, the number of $d$ with $(c, d) \in \mathcal{F}^{0cd}$ is at most $B(q, G, N, \epsilon)$ and so

$$\#\mathcal{F}^{0cd} \le N_c B(q, G, N, \epsilon).$$

By the same argument, the projections

$$x = [d - a], L = Sp([c - a], [d - a]), y_1[c - a], y_2 = [d - c]$$

$$x = [c - b], L = Sp([c - b], [d - b]), y_1 = [d - b], y_2[d - c]$$

give

$$\#\mathcal{F}^{acd} \le N_d B(q, G, N, \epsilon), \#\mathbb{F}^{bcd} \le N_c B(q, G, N, \epsilon)$$

respectively.

Thus

$$\#(\mathcal{F}^{0ac} \cup \mathcal{F}^{0ad} \cup \mathcal{F}^{0bc} \cup \mathcal{F}^{0bd} \cup \mathcal{F}^{abc} \cup \mathcal{F}^{abd} \cup \mathcal{F}^{0cd} \cup \mathcal{F}^{acd} \cup \mathcal{F}^{bcd})$$
$$\le (5N_c + 4N_d) B(q, G, N, \epsilon).$$

This union is the total number of triples that satisfy (1) and (2) but fail (3). So the total number of triples satisfying (1),(2),(3) is at least

$$N_c N_d - q^{n+4} - (5N_c + 4N_d) B(q, G, N, \epsilon).$$

To show this is nonzero, it suffices to show that

$$1 - \frac{q^{n+4}}{N_c N_d} - \left( \frac{5}{N_d} + \frac{4}{N_c} \right) B(q, G, N, \epsilon) > 0,$$

which by our lower bound on $N_c, N_d$ follows from our assumption

$$\frac{9 B(q, G, N, \epsilon)}{q^{n+1} \prod_{i \in B} (1 - 2q^{-\text{codim } V_i})} + \frac{q^{n+4}}{q^{2n+2} \prod_{i \in B} (1 - 2q^{-\text{codim } V_i})^2} < 1. \qquad \square$$

**3.3.26.** We are now ready to give the proof of (3.3.1).

*Proof of (3.3.1).* We let $f'$ be the map defined in (3.3.14). We refer in this proof to the diagram in Figure 3.4.

Assuming the inequality of (3.3.25), we can choose $(c, d) \in W$, and let $(\mathscr{P}, \mathscr{M}) = (\mathscr{P}(c, d), \mathscr{M}(c, d))$ be the resulting Desargues configuration. We have that all the lines in $\mathscr{M}$, except possibly for $(t_1, t_2, t_3)$, are taken to lines in $P_2$ under $f'$. Thus, in Figure 3.4, the dotted, dashed, dot-dashed, and non-bold solid lines are all taken to lines under $f'$. On the other hand, the images of the dotted lines intersect at $f'(t_1)$, the images of the dashed lines intersect at $f'(t_2)$, and the images of the dot-dashed lines intersect at $f'(t_3)$. By Desargues theorem, $f'(t_1)$, $f'(t_2)$, and $f'(t_2)$ are collinear and distinct, lying on the axis of perspectivity for the image Desargues configuration. Note that this also implies that $f'$ is injective. □

# Chapter Four

---

## Divisorial structures and definable linear systems

This chapter is devoted to studying the basic theory of the *divisorial structure* associated to an algebraic variety $X$. This structure consists of the Zariski topological space $|X|$ and the linear equivalence relation on Weil divisors. The key result is (4.1.14), which says that the divisorial structure uniquely characterizes the schemes associated to varieties of dimension at least 2 over infinite base fields (resp. dimension at least 3 over finite fields, with the additional assumption that the varieties are Cohen-Macaulay if the base field is $\mathbb{F}_2$), up to unique isomorphism. This Torelli-type theorem is one of the tehnical hearts of this work, the other being the construction of the divisorial structure of $X$ from $|X|$ in many circumstances (9.8.18).

Given a normal proper variety $X$ over a field $k$, if one knows the linear equivalence relation then one knows the sets $|D|$ that make up the linear systems. In fact, as we discuss in Section 4.2, one can deduce various properties of $D$ from $|D|$ and $|X|$, such as ampleness and basepoint-freeness. If we look at the set of elements of $|D|$ that contain a chosen subset of $X$, then we define a linear subspace of $|D|$. In Section 4.3 we consider when one can detect the lines among these definable linear subspaces, especially in the case that $D$ is ample and basepoint-free. By the theory of Chapter 3, we can then recover the projective space structure on $|D|$, which is a crucial step toward finding $X$ itself, as we will explain in Chapters 5 and 6.

## 4.1 DIVISORIAL STRUCTURES

In this section we introduce the key structure that will ultimately be the subject of our main reconstruction theorem.

**Notation 4.1.1.** For a Zariski topological space $Z$ we write $\mathrm{Div}(Z)$ for the set of divisors on $Z$. This is the free abelian group on the set of codimension 1 points of $Z$. When $X$ is a variety, we will write $\mathrm{Div}(X)$ for $\mathrm{Div}(|X|)$.

**Definition 4.1.2.** Let $K$ be a field. A normal, geometrically integral $K$-variety $X$ is *divisorially proper over $K$* if $H^0(X, \mathcal{O}_X) = K$ and for any reflexive sheaf $\mathcal{L}$ of rank 1 we have that $\Gamma(X, \mathcal{L})$ is finite-dimensional over $K$.

**Remark 4.1.3.** One could more generally define the notion of divisorially proper for a normal $K$-variety, but for our main results we will always need the additional assumptions on $X$, so we find the above definition more convenient.

**Lemma 4.1.4.** *If a normal $K$-variety $X$ is divisorially proper over $K$ and $U \subset X$ is an open subvariety such that $\mathrm{codim}(X \setminus U \subset X) \geq 2$ at every point, then $U$ is also divisorially proper over $K$.*

*Proof.* The variety $U$ is also geometrically integral, being schematically dense in $X$. Furthermore, the restriction map $K = H^0(X, \mathcal{O}_X) \to H^0(U, \mathcal{O}_U)$ is an isomorphism since $\mathrm{codim}(X \setminus U \subset X) \geq 2$ and $X$ is normal. Finally, any reflexive sheaf $\mathcal{L}$ of rank 1 on $U$ is the restriction of a reflexive sheaf $\mathcal{L}'$ of rank 1 on $X$, and Krull's theorem tells us that the restriction map

$$\Gamma(X, \mathcal{L}') \to \Gamma(U, \mathcal{L})$$

is an isomorphism of $K$-vector spaces.                                                                         $\square$

**4.1.5.** We define a category $\mathscr{DP}$, which we will refer to as the *category of divisorially proper varieties*, as follows.

The objects of $\mathscr{DP}$ consist of pairs $(K, X)$, where $K$ is a field and $X/K$ is a divisorially proper $K$-variety. We usually write simply $X$, instead of $(K, X)$, since the field $K$ can be recovered as $\Gamma(X, \mathcal{O}_X)$.

A morphism

$$(K, X) \to (L, Y)$$

in $\mathscr{DP}$ is a pair $(\varphi, j)$, where $\varphi \colon L \to K$ is an isomorphism of fields and $j \colon X \to Y$ is a $\varphi$-linear open immersion of varieties such that $Y \setminus j(X)$ has codimension at least 2 in $Y$.

**Remark 4.1.6.** This definition can be simplified using the language of schemes. In this language an object of $\mathscr{DP}$ is a scheme $X$ such that the following hold:

(1) $X$ is normal and $\kappa_X := \Gamma(X, \mathcal{O}_X)$ is a field. We refer to $\kappa_X$ as the *constant field of $X$*.

(2) The natural map $f \colon X \to \mathrm{Spec}(\kappa_X)$ is separated, of finite type, and has geometrically integral fibers.

(3) For every reflexive sheaf $\mathcal{L}$ of rank 1 the $\kappa_X$-vector space $\Gamma(X, \mathcal{L})$ is finite-dimensional.

Morphisms in $\mathscr{DP}$ are open immersions $j \colon X \to Y$ such that $Y \setminus j(X)$ has codimension at least 2.

**Definition 4.1.7.** A *divisorial structure* is a pair $(Z, \Lambda)$ with $Z$ a Zariski topological space and $\Lambda \subset \mathrm{Div}(Z)$ is a subgroup.

**Remark 4.1.8.** Any divisor $D \subset \mathrm{Div}(Z)$ can be written as a difference $D = D^+ - D^-$ of effective divisors $D^+, D^- \in \mathrm{Div}^+(Z)$. It follows that $\Lambda$ can also be specified by an equivalence relation on the monoid $\mathrm{Div}^+(Z)$, and conversely any equivalence relation on $\mathrm{Div}^+(Z)$ compatible with the monoid structure (a so-called congruence relation) defines a subgroup of $\mathrm{Div}(Z)$.

**Definition 4.1.9** (Restriction of a divisorial structure). Suppose $t := (Z, \Lambda)$ is a divisorial structure. Given an open subset $U \subset Z$, the *restriction of $t$ to $U$*, denoted by $t|_U$, is the divisorial structure $(U, \Lambda_U)$, where $\Lambda_U \subset \mathrm{Div}(U)$ is the image of $\Lambda$ under the restriction map $\mathrm{Div}(Z) \to \mathrm{Div}(U)$.

In other words, if we let $\mathrm{Div}(X) \to Q$ denote the quotient by $\Lambda$, we define $\Lambda_U$ by forming the pushout

$$
\begin{array}{ccc}
\mathrm{Div}(X) & \longrightarrow & Q \\
\downarrow & & \downarrow \\
\mathrm{Div}(U) & \longrightarrow & Q_U
\end{array}
$$

in the category of abelian groups, and $\Lambda_U$ is the kernel of $\mathrm{Div}(U) \to Q_U$.

**Definition 4.1.10** (Morphisms of divisorial structures). A *morphism of divisorial structures*

$$(Z, \Lambda) \to (Z', \Lambda')$$

is an open immersion of topological spaces $f \colon Z \to Z'$ such that

$$\mathrm{Div}(f) \colon \mathrm{Div}(Z) \to \mathrm{Div}(Z')$$

is a bijection and

$$\mathrm{Div}(f)(\Lambda) = \Lambda'.$$

**Notation 4.1.11.** We will write $\mathcal{T}$ for the category of divisorial structures.

**Definition 4.1.12.** The *divisorial structure* of an integral scheme $X$ is the pair

$$\tau(X) := (|X|, \Lambda_X),$$

where $|X|$ is the Zariski topological space of $X$ and

$$\Lambda_X \subset \mathrm{Div}(X)$$

is the subgroup of divisors rationally equivalent to 0.

**Remark 4.1.13.** The divisorial structure of an integral scheme $X$ can be obtained from the data of the triple

$$(|X|, \mathrm{Cl}(X), c \colon X^{(1)} \to \mathrm{Cl}(X)),$$

where $X^{(1)}$ is the set of codimension 1 points of $X$. Indeed, by the universal property of a free group on a set, giving the map $c$ is equivalent to giving a map of groups

$$\mathrm{Div}(X) \to \mathrm{Cl}(X),$$

and the kernel of this map is $\Lambda_X$. Conversely, from $\Lambda_X$ we obtain the class group as the quotient $\mathrm{Div}(X)/\Lambda_X$, and the map $c$ is induced by the natural map $X^{(1)} \to \mathrm{Div}(X)$.

Formation of the divisorial structure defines a functor

$$\mathscr{DP} \xrightarrow{\ \tau\ } \mathscr{T}. \tag{4.1.13.1}$$

One of the main results of this monograph is the following:

**Theorem 4.1.14.** *Let $(K, X)$ and $(L, Y)$ be objects of $\mathscr{DP}$ and assume one of the following conditions holds:*

(1) *$K$ is infinite and $X$ has dimension $\geq 2$.*

(2) *$K$ has cardinality $> 2$ and $X$ has dimension $\geq 3$.*

(3) *$K$ has cardinality $2$, $\dim(X) \geq 3$, and $X$ is Cohen-Macaulay.*

*Then the map*

$$\mathrm{Hom}_{\mathscr{DP}}((K, X), (L, Y)) \to \mathrm{Hom}_{\mathscr{T}}(\tau(K, X), \tau(L, Y))$$

*is bijective.*

The proof of (4.1.14) will be given in Chapter 5 (infinite field case) and Chapter 6 (finite field case) after some preliminary foundational work.

**Remark 4.1.15.** Note that if $(K, X)$ is a divisorially proper variety and $U \subset X$ is an open subset such that $X \setminus U$ has codimension $\geq 2$ in $X$, then $(K, U)$ is also a divisorially proper variety. In particular, if $(K, X)$ and $(L, Y)$ are divisorially proper varieties and

$$f : |X| \to |Y|$$

is a morphism of divisorial structures, then setting $Y'$ equal to the open sub-scheme of $Y$ given by the open subset $f(|X|) \subset |Y|$ we get a divisorially proper variety $(L, Y')$ and an isomorphism of divisorial structures $|X| \simeq |Y'|$. From this it follows that in order to prove (4.1.14) it suffices to show the following: if $(K, X)$ and $(L, Y)$ are divisorially proper varieties then any isomorphism of divisorial structures

$$\tau(K, X) \simeq \tau(L, Y)$$

is induced by a unique isomorphism $(K, X) \simeq (L, Y)$ in $\mathscr{DP}$. This is, in fact, the statement we show in Chapter 5 and Chapter 6.

## 4.2   REMARKS ON DIVISORS

In this section we gather a few facts about divisors on normal varieties. Our main purpose is to demonstrate that some basic features of such varieties—such as the maximal factorial open subscheme—can be characterized purely in terms of the divisorial structure.

Fix a field $k$. For a normal $k$-variety $X$ let

$$q \colon \operatorname{Div}(X) \to \operatorname{Cl}(X)$$

denote the quotient map to the class group. Given a divisor $D$ on $X$, upon identifying $|D|$ with the subset of effective divisors on $X$ that are linearly equivalent to $D$, we have a set-theoretic equality

$$|D| = q^{-1}(q(D)) \cap \operatorname{Div}^+(X).$$

In particular, the linear system is defined *as a set* by the divisorial structure.

There is a reflexive sheaf of rank 1 canonically associated to $D$ that we will write $\mathcal{O}(D)$. Members of $|D|$ are in bijection with sections $\mathcal{O} \to \mathcal{O}(D)$ in the usual way. Recall that $D$ is Cartier if and only if $\mathcal{O}(D)$ is an invertible sheaf on $X$.

**Lemma 4.2.1.** *Let $U \subset X$ be an open subscheme. Then the commutative diagram*

$$
\begin{array}{ccc}
\operatorname{Div}(X) & \longrightarrow & \operatorname{Cl}(X) \\
\downarrow & & \downarrow \\
\operatorname{Div}(U) & \longrightarrow & \operatorname{Cl}(U)
\end{array}
$$

*is a pushout diagram in the category of abelian groups.*

*Proof.* This follows from the observation the map on the kernels of the horizontal arrows is surjective. □

**Corollary 4.2.2.** *If $X$ is an integral scheme and $U \subset X$ is an open subscheme then the divisorial structure $\tau(U)$ is canonically isomorphic to the restriction $\tau(X)|_U$ (see (4.1.9)).*

*Proof.* By (4.2.1), we see that the induced relation on $\tau(X)|_U$ is precisely the relation for $\tau(U)$, giving the desired result. □

**Definition 4.2.3.** Given a variety $X$, there is a largest open subset that is factorial. We refer to this open subset as the *maximal factorial open subset*.

**Remark 4.2.4.** The existence of the maximal factorial open subset follows from the basic observation that if $U, V \subset X$ are factorial open subsets then $U \cup V$ is also factorial. Note, however, that we are *not* asserting that the set of points $x \in X$ for which the local ring $\mathcal{O}_{X,x}$ is factorial is equal to this open set. In some cases it is known that the set of such points is open, and hence equal to the maximal factorial open subset, but in general it is a subtle question (see [BGS11] for more discussion).

**Proposition 4.2.5.** *Let $X$ be a normal variety and let $D \subset X$ be a divisor.*

(1) *If $|D|$ is basepoint-free then $D$ is Cartier.*

(2) *If $D$ is ample then $D$ is $\mathbb{Q}$-Cartier.*

*Proof.* Since $X$ is quasi-compact, if $D$ is ample we know that $|nD|$ is basepoint-free for some $n$. Thus it suffices to prove the first statement. Given a point $x \in X$, choose $E \in |D|$ such that $x \notin E$. This gives some section $s \colon \mathcal{O} \to \mathcal{O}(D)$. Restricting to the local ring $R = \mathcal{O}_{X,x}$, we see that $s_x \colon R \to \mathcal{O}(D)_x$ is an isomorphism in codimension 1 (for otherwise $E$ would be supported at $x$). Since $\mathcal{O}(D)$ is reflexive, it follows that $s_x$ is an isomorphism, whence $\mathcal{O}(D)$ is invertible in a neighborhood of $x$. Since this holds at any $x \in X$, we conclude that $\mathcal{O}(D)$ is invertible, as desired. □

**Corollary 4.2.6.** *A normal variety $X$ is factorial if and only if it is covered by open subschemes $U \subset X$ with the property that every divisor class on $U$ is basepoint-free.*

*Proof.* If $X$ is factorial, then any affine open covering has the desired property, since the linear system of any Cartier divisor on an affine scheme is basepoint-free. On the other hand, if $X$ admits such a covering, then we know that every divisor class on $X$ is locally Cartier, whence it is Cartier. □

**Proposition 4.2.7.** *If $X$ is a normal $k$-variety then we can characterize the maximal factorial open subset of $X$ as the union of all open subsets $U \subset X$ such that every divisor class on $U$ is basepoint-free.*

*Proof.* This is an immediate consequence of (4.2.6). □

The preceding discussion implies that various properties of a variety $X$ and its divisors can be read off from the divisorial structure. We summarize this in the following.

**Proposition 4.2.8.** *Let $X$ be a normal variety and let*

$$\tau(X) = (|X|, \Lambda_X)$$

*be the associated divisorial structure. Then*

(1) *the property that $D \in \mathrm{Div}(X)$ has basepoint-free linear system $|D|$ depends only on $\tau(X)$;*

(2) *the property that $X$ is factorial depends only on $\tau(X)$;*

(3) *the maximal factorial open subset of $X$ depends only on $\tau(X)$;*

(4) *the condition that a divisor $D$ is ample depends only on $\tau(X)$.*

*Proof.* Let
$$q\colon \operatorname{Div}(X) \to \operatorname{Cl}(X)$$
denote the quotient map defined by $\Lambda_X$, so that for $D \in \operatorname{Div}(X)$ we have $|D| = q^{-1}(q(D)) \cap \operatorname{Div}^+(X)$. The condition that $|D|$ is basepoint-free is the statement that for every $x \in |X|$ there exists $E \in |D|$ such that $x \notin E$. Evidently this depends only on $\tau(X)$, proving (1).

Likewise the condition that a divisor $D$ is ample is the statement that the open sets defined by elements of $|nD|$ for $n \geq 0$ give a base for the topology on $|X|$. Again this clearly only depends on $\tau(X)$, proving (4).

Statement (2) follows from (4.2.6) and (4.2.1), which implies that the divisorial structure $\tau(U)$ for an open subset $U \subset X$ is determined by $|U| \subset |X|$ and $\tau(X)$.

Finally (3) follows from (4.2.7). □

**4.2.9.** The proofs of our main results will ultimately rely on reducing to the projective case. For the remainder of this section, we record some results about polarizations that we will need later.

**Lemma 4.2.10.** *Suppose we are given two divisorially proper varieties $X, Y \in \mathscr{DP}$ and an isomorphism $\varphi\colon \tau(X) \to \tau(Y)$ of the associated divisorial structures. If $X$ is polarizable and factorial, then so is $Y$ and the isomorphism*

$$\operatorname{Div}(X) \to \operatorname{Div}(Y)$$

*given by $\varphi$ preserves the classes of ample, basepoint-free, effective divisors.*

*Proof.* Since $X$ is factorial, all divisors are Cartier divisors. By (4.2.8), $Y$ is also factorial and polarizable, and the submonoids of ample basepoint-free divisors are preserved.

□

**Definition 4.2.11.** Suppose $X \in \mathscr{DP}$ is a divisorially proper variety. An open subscheme $U \subset X$ will be called *essential* if $\operatorname{codim}(X \setminus U \subset X) \geq 2$, $U$ is factorial, and $U$ is polarizable.

Note that if $U \subset X$ is essential, then the natural restriction map $\operatorname{Div}(X) \to \operatorname{Div}(U)$ is an isomorphism.

**Lemma 4.2.12.** *If $X$ is a normal $k$-variety then there is an open subvariety $U \subset X$ such that $\operatorname{codim}(X \setminus U \subset X) \geq 2$ and $U$ is quasi-projective. In particular, any $X \in \mathscr{DP}$ contains an essential open subset $U \subset X$.*

*Proof.* By Chow's lemma, there is a proper birational morphism $\pi\colon \widetilde{X} \to X$ with $\widetilde{X}$ quasi-projective. Since $X$ is normal, $\pi$ is an isomorphism in codimension 1. Thus, $\widetilde{X}$ and $X$ have a common open subset $U$ whose complement in $X$ has codimension at least 2, and which is quasi-projective. Passing to the maximal factorial open subset yields the second statement. □

**Lemma 4.2.13.** *Suppose* $X, Y \in \mathcal{DP}$ *are divisorially proper varieties and* $\varphi \colon \tau(X) \to \tau(Y)$ *is an isomorphism of divisorial structures. If* $U \subset X$ *is an essential open subset then* $\varphi(U) \subset Y$ *is an essential open subset and there is an induced isomorphism* $\tau(U) \xrightarrow{\sim} \tau(\varphi(U))$.

*Proof.* First note that since $\varphi$ induces a homeomorphism $|X| \to |Y|$, we have that $\mathrm{codim}(X \backslash U \subset X) = \mathrm{codim}(Y \backslash \varphi(U) \subset Y)$. In particular, if $U$ is divisorially proper then so is $\varphi(U)$ by (4.1.4). By (4.1.9), we have isomorphisms

$$\tau(X)|_U \xrightarrow{\sim} \tau(U)$$

and

$$\tau(Y)|_{\varphi(U)} \xrightarrow{\sim} \tau(\varphi(U)).$$

On the other hand, $\varphi$ induces an isomorphism $\tau(X)|_U \xrightarrow{\sim} \tau(Y)|_{\varphi(U)}$. The result thus follows from (4.2.10). □

## 4.3 DEFINABLE SUBSPACES IN LINEAR SYSTEMS

Fix a divisorially proper variety $X \in \mathcal{DP}$ with infinite constant field (this assumption is used in this section when we apply Bertini's theorem; see especially (4.3.13) below). Let $P := |D|$ be the linear system associated to an effective divisor $D$.

**Definition 4.3.1.** A subspace $V \subset P$ is *definable* if there is a subset $Z \subset X$ such that
$$V = V(Z) := \{E \in P \mid Z \subset E\}.$$

**Remark 4.3.2.** If $Z \subset X$ is a subset and $Z' \subset X$ is the closure of $Z$ then $V(Z) = V(Z')$. When considering definable subspaces it therefore suffices to consider subspaces defined by closed subsets.

**Remark 4.3.3.** Note that $V(Z)$ is the projective space associated to the kernel of the restriction map
$$H^0(X, \mathcal{O}_X(D)) \to H^0(Z_{\mathrm{red}}, \mathcal{O}_X(D)|_{Z_{\mathrm{red}}}),$$
where we write $Z_{\mathrm{red}} \subset X$ for the reduced subscheme associated to the subspace $Z \subset |X|$.

**4.3.4.** Let $\kappa_X$ denote the constant field of $X$. As we now explain, when the $\kappa_X$-

points in $X$ are Zariski dense, the structure of definable subspaces is significantly simpler than in the general case.

**Lemma 4.3.5.** *Assume that the $\kappa_X$-points are dense in $X$, and let $V = V(Z)$ be a nonempty definable subspace of a linear system $P$ on $X$. Then there is an ascending chain of closed subsets*

$$Z = Z_1 \subsetneq \cdots \subsetneq Z_n$$

*such that the induced chain*

$$V(Z) = V(Z_1) \supsetneq \cdots \supsetneq V(Z_n)$$

*is a full flag of linear subspaces ending in a point.*

*Proof.* By induction, it suffices to produce $Z_2 \supsetneq Z_1 = Z$ such that $V(Z_2) \subsetneq V(Z_1)$ has codimension 1.

For this, note that if $x \in X$ is a $\kappa_X$-point then either

$$V(Z \cup \{x\}) = V(Z)$$

or $V(Z \cup \{x\})$ has codimension 1 in $V(Z)$ since $x$ is $\kappa_X$-rational. Furthermore, equality happens if and only if $x \in E$ for all $E \in V(Z)$. It therefore suffices to observe that there exists a point $x \in X(\kappa_X)$ that does not lie in every element of $V(Z)$. This follows from fixing $E \in V(Z)$, noting that $E$ is a proper closed subset, and applying our assumption that the $\kappa_X$-points are Zariski dense. $\square$

**Corollary 4.3.6.** *The dimension of $P$ is equal to 1 more than the length of a maximal chain of definable subspaces.*

*Proof.* Take $Z = \emptyset$ in (4.3.5). $\square$

**Corollary 4.3.7.** *Assume that the $\kappa_X$-points are dense in $X$. For a basepoint-free linear system $P$ on $X$, the definable lines in $P$ are precisely those definable subspaces with more than one element that are minimal with respect to inclusions of definable subspaces.*

*Proof.* By (4.3.5), any definable set of higher dimension contains a definable line. $\square$

**Example 4.3.8.** The conclusions of (4.3.5) and (4.3.7) are false without the assumption that the $\kappa_X$-points are dense. For example, let $X \subset \mathbb{P}^2_{\mathbf{R}}$ be the conic given by

$$V(X^2 + Y^2 + Z^2) \subset \mathbb{P}^2_{\mathbf{R}},$$

so we have an isomorphism $\sigma : \mathbb{P}^1_{\mathbf{C}} \simeq X \otimes_{\mathbf{R}} \mathbf{C}$. If $\mathscr{L}$ is an ample invertible sheaf on $X$ then $\sigma^* \mathscr{L}_{\mathbf{C}} \simeq \mathcal{O}_{\mathbb{P}^1_{\mathbf{C}}}(2n)$ for $n > 0$. From this we see that each closed point $x \in X$ imposes a codimension 2 condition on $|\mathscr{L}|$, and since this projective space has dimension $2n$ we conclude that there are no definable lines in any ample linear system on $X$.

**4.3.9.** Detecting lines is more subtle over arbitrary fields. This is closely related to the counterexamples to (4.1.14) for curves discussed in Section 5.3.

**Lemma 4.3.10.** *Let $\mathscr{L}$ be a line bundle on $X$ with associated linear system $\mathbf{P}(V)$, where $V := H^0(X, \mathscr{L})$. Let $\ell \subset \mathbf{P}(V)$ be a line corresponding to a 2-dimensional subspace $T \subset V$. Let $Z' \subset X$ be the maximal reduced closed subscheme of the intersection of the zero loci of elements of $T$. Then $\ell$ is definable if and only if the dimension of the kernel*

$$K := \ker(H^0(X, \mathscr{L}) \to H^0(Z', \mathscr{L}|_{Z'}))$$

*is equal to 2.*

*Proof.* First suppose $\ell$ is definable, so we can write $\ell = V(Z)$ for some closed subset $Z \subset |X|$, which we view as a subscheme with the reduced structure. Then by definition $\ell$ is the projective subspace of $\mathbb{P}V$ associated to the kernel of the map

$$H^0(X, \mathscr{L}) \to H^0(Z, \mathscr{L}|_Z),$$

which must therefore equal $T$. In particular, we have $Z \subset Z'$, which implies that

$$T \subset K = \ker(H^0(X, \mathscr{L}) \to H^0(Z', \mathscr{L}|_{Z'}))$$
$$\cap$$
$$\ker(H^0(X, \mathscr{L}) \to H^0(Z, \mathscr{L}|_Z)) = T.$$

It follows that $K = T$, and, in particular, $K$ has dimension 2.

Conversely, if $K$ has dimension 2 then we have $T = K$ and $\ell = V(Z')$. $\square$

**Lemma 4.3.11.** *If $Y$ is a projective variety over a field $K$ and $D \subset Y$ is a Cartier divisor such that*

$$H^0(Y, \mathcal{O}_Y) \to H^0(D, \mathcal{O}_D)$$

*is surjective (for example, if $D$ is geometrically connected and geometrically reduced), then*

$$H^1(Y, \mathcal{O}_Y(-D)) \to H^1(Y, \mathcal{O}_Y)$$

*is injective.*

*Proof.* This follows from taking cohomology of the short exact sequence

$$0 \to \mathcal{O}_Y(-D) \to \mathcal{O}_Y \to \mathcal{O}_D \to 0. \qquad \square$$

**Lemma 4.3.12.** *Let $Y$ be an integral, projective variety over a field $K$ and let $\mathscr{L}$ be an ample invertible sheaf with sections $s_1, s_2 \in \Gamma(Y, \mathscr{L})$. Let $D_i$ be the zero* locus of $s_i$, and assume that $D_1 \cap D_2$ is geometrically reduced and of codimension 2. Then

(1) $D_1$ and $D_2$ are generically reduced;

(2) if $Y$ is geometrically integral and for $i = 1, 2$ the intersection $D_i \cap Y_{\mathrm{norm}}$ of $D_i$ with the normal locus of $Y$ is schematically dense in $D_i$, then

$$\dim_K H^0(Y, \mathscr{L}(-(D_1 \cap D_2))) = 2,$$

where $\mathscr{L}(-(D_1 \cap D_2))$ denotes the tensor product of $\mathscr{L}$ with the ideal sheaf of $D_1 \cap D_2$.

*Proof.* For (1) we show that $D_1$ is generically reduced (the result for $D_2$ follows by symmetry). Since $D_1 \cap D_2$ is reduced and of codimesion 2, $D_1 \cap D_2$ has a dense open subset that is regular. In a neighborhood of such a point $x \in D_1 \cap D_2$ the divisor $D_1$ is regular [Sta22, Tag 00NU], and therefore every irreducible component of $D_1$ which meets the regular locus of $D_1 \cap D_2$ is generically reduced. To complete the proof of (1) it suffices to note that every irreducible component of $D_1$ meets the regular locus of $D_1 \cap D_2$, since the non-regular locus of $D_1 \cap D_2$ has codimension $\geq 3$ in $Y$ and therefore codimension $\geq 2$ in each irreducible component of $D_1$. Since $D_2$ meets every irreducible component of $D_1$, since $\mathscr{L}$ is ample, this proves (1).

To prove (2) note that since $Y_{\mathrm{norm}}$ is $S_2$ the divisors $D_1 \cap Y_{\mathrm{norm}}$ and $D_2 \cap Y_{\mathrm{norm}}$ are $S_1$, and by (1) they are also $R_0$ and therefore $D_1 \cap Y_{\mathrm{norm}}$ and $D_2 \cap Y_{\mathrm{norm}}$ are reduced [Sta22, Tag 0344]. By our assumption that $D_i \cap Y_{\mathrm{norm}}$ is schematically dense in $D_i$ it follows that $D_1$ and $D_2$ are reduced. In fact, we claim that $D_1$ and $D_2$ are geometrically reduced. We give the proof for $D_1$. Since $D_1$ is reduced, for any dense open subset $j\colon U_1 \hookrightarrow D_1$ the map

$$\mathscr{O}_{D_1} \to j_* \mathscr{O}_{U_1}$$

is injective. The formation of this map commutes with passing to a finite extension of $k$, and therefore the same holds after making a finite extension of $k$. Combining this with (1) we get that $D_1$ is geometrically reduced, and the divisor $D_1$ is also geometrically connected since $\mathscr{L}$ is ample and $Y$ is geometrically integral. To get statement (2) from this, note that since $D_1$ and $D_2$ are reduced we have an exact sequence

$$0 \longrightarrow \mathscr{L}^{-1} \xrightarrow{(s_1, s_2)} \mathscr{O}_Y \oplus \mathscr{O}_Y \xrightarrow{(s_2, -s_1)} \mathscr{L}(-(D_1 \cap D_2)) \longrightarrow 0,$$

from which we get (2) by taking cohomology and applying (4.3.11). □

**Proposition 4.3.13.** *Let $\mathscr{O}_X(1)$ be a very ample invertible sheaf on $X$ with associated linear system $P = |\mathscr{O}_X(1)|$. Let $j\colon X \hookrightarrow \overline{X}$ be the compatification of $X$ provided by the given projective embedding. Let $\mathscr{O}_{\overline{X}}(1)$ be the line bundle on*

$\overline{X}$ *obtained from the embedding so* $|\mathcal{O}_X(1)| = |\mathcal{O}_{\overline{X}}(1)|$.

(1) *Let* $V \subset \mathrm{Gr}(1, P)$ *be the subset of lines* $\ell$ *spanned by elements* $D$ *and* $E$ *of* $P$ *on* $\overline{X}$ *for which* $D$ *is geometrically reduced,* $E$ *is geometrically integral, the intersection* $B := E \cap D \subset \overline{X}$ *is geometrically reduced and does not contain any components of* $D$ *or* $E$, *and the inclusions*

$$D \cap X \hookrightarrow D, \quad E \cap X \hookrightarrow E, \quad B \cap X \hookrightarrow B$$

*are all schematically dense. Then* $V$ *is a nonempty Zariski open subset of* $\mathrm{Gr}(1, P)$ *and every element of* $V$ *is definable.*

(2) *If* $D \in |\mathcal{O}_{\overline{X}}(1)| = P$ *is a geometrically reduced divisor in* $\overline{X}$ *for which* $D \cap X \subset D$ *is dense, then* $D$ *lies in the sweep of the maximal Zariski open subset of the definable locus in* $\mathrm{Gr}(1, P)$.

*Proof.* Let $D$ and $E$ be as in (1). Since $B \cap X \subset B$ is schematically dense, we have by (4.3.2)

$$V(B \cap X) = V(B).$$

By (4.3.10) it therefore suffices to show that $V(B) \subset P$ is a line. For this we apply (4.3.12) with $Y = \overline{X}$, $\mathscr{L} = \mathcal{O}_{\overline{X}}(1)$, $D_1 = E$, and $D_2 = D$. Note that $D_1 \cap D_2 = B$ is geometrically reduced by assumption. Furthermore, since $X$ is normal, being a divisorially proper variety, and $X \cap D \hookrightarrow D$ and $E \cap X \hookrightarrow E$ are schematically dense the conditions in (4.3.12 (2)) are satisfied and we conclude that $V(B)$ is a line. Finally note that the conditions on $D$ and $E$ are open conditions. Therefore to prove (1) it suffices to show that $V$ is nonempty, which follows from Bertini's theorem [FOV99, 3.4.10 and 3.4.14].

In fact, given geometrically reduced $D$ with $D \cap X \subset D$ dense, the set of $E$ such that $(D, E)$ satisfy the conditions in (1) is open and nonempty by [FOV99, 3.4.14]. From this statement (2) also follows.                                   $\square$

**Corollary 4.3.14.** *Assume that the dimensions of* $X$ *and* $Y$ *are* $\geq 2$ *and let* $\sigma : \tau(X) \to \tau(Y)$ *be an isomorphism of divisorial structures such that there are very ample invertible sheaves* $\mathcal{O}_X(1)$ *and* $\mathcal{O}_Y(1)$ *with* $\sigma$ *inducing a bijection*

$$s : |\mathcal{O}_X(1)| \to |\mathcal{O}_Y(1)|.$$

*Then the map* $s$ *satisfies the hypotheses of (3.2.1).*

*Proof.* Indeed, the locus of lines described in (4.3.13 (1)) suffices. Note that we know that the linear systems have dimension at least 2 because $X$ and $Y$ have dimension at least 2 and the linear systems are very ample.                     $\square$

**Example 4.3.15.** In general, the set of definable lines in $\mathrm{Gr}(1, P)$ is not open. An explicit example is the following.

Consider three $k$-points $A, B, C \in \mathbb{P}_k^2$, say $A = [0 : 0 : 1]$, $B = [0 : 1 : 0]$, and $C = [1 : 0 : 0]$. For a line $L \subset \mathbb{P}_k^2$ passing through $A$, let $T_L$ be the set of

$F \in H^0(\mathbb{P}_k^2, \mathcal{O}_{\mathbb{P}_k^2}(2))$ such that $V(F)$ passes through $A, B, C$ and is tangent to $L$ at $A$. Concretely, if $X$, $Y$, and $Z$ are the coordinates on $\mathbb{P}_k^2$ and $L$ is given by

$$\alpha X + \beta Y = 0,$$

then $T_L$ is given by

$$T_L = \{aXY + b(\alpha X + \beta Y)Z \mid a, b \in k\}.$$

In particular, $T_L$ gives a line $\ell_L$ in $|\mathcal{O}_{\mathbb{P}^2}(2)|$.

If $\alpha$ and $\beta$ are nonzero then $L$ does not pass through $B$ and $C$ and the set-theoretic base locus of $T_L$ is equal to $\{A, B, C\}$ and the space of degree 2 polynomials passing through these three points has dimension 3. Therefore for such $L$ the line $\ell_L$ is not definable.

However, for $L$ the lines $X = 0$ or $Y = 0$ the line $\ell_L$ is definable. Indeed, in this case the set-theoretic base locus of $\ell_L$ is given by the union of the line $L$ together with a third point not on the line, from which one sees that $T_L$ is definable.

Letting $\alpha$ and $\beta$ vary we obtain a 1-parameter family of lines $\mathbb{P}^1 \simeq \Sigma \subset \mathbb{G}\mathrm{r}(1, |\mathcal{O}_{\mathbb{P}_k^2}(2)|)$ whose general member is not definable but with two points giving definable lines. It follows that the definable locus is not open in this case.

# Chapter Five

---

## Reconstruction from divisorial structures: infinite fields

In this chapter we prove (4.1.14) in the case when the constant fields are infinite. Suppose $X$ and $Y$ are divisorially proper varieties of dimension at least 2 with infinite constant fields. We need to show that given an isomorphism $\varphi \colon \tau(X) \to \tau(Y)$ of divisorial structures, there is a unique isomorphism of schemes $f \colon X \to Y$ such that $\tau(f) = \varphi$.

We proceed in several steps. In Section 5.1, we show that it suffices to assume that $X$ is quasi-projective. Then, in Section 5.2 we handle the quasi-projective case. In this case, fixing an ample invertible sheaf $\mathscr{L}_X$ on $X$, the isomorphism $\varphi$ induces an invertible sheaf $\mathscr{L}_Y$ on $Y$, which is ample by (4.2.8). The map $\varphi$ then induces bijections

$$|\mathscr{L}_X^{\otimes n}|^{\text{set}} \to |\mathscr{L}_Y^{\otimes n}|^{\text{set}}$$

for all $n$, and our challenge is to show that these are algebraic maps that can be lifted to isomorphisms between the homogeneous coordinate rings of $X$ and $Y$. This is achieved by showing that these bijections preserve definable lines, since in dimension at least 2 a general pencil in a basepoint-free ample linear system is determined by its topological base locus.

The fundamental theorem of definable projective geometry (3.1.5) then algebraizes the bijections of linear systems. One must then check that the linearizations given by (3.1.5) on the graded components are all linear over the same isomorphism $\kappa_X \to \kappa_Y$ of constant fields. This uses the fact that these spaces are related by addition of divisors.

Having created the isomorphisms of graded components, we show starting in (5.2.7) that these isomorphisms respect the graded components of the ideals of $X$ and $Y$ in their respective projective embeddings, completing the proof.

Finally, in Section 5.3, we show that over non-algebraically closed base fields, (4.1.14) admits counterexamples. While one might naïvely think that the use of model theory in [Zil14] is responsible for the assumption of an algebraically closed base field, these counterexamples show that something deeper is happening. It remains somewhat mysterious to the authors what the ultimate truth should be in dimension 1, and how precisely the model-theoretic techniques employed in [Zil14] are related to the purely algebro-geometric techniques used in the present work.

## 5.1 REDUCTION TO THE QUASI-PROJECTIVE CASE

**Proposition 5.1.1.** *In order to prove (4.1.14), it suffices to prove it assuming that $X$ is quasi-projective.*

*Proof.* By (4.2.13), there are essential open subsets $U \subset X$ and $V \subset Y$ such that $V = \varphi(U)$ and $\tau$ induces an isomorphism $\tau(U) \to \tau(V)$. If we know the result in the quasi-projective case then the homeomorphism $|U| \to |V|$ induced by $\varphi$ extends to an algebraic isomorphism $f_U \colon U \to V$ such that $\tau(f_U) = \varphi|_U$. By (5.1.4) below, $f$ extends uniquely to an isomorphism of schemes $f \colon X \to Y$ such that $\tau(f) = \varphi$. $\qquad\square$

We include some straightforward lemmas, culminating in (5.1.4), showing that an isomorphism between open subsets of varieties that admits a topological extension has an algebraic extension.

**Lemma 5.1.2.** *If $X$ is a variety over a field $K$ then for any point $x \in X$ we have that*
$$\overline{\{x\}} = \bigcap \overline{\{y\}},$$
*the intersection taken over all points $y \in X$ of codimension at most $1$ such that $x \in \overline{\{y\}}$.*

*Proof.* It suffices to show that, for all $z \in X$ with $z \notin \overline{\{x\}}$, there exists a codimension $1$ point $y \in X$ such that $x \in \overline{\{y\}}$ but $z \notin \overline{\{y\}}$. Choose an affine open $\mathrm{Spec}(A)$ containing $z$. If $x \in \mathrm{Spec}(A)$ then, since $z \notin \overline{\{x\}}$, there must be $f \in A$ vanishing at $x$ but not $z$. By the Hauptidealsatz, the vanishing locus of $f$ is a union of codimension $1$ points, one of which must contain $x$, but none of which contain $z$. If $x \notin \mathrm{Spec}(A)$ then the complement of $\mathrm{Spec}(A)$ is a union of codimension $1$ points, one of which must contain $x$, but none of which contain $z$. $\qquad\square$

**Lemma 5.1.3.** *Suppose $f, g \colon |X| \to |Y|$ are homeomorphisms of the Zariski topological spaces of two normal varieties. Given an open subset $U \subset |X|$ containing all points of codimension $1$, if $f|_U = g|_U$ then $f = g$.*

*Proof.* By (5.1.2), we can characterize any point $x \in X$ as the unique generic point of an irreducible intersection of closures of codimension $\leq 1$ points. But $f$ and $g$ establish the same bijection on the sets of points of codimension $\leq 1$, and, since they are homeomorphisms, therefore the same bijections on the closures of those points. The result follows. $\qquad\square$

**Lemma 5.1.4.** *Suppose $X$ and $Y$ are normal varieties, $U \subset X$ and $V \subset Y$ are dense open subvarieties with complements of codimension at least $2$. Suppose $f \colon |X| \to |Y|$ is a homeomorphism of Zariski topological spaces such that $f(U) = V$ and $f|_U$ extends to an isomorphism $\widetilde{f_U} \colon U \to V$ of schemes. Then $\widetilde{f_U}$ extends to a unique isomorphism of schemes $\widetilde{f} \colon X \to Y$ whose associated morphism of topological spaces is $f$.*

*Proof.* Let us first show that $\widetilde{f}_U$ extends to a morphism of schemes $\widetilde{f}\colon X \to Y$. If $W_1, W_2 \subset X$ are two open subsets and $\widetilde{f}_{W_i}\colon W_i \to Y$ $(i = 1, 2)$ are morphisms of schemes such that $\widetilde{f}_{W_i}$ and $\widetilde{f}_U$ agree on $W_i \cap U$, then since $Y$ is separated the morphisms $\widetilde{f}_{W_1}$ and $\widetilde{f}_{W_2}$ agree on $W_1 \cap W_2$. To extend $\widetilde{f}_U$ it therefore suffices to show that $\widetilde{f}_U$ extends locally on $X$. In particular, by covering $X$ by open subsets of the form $f^{-1}(\mathrm{Spec}(A))$ for affines $\mathrm{Spec}(A) \subset Y$, we are reduced to proving the existence of an extension in the case when $Y = \mathrm{Spec}(A)$ is affine. In this case, to give a morphism of schemes $X \to \mathrm{Spec}\, A$, it suffices to give a morphism of rings $A \to \Gamma(X, \mathcal{O}_X)$. By Krull's theorem, $\Gamma(U, \mathcal{O}_X) = \Gamma(X, \mathcal{O}_X)$. Thus, the morphism $\widetilde{f}_U\colon U \to \mathrm{Spec}\, A$ extends uniquely to a morphism $\widetilde{f}\colon X \to \mathrm{Spec}\, A$, and we get the desired extension $\widetilde{f}$.

Applying the same argument to the inverse of $f$, and using that $X$ is separated, we see that in fact $\widetilde{f}$ is an isomorphism. In particular, its associated map of topological spaces is a homeomorphism and agrees with $f$ on $|U|$. We conclude by (5.1.3) that $|\widetilde{f}| = f$. $\qquad\square$

## 5.2   THE QUASI-PROJECTIVE CASE

We now assume that $X$ is quasi-projective and the ground field $k$ is infinite. Let $\mathcal{O}_X(1)$ denote a very ample invertible sheaf on $X$ and for $m \geq 1$ let

$$\Sigma\colon |\mathcal{O}_X(1)|^{\times m} \to |\mathcal{O}_X(m)|$$

denote the addition map on divisors. So a point of $|\mathcal{O}_X(1)|^{\times m}$ is given by a collection of divisors $\underline{D} = (D_1, \dots, D_m)$ and $\Sigma(\underline{D})$ corresponds to the divisor $D_1 + \cdots + D_m$.

**Lemma 5.2.1.** *For a general point $\underline{D}$ of $|\mathcal{O}_X(1)|^{\times m}$ the point $\Sigma(\underline{D}) \in |\mathcal{O}_X(m)|$ lies in the sweep of the maximal Zariski open subset of the set of definable lines in $\mathrm{Gr}(1, |\mathcal{O}_X(m)|)$.*

*Proof.* Let $\overline{X}$ be the projective closure of $X$ in the embedding given by $\mathcal{O}_X(1)$. Note that $\overline{X}$ is also geometrically integral. Indeed, if $j\colon X \hookrightarrow \overline{X}$ is the inclusion then the map $\mathcal{O}_{\overline{X}} \to j_*\mathcal{O}_X$ is injective, and remains injective after base field extension. Since $X$ is geometrically integral, it follows that $\overline{X}$ is as well.

By Bertini's theorem [FOV99, 3.4.14], for a general choice of

$$\underline{D} \in |\mathcal{O}_{\overline{X}}(1)|^{\times m} = |\mathcal{O}_X(1)|^{\times m}$$

the point $\Sigma(\underline{D}) \in |\mathcal{O}_{\overline{X}}(m)|$ satisfies the conditions on $D$ in (4.3.13 (2)) and the result follows. $\qquad\square$

**Lemma 5.2.2.** *Let $X$ be a $k$-variety and $\mathcal{O}_X(1)$ a very ample invertible sheaf* *on X. Given a closed point $z \in X$, we have that*

$$\{z\} = \bigcap |D| \subset |X|, \qquad (5.2.2.1)$$

*the intersection taken over all irreducible divisors $D$ in $|\mathcal{O}_X(m)|$ containing $z$, for some $m$.*

*Proof.* This follows from Bertini's theorem. Precisely, by [Sta22, Tag 0FD5] (applied to the projective closure of $X$ if $X$ is not proper) there exists an integer $m > 0$ such that if

$$V_m := \mathrm{Ker}(H^0(X, \mathcal{O}_X(m)) \to \mathcal{O}_X(m)(z))$$

then the map

$$X \setminus \{z\} \to \mathbb{P}(V_m)$$

is an immersion. We then get the result by cutting with hyperplanes. $\square$

**Corollary 5.2.3.** (1) *With notation and assumptions as in (5.2.2), let $z \in X$ be a closed point and for each $m$ let $U_m \subset V_m$ be an open dense subset of the space $V_m$ of divisors in $|\mathcal{O}_X(m)|$ passing through $z$. Then for some $m$ there exists an integer $r > 0$ and irreducible divisors $D_1, \ldots, D_r \in U_m \subset |\mathcal{O}_X(m)|$ such that*

$$\{z\} = D_1 \cap \cdots \cap D_r.$$

(2) *If $X$ is a divisorially proper variety of dimension at least 2 with constant field $k$ and very ample invertible sheaf $\mathcal{O}_X(1)$, then for any closed point $z \in X$ there exists an integer $m > 0$ and divisors $D_1, \ldots, D_r \in |\mathcal{O}_X(m)|$ in the sweep of the maximal open subset of the definable lines in $|\mathcal{O}_X(m)|$ such that*

$$\{z\} = D_1 \cap \cdots \cap D_r.$$

*Proof.* (1) By (5.2.2) there exists integers $r, m$ and irreducible divisors

$$D_1, \ldots, D_r \in |\mathcal{O}_X(m)|$$

such that

$$\{z\} = D_1 \cap \cdots \cap D_r.$$

Now observe that the locus of tuples $(D_1, \ldots, D_r)$ is a Zariski open subset of $|\mathcal{O}_X(m)|^r$ and therefore has nonempty intersection with $U_m^r$.

To prove (2), it suffices by (1) and (4.3.13 (2)) to show that for $m$ sufficiently large there is a dense open subset $U \subset V_m$ of geometrically reduced and irreducible divisors. For this, let $\overline{X}$ be the closure of $X$ in the projective embedding provided by $\mathcal{O}_X(1)$, and let $\mathcal{O}_{\overline{X}}(1)$ be the canonical extension of $\mathcal{O}_X(1)$ to $\overline{X}$. Since a divisorially proper variety is geometrically reduced (see (4.1.6)) and $X \hookrightarrow \overline{X}$ is schematically dense, the projective variety $\overline{X}$ is geometrically reduced. Now apply (11.1.16) to $\overline{X}$, with $Z_i = X$ (with a single index $i$) and $W = \{z\}$.                                                                                                                    □

**Proposition 5.2.4.** *Suppose $X$ and $Y$ are divisorially proper varieties of dimension at least 2 with infinite constant fields, and assume that $X$ is polarizable. Given an isomorphism $\varphi\colon \tau(X) \to \tau(Y)$, the associated homeomorphism $|X| \to |Y|$ extends to an isomorphism $X \to Y$ in $\mathfrak{DV}$.*

*Proof.* Let $D$ be an ample basepoint-free divisor with $\mathcal{O}_X(D) = \mathcal{O}_X(1)$ and let $\mathcal{O}_Y(1)$ denote $\mathcal{O}_Y(\varphi(D))$. After possibly taking a power of our choice of polarization, we may assume that $\mathcal{O}_X(1)$ and $\mathcal{O}_Y(1)$ are very ample. Note that we are not asserting that we can detect very ampleness from $\tau(X)$ and $\tau(Y)$, just that we know that such a multiple must exist, so we are free to choose one.

By (4.3.13 (1)), for each $m > 0$, the sets of definable lines in the Grassmannians of $|\mathcal{O}_X(m)|$ and $|\mathcal{O}_Y(m)|$ contain dense Zariski open sets, and thus by (3.1.5), there is an isomorphism $\sigma_m\colon \kappa_X \to \kappa_Y$ and a $\sigma_m$-linear isomorphism $\gamma_m\colon |\mathcal{O}_X(m)| \to |\mathcal{O}_Y(m)|$ that agrees with $\varphi$ on a dense Zariski open subset $U \subset |\mathcal{O}_X(m)|$.

Consider the diagram of addition maps

$$
\begin{array}{ccc}
|\mathcal{O}_X(m)| & \longrightarrow & |\mathcal{O}_Y(m)| \\
\Sigma_X \uparrow & & \Sigma_Y \uparrow \\
|\mathcal{O}_X(1)|^{\times m} & \longrightarrow & |\mathcal{O}_Y(1)|^{\times m}.
\end{array}
$$

Since a general sum of divisors in $\mathcal{O}(1)$ lies in the sweep of the maximal open subset of the definable points by (5.2.1), we see that the associated diagram of sets

$$
\begin{array}{ccc}
|\mathcal{O}_X(m)| & \xrightarrow{\gamma_m} & |\mathcal{O}_Y(m)| \\
\Sigma_X \uparrow & & \Sigma_Y \uparrow \\
|\mathcal{O}_X(1)|^{\times m} & \xrightarrow{\gamma_1^{\times m}} & |\mathcal{O}_Y(1)|^{\times m}
\end{array}
\qquad (5.2.4.1)
$$

commutes over a dense open set of points. Because the maps appearing are maps on point sets arising from maps of integral separated schemes, the diagram commutes. (The vertical maps preserve the base field and thus are even maps of varieties, but the horizontal maps may not.)

**Lemma 5.2.5.** *The two isomorphisms of fields $\sigma_1, \sigma_m\colon \kappa_X \to \kappa_Y$ are equal.*

**Remark 5.2.6.** The basic reason why this is nontrivial is that the map $\Sigma$ does not preserve the linear structure.

*Proof.* Let $U_1 \subset |\mathcal{O}_X(1)|$ (resp. $U_m \subset |\mathcal{O}_X(m)|$) be the sweep of the maximal open subset in the set of definable lines in $|\mathcal{O}_X(1)|$ (resp. $|\mathcal{O}_X(m)|$). Then $U_1^{\times m} \subset$

$|\mathcal{O}_X(1)|^{\times m}$ is a nonempty Zariski open subset, and therefore

$$V := (\Sigma_X)^{-1}(U_m) \cap U_1^{\times m} \subset |\mathcal{O}_X(1)|^{\times m}$$

is a Zariski open subset of $|\mathcal{O}_X(1)|^{\times m}$. We can therefore find points

$$P, Q, R, P_2, \ldots, P_m \in |\mathcal{O}_X(1)|$$

such that the three points of $|\mathcal{O}_X(1)|^{\times m}$ given by

$$(P, P_2, \ldots, P_m), \quad (Q, P_2, \ldots, P_m), \quad (R, P_2, \ldots, P_m)$$

lie in $V$ and $P, Q, R \in |\mathcal{O}_X(1)|$ are collinear. Since $\gamma_1$ and $\gamma_m$ agree with the maps defined by $\varphi$ on $U_1$ and $U_m$, it follows that we have equalities

$$(\gamma_m \circ \Sigma_X)(P, P_2, \ldots, P_m) = (\Sigma_Y \circ \gamma_1^{\times m})(P, P_2, \ldots, P_m) =: \overline{P}$$
$$(\gamma_m \circ \Sigma_X)(Q, P_2, \ldots, P_m) = (\Sigma_Y \circ \gamma_1^{\times m})(Q, P_2, \ldots, P_m) =: \overline{Q}$$
$$(\gamma_m \circ \Sigma_X)(R, P_2, \ldots, P_m) = (\Sigma_Y \circ \gamma_1^{\times m})(R, P_2, \ldots, P_m) =: \overline{R}$$

in $|\mathcal{O}_Y(m)|$. Let $\overline{L} \subset |\mathcal{O}_Y(m)|$ be the line through $\overline{P}$ and $\overline{Q}$, and let $L \subset |\mathcal{O}_X(1)|$ be the line through $P$ and $Q$. Then

$$\Sigma_X(L \times \{P_2\} \times \cdots \{P_m\})$$

is the line in $|\mathcal{O}_X(m)|$ through the two points

$$\Sigma_X(P, P_2, \ldots, P_m), \quad \Sigma_X(Q, P_2, \ldots, P_m).$$

Since $\gamma_m$ takes lines to lines, it follows that

$$(\gamma_X \circ \Sigma_X)(L \times \{P_2\} \times \cdots \{P_m\}) = \overline{L}.$$

Similarly, since $\gamma_1$ takes lines to lines and agrees on $U_1$ with the map defined by $\varphi$, we find that

$$(\Sigma_Y \circ \gamma_1^{\times m})(L \times \{P_2\} \times \cdots \{P_m\}) = \overline{L}.$$

Since $\gamma_m \circ \Sigma_X$ and $\Sigma_Y \circ \gamma_1^{\times m}$ agree on a dense open subset of $L$, viewed as embedded in $|\mathcal{O}_X(1)|^{\times m}$ via the identification

$$L \simeq L \times \{P_2\} \times \cdots \times \{P_m\},$$

we conclude that the two compositions

$$\kappa_X \xrightarrow{\;\alpha\;} L \subset |\mathcal{O}_X(1)|^{\times m} \xrightarrow{\;\gamma_m \circ \Sigma_X\;} \overline{L} \xrightarrow{\;\beta^{-1}\;} \kappa_Y$$

and

$$\kappa_X \xrightarrow{\ \alpha\ } L \subset |\mathcal{O}_X(1)|^{\times m} \xrightarrow{\ \Sigma_Y \circ \gamma_1^{\times m}\ } \overline{L} \xrightarrow{\ \beta^{-1}\ } \kappa_Y$$

agree on all but finitely many elements of $\kappa_X$, where $\alpha : \kappa_X \simeq L \setminus \{R\}$ (resp. $\beta : \kappa_Y \simeq \overline{L} \setminus \{\overline{R}\}$) is the isomorphism obtained as in the proof of (3.1.5) using the three points $P$, $Q$, $R$ (resp. $\overline{P}$, $\overline{Q}$, $\overline{R}$). Now the first of these maps is the map $\sigma_m$ and the second is $\sigma_1$. We conclude that $\sigma_1(a) = \sigma_m(a)$ for all but finitely many elements $a \in \kappa_X$, which implies that $\sigma_1 = \sigma_m$.  $\square$

In the rest of the proof we write $\sigma \colon \kappa_X \to \kappa_Y$ for the isomorphism $\sigma_m = \sigma_1$.

**5.2.7.** Next observe that the diagram of schemes

$$
\begin{array}{ccc}
|\mathcal{O}_X(m)|^{\mathrm{var}} & \xrightarrow{\ \gamma_m\ } & |\mathcal{O}_Y(m)|^{\mathrm{var}} \\[4pt]
\Sigma_X \big\uparrow & & \Sigma_Y \big\uparrow \\[4pt]
(|\mathcal{O}_X(1)|^{\mathrm{var}})^{\times m} & \xrightarrow{\ \gamma_1^{\times m}\ } & (|\mathcal{O}_Y(1)|^{\mathrm{var}})^{\times m}
\end{array}
\qquad (5.2.7.1)
$$

(5.2.4.1) commutes, since the two morphisms obtained by going around the different directions of the diagram are semilinear with respect to the same field isomorphism and agree on a dense set of points.

Consider the embeddings

$$\nu_X \colon X \hookrightarrow |\mathcal{O}_X(1)|^{\vee}$$

and

$$\nu_Y \colon Y \hookrightarrow |\mathcal{O}_Y(1)|^{\vee}$$

and let $\overline{X}$ (resp. $\overline{Y}$) be the scheme-theoretic closure of $\nu_X(X)$ (resp. $\nu_Y(Y)$). Let $S_X$ (resp. $S_Y$) be the symmetric algebra on $\Gamma(X, \mathcal{O}_X(1))$ (resp. $\Gamma(Y, \mathcal{O}_Y(1))$), so $\overline{X}$ (resp. $\overline{Y}$) is given by a graded ideal $I_{\overline{X}} \subset S_X$ (resp. $I_{\overline{Y}} \subset S_Y$).
Choosing a lift

$$\widetilde{\gamma}_1 \colon \Gamma(X, \mathcal{O}_X(1)) \to \Gamma(Y, \mathcal{O}_Y(1))$$

yields an induced $\sigma$-linear isomorphism of graded rings

$$\gamma^{\natural} \colon S_X \to S_Y$$

that is uniquely defined up to scalars. We claim that $\gamma^{\natural}(I_{\overline{X}}) = I_{\overline{Y}}$.

For this, consider the diagram

$$
\begin{array}{ccc}
\Gamma(X, \mathcal{O}_X(m)) & \xrightarrow{\;\;\tilde{\gamma}_m\;\;} & \Gamma(Y, \mathcal{O}_Y(m)) \\
{\scriptstyle p_X}\big\uparrow & & \big\uparrow{\scriptstyle p_Y} \\
S_X^m = \Gamma(X, \mathcal{O}_X(1))^{\otimes m} & \xrightarrow{\;\;\gamma_m^{\natural}\;\;} & \Gamma(Y, \mathcal{O}_Y(1))^{\otimes m} = S_Y^m \\
\big\uparrow & & \big\uparrow \\
\Gamma(X, \mathcal{O}_X(1))^{\times m} & \xrightarrow{\;\;\tilde{\gamma}^{\times m}\;\;} & \Gamma(Y, \mathcal{O}_Y(1))^{\times m}
\end{array}
$$

arising as follows. The vertical arrows are the natural multiplication maps, and the induced linear maps from the universal property of $\otimes$. The arrow $\tilde{\gamma}_m$ is a lift of $\gamma_m$. By the commutativity of diagram (5.2.7.1), we see that this diagram commutes (up to suitably scaling $\tilde{\gamma}_m$), which implies that $\gamma_m^{\natural}(I_{\overline{X},m}) = I_{\overline{Y},m}$, as desired.

In summary, we have shown that if

$$
A_{\overline{X}} \subset \oplus_{m \geq 0} \Gamma(X, \mathcal{O}_X(m)) \quad (\text{resp. } A_{\overline{Y}} \subset \oplus_{m \geq 0} \Gamma(Y, \mathcal{O}_Y(m)))
$$

denotes the subring generated by $\Gamma(X, \mathcal{O}_X(1))$ (resp. $\Gamma(Y, \mathcal{O}_Y(1))$), then we have an isomorphism of graded rings

$$
\tilde{\gamma} \colon A_{\overline{X}} \to A_{\overline{Y}}
$$

such that the isomorphism induced by $\tilde{\gamma}$ in degree $m$

$$
|\mathcal{O}_{\overline{X}}(m)| \to |\mathcal{O}_{\overline{Y}}(m)|
$$

fits into a commutative diagram

$$
\begin{array}{ccc}
|\mathcal{O}_{\overline{X}}(m)| & \longrightarrow & |\mathcal{O}_{\overline{Y}}(m)| \\
\big\downarrow & & \big\downarrow \\
|\mathcal{O}_X(m)| & \xrightarrow{\;\;\tilde{\gamma}_m\;\;} & |\mathcal{O}_Y(m)|,
\end{array}
$$

where the vertical maps are the restriction maps.

In other words, if we let

$$
f \colon \overline{X} \to \overline{Y}
$$

be the isomorphism given by $\tilde{\gamma}$, then the diagram

$$
\begin{array}{ccc}
\overline{X} & \xrightarrow{\ \ f\ \ } & \overline{Y} \\
\wr\downarrow & & \downarrow\wr \\
\mathrm{Proj}(A_{\overline{X}}) & \xrightarrow{\ \tilde{\gamma}\ } & \mathrm{Proj}(A_{\overline{Y}}) \\
\uparrow & & \uparrow \\
|\mathbb{O}_X(1)|^{\vee} & \xrightarrow{\ \gamma_1^{\vee}\ } & |\mathbb{O}_Y(1)|^{\vee}
\end{array}
$$

commutes. The commutativity of the top square in this diagram implies that if $D \subset \overline{X}$ is an effective divisor in $|\mathbb{O}_{\overline{X}}(m)|$ then the image of the divisor $f(D) \subset \overline{Y}$ in $|\mathbb{O}_Y(m)|$ (i.e., the restriction $f(D)|_Y$) is the divisor $\gamma_m(D)$. In particular, if $D$ is in the sweep of the maximal open subset of the definable lines with generic point $\eta_D \in X$ then $f(\eta_D) = \varphi(\eta_D)$.

By (5.2.3) we conclude that $f$ acts the same as $\varphi$ on every closed point of $X$. Since $|X|$ is a Zariski topological space, it follows that $\varphi$ and $f$ have the same action on $|X|$. Thus, $\tau$ is fully faithful.     □

This completes the proof of (4.1.14) in the case of infinite constant fields.     □

## 5.3   COUNTEREXAMPLES IN DIMENSION 1

In this section we provide counterexamples to (4.1.14) for schemes of dimension 1 over arbitrary fields. This shows that the assumption of algebraically closed base fields in [Zil14] is necessary and not simply an artifact of the proof.

**5.3.1.** Let $K$ be a field and let $C/K$ be a smooth, projective, geometrically connected curve over $K$ with $\mathrm{Pic}(C) \simeq \mathbb{Z}$ (note that in this case there is a unique such isomorphism sending an ample class to a positive integer). In this case the data of rational equivalence on divisors is captured by the function

$$
d : \{\text{closed points of } C\} \to \mathbb{N}
$$

sending a closed point $c \in C$ to the class of the corresponding divisor.

Given a second such pair $(K', C')$ with associated function

$$
d' : \{\text{closed points of } C'\} \to \mathbb{N}
$$

we find that the corresponding divisorial structures $\tau(C)$ and $\tau(C')$ are isomorphic if and only if for all $n \in \mathbb{N}$ the sets

$$
\{c \in C \,|\, d(c) = n\}, \quad \{c' \in C' \,|\, d'(c') = n\}
$$

have the same cardinality. This gives rise to non-isomorphic fields and curves with isomorphic divisorial structures. Specifically, the following gives examples of non-isomorphic curves for any two different genera with isomorphic divisorial structures:

**Proposition 5.3.2.** *Fix an infinite field $K$. For an integer $g > 1$ let $C_g \to \operatorname{Spec} K_g$ be the generic curve of genus $g$ (so that $K_g$ is the function field of $\mathcal{M}_g$—the moduli stack over $K$ classifying genus $g$ curves). Then*

(1) $\operatorname{Pic}(C_g) = \mathbb{Z} \cdot [K_{C_g}]$

(2) *if $d$ is the associated function then for all $n > 0$ the cardinality of*

$$S_g(n) := \{c \in C_g \mid d(c) = n\}$$

*is equal to the cardinality of $K$.*

*Proof.* Statement (1) is the Franchetta Conjecture (e.g., [Sch03, Theorem 5.1]). For (2), consider for each $n$ the linear system $|nK_{C_g}|$. By Riemann-Roch, this has projective dimension $(2n-1)g - 2n$ for $n > 1$ and dimension $g-1$ for $n = 1$. For each pair of positive integers $a$ and $b$, the natural map

$$|aK_{C_g}| \times |bK_{C_g}| \to |(a+b)K_{C_g}|$$

has proper closed image (by a simple dimension count). We conclude that for each $a \in \mathbb{N}$, there is a Zariski open $U_a \subset |aK_{C_g}|$ whose points correspond precisely to $S_g(a)$. This shows that $S_g(a)$ has the same cardinality as $K_g$, which is also the cardinality of $K$ since $K_g$ is a finitely generated field extension of $K$. □

A similar counterexample exists for curves of genus 1 over finite fields.

**Proposition 5.3.3.** *Fix a finite field $\mathbf{F}_q$. Let $E_1$ and $E_2$ be two elliptic curves over $\mathbf{F}_q$ such that there is a group isomorphism $E_1(\mathbf{F}_q) \cong E_2(\mathbf{F}_q)$. Then there is an isomorphism between the divisorial structures $\tau(E_1) \cong \tau(E_2)$.*

It is easy to find examples of two non-isomorphic elliptic curves over the same finite field with isomorphic groups of rational points. For example, by the Hasse bound, the number of possible isomorphism classes of groups of rational points of elliptic curves over $\mathbf{F}_q$ is less than the number of possible $j$ invariants for large $q$.

*Proof.* We can extend any isomorphism $E_1(\mathbf{F}_q) \to E_2(\mathbf{F}_q)$ to an isomorphism of their Picard groups $f: \operatorname{Cl}(E_1) \to \operatorname{Cl}(E_2)$ preserving the degree. To do this, we can send a fixed degree 1 divisor on $E_1$ to any fixed degree 1 divisor on $E_2$. Fix such an $f$.

There is a norm map $E_i(\mathbf{F}_{q^n}) \to E_i(\mathbf{F}_q)$ that sends $x$ to $x + \operatorname{Frob}_q(x) + \operatorname{Frob}_q^2(x) + \cdots + \operatorname{Frob}_q^{n-1}(x)$. This map is a group homomorphism, and it is surjective: Indeed, it is a nonconstant morphism of elliptic curves, thus surjective on geometric points, and for $y \in E_i(\mathbf{F}_q)$, we have $\mathrm{Frob}_q(y) - y = 0$, so $x \in E(\overline{\mathbf{F}_q})$ with norm $y$ satisfies $\mathrm{Frob}_q^n(x) - x = 0$ and thus $x \in E_i(\mathbf{F}_{q^n})$. In particular, the number of elements of $E_i(\mathbf{F}_{q^n})$ with any given norm is $\#E_i(\mathbf{F}_{q^n})/\#E_i(\mathbf{F}_q)$.

Let us now calculate, for $D \in \mathrm{Cl}(E_i)$, the number of closed points in $E_i$ with class $D$. Let $n$ be the degree of $D$. If $n \leq 0$, then there are no such closed points. For $n > 1$, for any $k$ dividing $n$, every closed point $x \in E_i$ of degree $n/k$ gives an orbit of $n/k$ points in $E_i(\mathbf{F}_{q^n})$, which has norm $D - n[0]$ if $D$ is $k$ times the class of $x$. Every point of $E(\mathbf{F}_{q^n})$ arises from exactly one closed point this way. Thus, we have the inclusion-exclusion

$$\#\{x \in |E_i| \mid [x] = D\} = \frac{1}{n} \sum_{m \mid n} \mu(m) \#\{D' \in \mathrm{Cl}(E_i) \mid mD' = D\} \frac{\#E_i(\mathbf{F}_{q^{n/m}})}{\#E_i(\mathbf{F}_q)}.$$

Because $E_1$ and $E_2$ have the same number of points and lie over the same finite field, they have the same $L$-function, and thus $\#E_1(\mathbf{F}_{q^n}) = \#E_2(\mathbf{F}_{q^n})$ for all $n$.

It follows that, for each $D$ in $\mathrm{Cl}(E_1)$, the number of closed points in $E_1$ with class $D$ is equal to the number of closed points in $E_2$ with class $f(D)$, because the group isomorphism $f$ guarantees that the number of $D'$ satisfying $mD' = D$ is equal to the number of $D'$ satisfying $mD' = f(D)$.

Thus, we can choose for each $D$ a bijection between the closed points of $E_1$ with class $D$ and the closed points of $E_2$ with class $f(D)$. Combining all these bijections, we get a bijection between the closed points of $E_1$ and $E_2$ such that the induced diagram

$$\begin{array}{ccc} \mathrm{Div}(E_1) & \longrightarrow & \mathrm{Cl}(E_1) \\ \downarrow & & \downarrow \\ \mathrm{Div}(E_2) & \longrightarrow & \mathrm{Cl}(E_2) \end{array}$$

commutes. Sending the generic point of the $E_1$ to the generic point of $E_2$, we obtain an isomorphism on topological spaces that respects the divisorial structure, as desired. $\qquad\square$

# Chapter Six

---

## Reconstruction from divisorial structures: finite fields

In this chapter we prove (4.1.14) in the case that $K$ is finite.

The proof follows a similar strategy to that of the case over infinite fields in Chapter 5: the isomorphism $\varphi$ induces a bijection of sets between the projectivizations of the graded pieces of the homogeneous coordinate rings of the varieties (with respect to suitable ample line bundles). The challenge is then to show that these bijections are suitably linear and can be lifted to an isomorphism of graded rings. The key technical ingredients are the Bertini-Poonen theorem, generalized to complete intersections in [BK12] and reviewed in Section 6.1 below, and the probabilistic fundamental theorem of projective geometry, (3.3.1).

The starting point is the idea that, over a finite field, proving that a property holds over an open set is often insufficient (as the open set could avoid all the finite-field-valued points) and instead one wants to lower bound the proportion of points on which the property holds. Furthermore, for properties of a hypersurface, it is easier to estimate the limit of this proportion as the degree of the hypesurface goes to $\infty$. For example, the original Bertini-Poonen theorem relates to the proportion of hypersurfaces that are smooth among all hypersurfaces of a given degree, in the large degree limit.

In order to apply the probabilistic fundamental theorem of projective geometry (3.3.1), we need a lower bound for the proportion of lines that are definable that is close to 1. Thus, we give a geometric criterion (6.2.3) for the line through two hypersurfaces to be definable, similar to that used earlier in (4.3.13). In (6.1.6) we prove a general Bertini-Poonen-type result that we use in (6.2.3) to obtain the desired lower bound for the proportion of definable lines in (6.2.3).

A key caution is that the proportion of hypersurfaces singular at a given closed point is positive, independent of the degree. Thus, our criterion (4.3.13) cannot assume that either hypersurface is smooth, even if $X$ is smooth. Instead we need to make do with assumptions weaker than this, like generic smoothness. If $X$ is Cohen-Macaulay, we can use that generically smooth Cohen-Macaulay schemes are reduced, but if not, we need to instead assume that the hypesurfaces avoid the non-$S_3$ points of $X$. This requires formulating (3.3.1), as we did, to involve only lines through points avoiding certain linear subspaces, and this requires a $q > 2$ assumption unless $X$ is Cohen-Macaulay.

Once the proportion of lines that are definable is proven to converge to 1 and the probabilistic fundamental theorem is applied, we obtain for each $n$ a semilinear isomorphism between the spaces of global sections of $\mathcal{O}_X(n)$ and $\mathcal{O}_Y(n)$. The main remaining difficulty is to prove these are compatible with the multiplication structure on the section rings. The two different multiplication structures for $X$ and $Y$ give two different bilinear maps, so we must check that two bilinear maps agree (up to scaling) by first showing that they agree (up to scaling) on a high proportion of pairs of vectors, and then concluding by an elementary argument that they agree (up to scaling) everywhere.

## 6.1    THE BERTINI-POONEN THEOREM

In fact, we will not need the main results of [Poo04, BK12], but only a certain key lemma. Poonen's argument, and its variant due to Bucur and Kedlaya, proceeds by treating points of small, medium, and large degrees separately. For our purposes, we will need only their results about points of large degree. We introduce some notation so that this result can be stated.

**6.1.1.** (The large degree estimate) Let $\mathbf{F}$ be a finite field with $q$ elements and let $r$ and $n$ be positive integers. Let $X/\mathbf{F}$ be a smooth, quasi-projective variety over $\mathbf{F}$ of dimension $m \geq r$ (slightly more generally we could consider here a quasi-projective $\mathbf{F}$-scheme of equidimension $m \geq r$) equipped with an embedding

$$X \hookrightarrow \mathbb{P}^n$$

defining an invertible sheaf $\mathcal{O}_X(1)$ on $X$.
   Let $S$ denote the polynomial ring in $n+1$ variables over $\mathbf{F}$ and let $S_d \subset S$ denote the degree $d$ elements in this ring, so we have a ring homomorphism

$$S \to \Gamma_*(X, \mathcal{O}_X(1))$$

restricting to a map

$$S_d \to \Gamma(X, \mathcal{O}_X(d))$$

of vector spaces.
   Fix functions

$$g_i \colon \mathbb{N} \to \mathbb{N}$$

for $i = 2, \ldots, r$ such that there exists an integer $w > 0$ for which

$$d \leq g_i(d) \leq wd$$

for all $d \in \mathbb{N}$ and all $i$. For notational reasons it will be convenient to also write $g_1 \colon \mathbb{N} \to \mathbb{N}$ for the identity function.

**6.1.2.** Let $\mathcal{S}$ denote the product $\prod_{j=1}^r S$ and let $\mathcal{S}_d$ denote the subset

$$S_d \times S_{g_2(d)} \times \cdots \times S_{g_r(d)} \subset \mathcal{S}.$$

For a section $f \in S_d$ let $H_{X,f} \subset X$ be the closed subscheme defined by the image of $f$ in $\Gamma(X, \mathcal{O}_X(d))$, and for $\underline{f} = (f_1, \ldots, f_r) \in \mathcal{S}_d$ let

$$X_{\underline{f}} := \bigcap_{i=1}^{r} H_{X, f_i}. \tag{6.1.2.1}$$

For an integer $d$ let $W_d \subset \mathcal{S}_d$ denote the subset of vectors $\underline{f}$ such that the intersection $X_{\underline{f}}$ is smooth of dimension $m - r$ at all closed points $P$ of degree $> d/(m+1)$ in this intersection, and define

$$e_d := 1 - \frac{\#W_d}{\#\mathcal{S}_d}.$$

**Lemma 6.1.3.** *There exists a constant $C$, depending on $n$, $r$, $m$, $w$, and the degree of $\overline{X} \subset \mathbb{P}^n$, such that*

$$e_d \leq Cd^m q^{-\min\{d/(m+1), d/p\}}.$$

*In particular,*

$$\lim_{d \to \infty} e_d = 0.$$

*Proof.* This is [BK12, 2.7]. □

We recall some additional useful notation from [Poo04], which we will use in stating consequences of (6.1.3).

**Notation 6.1.4.** For any subset $A$ of a set $B$, write

$$\mu_B(A) = \frac{\#(A)}{\#(B)}$$

for the proportion of elements of $B$ that lie in $A$.
    For a subset $\mathcal{P} \subset \mathcal{S}$ write

$$\mu(\mathcal{P}) := \lim_{d \to \infty} \mu_{\mathcal{S}_d}(\mathcal{S}_d \cap \mathcal{P}) = \lim_{d \to \infty} \frac{\#(\mathcal{S}_d \cap \mathcal{P})}{\#\mathcal{S}_d}$$

and

$$\overline{\mu}(\mathcal{P}) := \limsup_{d \to \infty} \mu_{\mathcal{S}_d}(\mathcal{S}_d \cap \mathcal{P}) = \limsup_{d \to \infty} \frac{\#(\mathcal{S}_d \cap \mathcal{P})}{\#\mathcal{S}_d}.$$

**6.1.5. (Variants)** As in (6.1.1), let $\mathbf{F}$ be a finite field and let $X \subset \mathbb{P}^n$ be a quasi-projective variety of dimension $m > r$.
    Let $H_d \subset \mathcal{S}_d$ be the subset of elements $(f_1, \ldots, f_r)$ such that for every subset $R \subset \{1, \ldots, r\}$ the scheme-theoretic intersection

$$X_R := \bigcap_{i \in R} X_{f_i}$$

is generically smooth of dimension $m - \#R$. Let $H \subset \mathcal{S}$ denote the union of the $H_d$.

**Theorem 6.1.6.** *We have* $\boldsymbol{\mu}(H) = 1$.

*Proof.* For a given $R$, let $H_{R,d} \subset \mathcal{S}_d$ be the subset of those vectors for which the intersection $X_R$ is generically smooth of dimension $m - \#R$, and let $H_R$ denote the union of the $H_{R,d}$. Then

$$1 - \frac{\#H_d}{\#\mathcal{S}_d} \leq \sum_R \left( 1 - \frac{\#H_{R,d}}{\#\mathcal{S}_d} \right).$$

It therefore suffices to show that $\boldsymbol{\mu}(H_R) = 1$. Furthermore, this case reduces immediately to the case when $R = \{1, \ldots, r\}$, which we assume henceforth.
　　Let $\overline{X} \subset \mathbb{P}^n$ be the closure of $X$ with the reduced structure, and fix a finite stratification $\overline{X} = \{Y_i\}_{i \in I}$ with each $Y_i$ a smooth locally closed subvariety of $\overline{X}$, and one of the strata $Y_0$ equal to the smooth locus of $X$. If we further arrange that each $Y_{i,R} \subset Y_i$ has the expected dimension then it follows that the inclusion

$$X_R \cap X_0 \hookrightarrow X_R$$

is dense.
　　For an integer $s$ let $E_{Y_i,d}^{(s)} \subset \mathcal{S}_d$ denote the subset of those vectors $(f_1, \ldots, f_r)$ for which the intersections $X_{i,f}$ are smooth of the expected dimension at all points $P$ of degree $\geq s$. Let $E_d^{(s)}$ denote the intersection of the $E_{Y_i,d}^{(s)}$.
　　Observe that since we assumed that $r < m$, the closed points of degree $\geq s$ are dense in any irreducible component of $X_{0,f}$. In particular, we have $E_d^{(s)} \subset H_d$. Let $E^{(s)}$ denote the union of the $E_d^{(s)}$. Taking $s = \lfloor \frac{d}{m+1} + 1 \rfloor$, we have

$$E_{Y_i,d}^{(s)} = W_{Y_i,d},$$

where $W_{Y_i,d}$ is defined as in (6.1.2) applied to $Y_i$.
　　By this and (6.1.3) we have that

$$\lim_{d \to \infty} \frac{\sum_i \#(\mathcal{S}_d \setminus E_{Y_i,d})}{\#\mathcal{S}_d} = 0.$$

We conclude that

$$\lim_{d \to \infty} \frac{\#E_d^{(s)}}{\#\mathcal{S}_d} = 1,$$

and it follows that

$$\boldsymbol{\mu}(H_R) = 1,$$

as desired.

$\square$

## 6.2   PREPARATORY LEMMAS

We continue with the setup of (6.1.1).

**Lemma 6.2.1.** *Let $k$ be a field, let $\overline{D}/k$ be a geometrically irreducible, proper $k$-variety, and let $D \subset \overline{D}$ be a dense open subvariety with $D$ geometrically reduced and $\mathrm{codim}(\overline{D} - D, \overline{D}) \geq 2$. Then $H^0(D, \mathcal{O}_D) = k$.*

*Proof.* We may without loss of generality assume that $\overline{D}$ is reduced, and furthermore, by replacing $\overline{D}$ by its normalization, that $\overline{D}$ is normal. Since $\overline{D}$ is geometrically reduced and irreducible, it follows that $H^0(\overline{D}, \mathcal{O}_{\overline{D}}) = H^0(D, \mathcal{O}_D) = k$.   □

**Lemma 6.2.2.** *Let $X$ be a divisorially proper variety over a perfect field $k$, and let $\mathcal{O}_X(1)$ be a very ample invertible sheaf with associated linear system $P$. Let*

$$j \colon X \hookrightarrow \overline{X}$$

*be the compactification of $X$ provided by the given projective embedding, so $X$ is schematically dense in $\overline{X}$. Fix a finite stratification $\{Y_i\}_{i \in I}$ of $\overline{X}$ with each $Y_i$ smooth and equidimensional, and $Y_i \hookrightarrow \overline{X}$ locally closed. Let*

$$F, G \in H^0(X, \mathcal{O}_X(1)) = H^0(\overline{X}, \mathcal{O}_{\overline{X}}(1))$$

*be two linearly independent sections. Let $\overline{D}_F$ (resp. $\overline{D}_G$) be the zero locus in $\overline{X}$ of $F$ (resp. $G$) and set $D_F := \overline{D}_F \cap X$ (resp. $D_G := \overline{D}_G \cap X$), and assume they satisfy the following:*

(1) *$\overline{D}_F$ is geometrically irreducible.*

(2) *The intersection $\overline{D}_F \cap Y_i$ has dimension $\dim(Y_i) - 1$ for all $i$, and the intersection $\overline{D}_F \cap \overline{D}_G \cap Y_i$ has dimension $\dim(Y_i) - 2$ for all $i$ (here we make the convention that the empty variety has dimension $-1$ as well as $-2$).*

(3) *$D_F$ and the intersection $D_F \cap D_G$ are generically smooth.*

(4) *$X$ is geometrically $S_3$ at every point of $D_F$.*

*Then $F$ and $G$ span a definable line.*

*Proof.* Assumption (2) implies that $D_F \subset \overline{D}_F$ is dense. Furthermore, $D_F$ is $S_2$ because it is a hypersurface in an $S_3$ variety, and because it is generically smooth is $R_0$, hence it is reduced. Both $S_2$ and $R_0$ hold geometrically, so it is geometrically reduced. Similarly, $D_F \cap D_G$ is geometrically $R_0$ and $S_1$ and thus geometrically reduced, and $D_F \cap D_G \subset \overline{D}_F \cap \overline{D}_G$ is dense. We need to show that the kernel of the restriction map

$$H^0(X, \mathcal{O}_X(1)) \to H^0(D_F \cap D_G, \mathcal{O}_{D_F \cap D_G}(1))$$

is the span of $F$ and $G$.

To this end, let $W$ denote $(\overline{D}_F \cap \overline{D}_G) \setminus (D_F \cap D_G)$. By assumption (2), $W$ has codimension at least 2 in $\overline{D}_F$, and we have a closed immersion

$$D_F \cap D_G \hookrightarrow \overline{D}_F \setminus W.$$

From this we therefore get an exact sequence

$$0 \to H^0(\overline{D}_{F,\mathrm{red}} \setminus W, \mathcal{O}_{\overline{D}_{F,\mathrm{red}}}) \xrightarrow{\cdot} H^0(\overline{D}_{F,\mathrm{red}} \setminus W, \mathcal{O}_{\overline{D}_{F,\mathrm{red}}}(1))$$

$$\downarrow$$

$$H^0(D_F \cap D_G, \mathcal{O}_{D_F \cap D_G}(1)).$$

From the commutative diagram

$$
\begin{array}{ccc}
H^0(\overline{X}, \mathcal{O}_{\overline{X}}(1)) & \xrightarrow{\simeq} & H^0(X, \mathcal{O}_X(1)) \\
\downarrow & & \downarrow \\
H^0(\overline{D}_{F,\mathrm{red}} \setminus W, \mathcal{O}_{\overline{D}_{F,\mathrm{red}} \setminus W}(1)) & \longrightarrow & H^0(D_F, \mathcal{O}_{D_F}(1))
\end{array}
$$

we see that the kernel of the map

$$H^0(\overline{X}, \mathcal{O}_{\overline{X}}(1)) \to H^0(\overline{D}_{F,\mathrm{red}} \setminus W, \mathcal{O}_{\overline{D}_{F,\mathrm{red}} \setminus W}(1))$$

is 1-dimensional generated by $F$. From this and the argument used in (4.3.12) we see that to prove the proposition it suffices to show that the dimension of $H^0(\overline{D}_{F,\mathrm{red}} - W, \mathcal{O}_{\overline{D}_{F,\mathrm{red}}})$ is 1. This follows from (6.2.1). $\qquad\square$

**Lemma 6.2.3.** *Let $X/k$ be a quasi-projective, divisorially proper variety of dimension at least 3, and let $\mathcal{O}_X(1)$ be a very ample line bundle on $X$. Let $B$ be the set of points of $X$ that are not $S_3$. For each $i \in B$, let $V_{i,n} \subset \mathbb{P}H^0(\overline{X}, \mathcal{O}_{\overline{X}}(n))$ be the set of hypersurfaces containing $i$. Let $X \hookrightarrow \mathbb{P}$ be the embedding into projective space provided by $\mathcal{O}_X(1)$ and let $\overline{X} \subset \mathbb{P}$ be the closure of $X$, with the reduced subscheme structure. Let $\mathscr{H}_n$ be the set of lines in $\mathbb{P}H^0(\overline{X}, \mathcal{O}_{\overline{X}}(n))$ and let $\mathscr{H}_n^{\mathrm{def},B} \subset \mathscr{H}_n$ be the subset of lines which are either definable as lines in $\mathbb{P}H^0(X, \mathcal{O}_X(n))$ or contained in $\bigcup_{i \in B} V_{i,n}$. Then*

$$\lim_{n \to \infty} \boldsymbol{\mu}_{\mathscr{H}_n}(\mathscr{H}_n^{\mathrm{def},B}) = 1.$$

*Proof.* Fix a finite stratification $\{Y_i\}_{i \in I}$ of $\overline{X}$ into locally closed smooth subvarieties.

Let $\mathscr{P}_n$ denote the set of pairs of linearly independent elements

$$f_1, f_2 \in \Gamma(\overline{X}, \mathcal{O}_{\overline{X}}(n)),$$

and let $\mathscr{P}'_n \subset \mathscr{P}_n$ denote the subset of pairs $(f_1, f_2)$ for which the associated divisors $\overline{D}_{a_1,a_2} := V(a_1 f_1 + a_2 f_2) \cap \overline{X}$ have the following properties:

(1) $D_{a_1,a_2}$ is geometrically irreducible for all $(a_1 : a_2) \in \mathbb{P}^1(k)$;

(2) The double intersection $\overline{D}_1 \cap \overline{D}_2 \cap Y_i$ and the $D_{a_1,a_2}$, for all $(a_1 : a_2) \in \mathbb{P}^1(k)$, have the expected dimension;

(3) The double intersection $\overline{D}_1 \cap \overline{D}_2 \cap X$ and the $\overline{D}_{a_1,a_2} \cap X$, for all $(a_1 : a_2) \in \mathbb{P}^1(k)$, are generically smooth.

There is a map
$$Sp: \mathscr{P}_n \to \mathscr{H}_n$$
sending a pair $(f_1, f_2)$ to the line spanned by $f_1$ and $f_2$. By (6.2.2) the image of $\mathscr{P}'_n$ is contained in $\mathscr{H}_n^{\mathrm{def},B}$. Indeed, if the line spanned by $f_1$ and $f_2$ is contained in $\bigcup_i V_i$ then we are done, and otherwise some $a_1 f_1 + b f_2$ does not intersect the non-$S_3$ locus $B$. We apply (6.2.2) to $D_{af_1 + bf_2}$ and to any other divisor in the pencil.

Because the fiber of $Sp$ over each line in $\mathscr{H}_n$ has the same size, and the image of $\mathscr{P}'_n$ is contained in $\mathscr{H}_n^{\mathrm{def}}$, we have
$$\mu_{\mathscr{P}_n}(\mathscr{P}'_n) \leq \mu_{\mathscr{H}_n}(\mathscr{H}_n^{\mathrm{def},B}),$$
and it suffices to show that
$$\lim_{n \to \infty} \mu_{\mathscr{P}_n}(\mathscr{P}'_n) = 1.$$

This follows from (6.1.6) and [CP16, Theorem 1.1]. (Note that, if the condition that a divisor is geometrically irreducible and generically smooth has density 1, then the condition that $q+1$ divisors are geometrically irreducible and generically smooth has density 1.) □

**6.2.4.** For integers $n_1, n_2$ with $n_1 \leq n_2 \leq 2n_1$ consider the subset
$$\mathscr{T}_{n_1,n_2} \subset S_{n_1} \oplus S_{n_2} \oplus S_{n_1+n_2}$$
whose elements are triples $(f_1, f_2, f_3)$ for which either the elements $f_1 f_2$ and $f_3$ span a definable line in $\mathbb{P}\Gamma(X, \mathcal{O}_X(n_1 + n_2))$ or $f_3$ vanishes at a point of $B$.

**Lemma 6.2.5.** *For any function $g: \mathbb{N} \to \mathbb{N}$ such that $n \leq g(n) \leq 2n$ for all $n$, we have*
$$\lim_{n \to \infty} \frac{\#\mathscr{T}_{n,g(n)}}{\#(S_n \oplus S_{g(n)} \oplus S_{n+g(n)})} = 1.$$

*Proof.* Fix a stratification $\{Y_i\}_{i \in I}$ of $\overline{X}$ into smooth subvarieties.

By (6.2.2) the set $\mathscr{T}_{n,g(n)}$ contains the set $\mathscr{T}'_{n,g(n)}$ of triples $(f_1, f_2, f_3) \in S_n \oplus S_{g(n)} \oplus S_{n+g(n)}$ satisfying the condition that the zero locus of each $f_i$ in

$\overline{X}$ is irreducible, for all $R \subset \{1, 2, 3\}$ the intersection $X_R$ is generically smooth, the intersections $\overline{X}_R \cap Y_i$ have the expected dimension for all $i$, and the zero locus of $f_3$ does not intersect the non-$S_3$ points of $X$. The result then follows from (6.1.6) and [CP16, Theorem 1.1]. □

## 6.3 RECONSTRUCTION OVER FINITE FIELDS

By the same argument as in (5.1.1), which did not require any assumption on the ground field, it suffices to prove (4.1.14) in the case when $X$ and $Y$ are quasi-projective.

**6.3.1.** Fix $\epsilon > 0$. In the course of the proof we will make various assumptions on $\epsilon$ being sufficiently small. As there are only finitely many steps, this is a harmless practice.

Fix an ample invertible sheaf $\mathcal{O}_X(1)$ on $X$ represented by an effective divisor $D$. By (4.2.8) the property of being ample depends only on the divisorial structure, and therefore $\varphi(D)$ defines an ample invertible sheaf on $Y$, which we denote by $\mathcal{O}_Y(1)$. After replacing $\mathcal{O}_X(1)$ by $\mathcal{O}_X(n)$ for sufficiently large $n$ we may assume that $\mathcal{O}_X(1)$ and $\mathcal{O}_Y(1)$ are very ample.

Note that, because $X$ and $Y$ are normal, they are $S_2$, and so their loci of non-$S_3$ points consist of only finitely many points (see (9.2.3)). Hence, for $n$ sufficiently large, a positive proportion of hypersurfaces of degree $n$ do not intersect the non-$S_3$ points. Thus, by choosing $n$ sufficiently large we may assume by (6.2.3) that there exist definable lines in $\mathbf{P}(\Gamma(X, \mathcal{O}_X(1)))$ and $\mathbf{P}(\Gamma(Y, \mathcal{O}_Y(1)))$. Since the number of elements in a definable line is one more than the size of the base field, we see that the finite fields $\kappa_X$ and $\kappa_Y$ are isomorphic to the same finite field $\mathbf{F}$, and in particular have the same number of elements, which we will denote by $q$.

**6.3.2.** Let $\overline{X} \subset |\mathcal{O}_X(1)|^\vee$ (resp. $\overline{Y} \subset |\mathcal{O}_Y(1)|^\vee$) be the scheme-theoretic closure of $X$ (resp. $Y$). Define graded rings

$$A_{\overline{X}} := \oplus_{n \geq 0} \Gamma(\overline{X}, \mathcal{O}_{\overline{X}}(n)), \quad A_{\overline{Y}} := \oplus_{n \geq 0} \Gamma(\overline{Y}, \mathcal{O}_{\overline{Y}}(n)),$$

$$A_X := \oplus_{n \geq 0} \Gamma(X, \mathcal{O}_X(n)), \quad A_Y := \oplus_{n \geq 0} \Gamma(Y, \mathcal{O}_Y(n)),$$

so $A_{\overline{X}} \subset A_X$ and $A_{\overline{Y}} \subset A_Y$. For $m > 0$ and any of these graded rings $A$ write $A(m)$ for the subring $A(m) \subset A$ given by

$$A(m) := \oplus_{n \geq 0} A^{nm}.$$

Write $|A^n_{\overline{X}}| \subset |nD|$ for $\mathbf{P}(\Gamma(\overline{X}, \mathcal{O}_{\overline{X}}(n))) \subset \mathbf{P}(\Gamma(X, \mathcal{O}_X(n)))$, and similarly for $|A^n_{\overline{Y}}|$. By (6.2.3), for $n$ sufficiently large the proportion of lines in the linear system $|A^n_{\overline{X}}|$ that are both not definable and not contained in $\bigcup_{i \in B} V_i$ is at most $\epsilon$. By (3.3.3) we may choose $\epsilon$ sufficiently small, and thereafter $n$ sufficiently large so that (3.3.1) applies to the map

$$|A_{\overline{X}}^n| \hookrightarrow \mathbb{P}(\Gamma(X, \mathcal{O}_X(n))) \to \mathbb{P}(\Gamma(Y, \mathcal{O}_Y(n))).$$

The assumption that $q > 2$ or $B$ is empty is satisfied because, in the case $q = 2$, we assume that $X$ is Cohen-Macaulay, therefore $S_k$ for any $k$, so in particular $B$ is empty. We therefore find an integer $n_0$ such that for each $n \geq n_0$ we get an isomorphism of fields

$$\sigma_n \colon \kappa_X \to \kappa_Y$$

and a $\sigma_n$-linear map

$$\gamma_n \colon A_{\overline{X}}^n \to \Gamma(Y, \mathcal{O}_Y(n))$$

such that the induced morphism of projective spaces

$$f'_n \colon \mathbf{P}(\Gamma(\overline{X}, \mathcal{O}_{\overline{X}}(n))) \to \mathbf{P}(\Gamma(Y, \mathcal{O}_Y(n)))$$

agrees with the map

$$f_n \colon |A_{\overline{X}}^n| \to |\mathcal{O}_Y(n)|$$

defined by $\varphi$ on a proportion of points

$$1 - \delta, \tag{6.3.2.1}$$

where $\delta > 0$ is any constant that we choose.

**6.3.3.** Next we prove that the $f'_n$ are close to multiplicative.

**Claim 6.3.4.** *We may take $n_0$ sufficiently large such that, for any $n_1 \geq n_0$ and for $n_2 = n_1$ or $n_2 = n_1 + 1$, we have*

$$f'_{n_1}(s_1) f'_{n_2}(s_2) = f'_{n_1 + n_2}(s_1 s_2)$$

*for a proportion $1 - 3\delta$ of pairs*

$$(s_1, s_2) \in A_{\overline{X}}^{n_1} \times A_{\overline{X}}^{n_2}.$$

*Proof.* Suppose

$$f'_{n_1}(s_1) f'_{n_2}(s_2) \neq f'_{n_1 + n_2}(s_1 s_2).$$

Then either $f_{n_1}(s_1) \neq f'_{n_1}(s_1)$, $f_{n_2}(s_2) \neq f'_{n_2}(s_2)$, or

$$f_{n_1}(s_1) f_{n_2}(s_2) \neq f'_{n_1 + n_2}(s_1 s_2).$$

The first two occur with probability at most $\delta$, so it suffices to prove the third occurs with probability at most $\delta$. If

$$f_{n_1}(s_1) f_{n_2}(s_2) \neq f'_{n_1 + n_2}(s_1 s_2),$$

then by (3.3.12) the number of pairs $s_3, s_4$ with $s_1 s_2, s_3, s_4$ collinear, $s_3 \notin$

$\bigcup_{i \in B} V_i$, and $f(s_1 s_2), f(s_3), f(s_4)$ collinear is at most

$$A(q, N_{n_1+n_2}, G_{n_1+n_2}, \epsilon) + q(q+1).$$

(Here $A(q, N_{n_1+n_2}, G_{n_1+n_2}, \epsilon)$ is defined in (3.3.0.1).) If $s_1 s_2$ and $s_3$ span a definable line then all $q - 1$ choices of $s_4$ satisfy this condition, so the number of $s_3$ with $s_3 \notin \bigcup_{i \in B} V_i$ and $Sp(s_1 s_2, s_3)$ definable is

$$\leq \frac{A(q, N_{n_1+n_2}, G_{n_1+n_2}, \epsilon) + q(q+1)}{q - 1}.$$

Thus for each $s_1, s_2$ satisfying this last condition, the number of $s_3$ with $s_3 \notin \bigcup_{i \in B} V_i$ and $Sp(s_1 s_2, s_3)$ not definable is at most

$$G_{n_1+n_2} - 1 - \frac{A(q, N_{n_1+n_2}, G_{n_1+n_2}, \epsilon) + q(q+1)}{q - 1}.$$

Taking $n_1, n_2$ sufficiently large and $\epsilon$ sufficiently small, this is $\geq c N_{n_1+n_2}$, where $c$ is some nonzero constant, for instance, $\frac{1}{2} \prod_{i \in B} (1 - q^{\deg i})$, taking $\deg i$ to be the degree of the closed point $i$ or $\infty$ if $i$ is not a closed point.

On the other hand, by (6.2.5), the number of triples $s - 1, s_2, s_3$ with $s_3 \notin \bigcup_{i \in B} V_i$ and $Sp(s_1, s_2, s_3)$ not definable is at most $c \delta N_{n_1} N_{n_2} N_{n_1+n_2}$ for $n_1$ sufficiently large and $n_2 = n_1$ or $n_1 + 1$. So the number of pairs $s_1, s_2$ with

$$f_{n_1}(s_1) f_{n_2}(s_2) \neq f'_{n_1+n_2}(s_1 s_2)$$

is at most $\delta N_{n_1} N_{n_2}$ and so the proportion of such pairs is at most $\delta$, as desired.

$\square$

**6.3.5.** Next we show that the $\gamma_n$ are close to multiplicative.

**Claim 6.3.6.** *For $n_0$ sufficiently large as in (6.3.4), for any $n_1$ and $n_2 = n_1$ or $n_2 = n_1 + 1$, there exists a constant $c_{n_1, n_2}$ such that*

$$\gamma_{n_1}(s_1) \gamma_{n_2}(s_2) = c_{n_1, n_2} \gamma_{n_1+n_2}(s_1 s_2) \tag{6.3.6.1}$$

*for all pairs $(s_1, s_2)$.*

*Proof.* Let $W_n$ denote $A_X^n$, viewed as a vector space over the prime field $\mathbf{F}_p$. We then have two bilinear forms

$$b_X, b_Y : W_{n_1} \times W_{n_2} \to \Gamma(Y, \mathcal{O}_Y(n_1 + n_2))$$

given by

$$b_X(s_1, s_2) := \gamma_{n_1}(s_1) \gamma_{n_2}(s_2), \quad b_Y(s_1, s_2) := \gamma_{n_1+n_2}(s_1 s_2).$$

These forms have the property that they agree up to a scalar for a proportion of $1 - 3\delta$ of pairs $(s_1, s_2)$.

Given $s_1$, let $Y_{s_1}$ be the set of $s_2$ such that $b_X(s_1, s_2)$ is a scalar multiple of $b_Y(s_1, s_2)$. Let $p(s_1) = \#Y_{s_1}/\#W_{n_2}$. It follows from the above remarks that we have

$$\#\{s_1|p(s_1) \leq 1 - \sqrt{3\delta}\} \cdot \sqrt{3\delta} \leq 3\delta \cdot \#W n_1.$$

Thus, for a proportion of $1 - \sqrt{3\delta}$ of elements $s_1$ the two forms $b_X(s_1, s_2)$ and $b_Y(s_1, s_2)$ agree up to a scalar for a proportion of $1 - \sqrt{3\delta}$ of elements $s_2$.

Fix $s_1$ for which $p(s_1) \leq 1 - \sqrt{3\delta}$. Each of the maps

$$b_X(s_1, -), b_Y(s_1, -) : W_{n_2} \to \Gamma(Y, \mathcal{O}_Y(n_1 + n_2))$$

is injective, which implies that

$$\mathrm{rank}(b_X(s_1, -) - \alpha b_Y(s_1, -)) + \mathrm{rank}(b_X(s_1, -) - \alpha' b_Y(s_1, -)) \geq \dim(W_{n_2})$$

for any distinct elements $\alpha$, $\alpha'$. It follows that there is at most one $\alpha \neq 0$ for which the rank of $b_X(s_1, -) - \alpha b_Y(s_1, -)$ is less than or equal to $\dim(W_{n_2})/2$.

Suppose that in fact we have

$$\mathrm{rank}(b_X(s_1, -) - \alpha b_Y(s_1, -)) \geq \dim(W_{n_2})/2$$

for all $\alpha$. Then the proportion of $s_2$ for which $b_X(s_1, s_2)$ is a scalar multiple of $b_Y(s_1, s_2)$ is at most

$$\frac{q - 1}{q^{\dim(W_{n_2})/2}},$$

and we obtain the inequality

$$\frac{q - 1}{q^{\dim(W_{n_2})/2}} \geq 1 - \sqrt{3\delta}.$$

For $n$ chosen sufficiently large relative to $\delta$ this is a contradiction. We conclude that there exists exactly one scalar $\alpha_0$ such that

$$\mathrm{rank}(b_X(s_1, -) - \alpha_0 b_Y(s_2, -)) < \dim(W n_2)/2.$$

Now in this case we find that the proportion of $s_2$ for which $b_X(s_1, s_2)$ is a scalar multiple of $b_Y(s_1, s_2)$ is at most

$$\frac{(q - 2)}{q^{\dim(W n_2)/2}} + \frac{1}{p^{r_0}},$$

where $r_0$ is the rank of $b_X(s_1, -) - \alpha_0 b_Y(s_1, -)$. For $\epsilon$ suitably small we see that this implies that in fact $r_0 = 0$ and $b_X(s_1, s_2) = \alpha_0 b_Y(s_1, s_2)$ for all $s_2$.

Note that this argument is symmetric in $s_1$ and $s_2$. That is, for a fixed $s_2$ subject to the condition that $b_X(s_1, s_2)$ is a scalar multiple of $b_Y(s_1, s_2)$, we find that there exists a constant $\beta$ such that

$$b_X(s_1, s_2) = \beta b_Y(s_1, s_2)$$

for all $s_1$. From this it follows that in fact the constant $\alpha_0$ in the previous paragraph is independent of the choice of $s_1$. Furthermore, using the bilinearity we find that there exists a constant $c_{n_1,n_2}$ such that

$$b_X(s_1, s_2) = c_{n_1,n_2} b_Y(s_1, s_2)$$

for all pairs $(s_1, s_2)$. In other words, we have the equality (6.3.6.1)                  □

**Claim 6.3.7.** *For $n_2$ sufficiently large as in (6.3.4), and for every $n \geq n_0$ and integer $m \geq 1$ there exists a constant $c_m$ such that for all sections*

$$s_1, \ldots, s_m \in A_{\overline{X}}^n$$

*we have*

$$\gamma_{nm}(s_1 \cdots s_m) = c_m \gamma_n(s_1) \cdots \gamma_n(s_m).$$

*Proof.* This we show by induction, the case $m = 1$ being vacuous. For the inductive step write $m = a + b$ for positive integers $a$ and $b$ with $a = b = m/2$ if $m$ is even, and $a = (m-1)/2$ and $b = (m+1)/2$ if $m$ is odd. Then by the above discussion there exists a constant $c_{a,b}$ such that

$$\gamma_{nm}(s_1 \cdots s_m) = c_{a,b}\gamma_{na}\left(\prod_{i=1}^{a} s_i\right)\gamma_{bn}\left(\prod_{j=1}^{b} s_{a+j}\right).$$

By our inductive hypothesis this equals

$$c_{a,b}c_a c_b \gamma_n(s_1) \cdots \gamma_n(s_m),$$

so we can take $c_m = c_{a,b}c_a c_b$.                                                   □

**6.3.8.** In particular, after possibly choosing $n_0$ even bigger so that $A_{\overline{X}}(n_0)$ is generated by $A_{\overline{X}}^{n_0}$ we get an injective ring homomorphism

$$\rho_{\overline{X},n_0} : A_{\overline{X}}(n_0) \to A_Y(n_0)$$

given in degree $mn_0$ by $\gamma_{mn_0}/c_m$.

**6.3.9.** The map $\rho_{\overline{X},n_0}$ defines a rational map

$$\lambda : Y -- \blacktriangleright \overline{X}.$$

Let $Y^\circ \subset Y$ be the maximal open subset over which $\lambda$ is defined and the map $\rho_{\overline{X},n_0}$ induces an isomorphism $\lambda^* \mathcal{O}_{\overline{X}}(n_0) \simeq \mathcal{O}_Y(n_0)$. We claim that the two maps of topological spaces

$$|\lambda|, \varphi^{-1} \colon |Y^\circ| \to |\overline{X}| \tag{6.3.9.1}$$

agree, where we write $\varphi^{-1}$ also for the composition

$$|Y| \xrightarrow{\varphi^{-1}} |X| \lhook\joinrel\longrightarrow |\overline{X}|.$$

To prove this it suffices to show that these two maps agree on all closed points. Suppose to the contrary that we have a closed point $y \in Y^\circ$ such that $\lambda(y) \neq \varphi^{-1}(y)$. Consider the subset

$$T_m \subset A_{\overline{X}}^{n_0 m}$$

of sections $g \in \Gamma(\overline{X}, \mathcal{O}_{\overline{X}}(n_0 m))$ whose zero locus contains both $\lambda(y)$ and $\varphi^{-1}(y)$. Now any section $g$ whose zero locus contains $\lambda(y)$ and for which $f_{n_0 m}(g) = f'_{n_0 m}(g)$ lies in $T_m$ by definition of $\lambda$ and $f_n$. It follows that

$$\frac{\#T_m}{\#A_{\overline{X}}^{n_0 m}} \geq \frac{1}{q^{\deg(\lambda(y))}} - \tilde{\epsilon}. \tag{6.3.9.2}$$

On the other hand, for $m$ sufficiently big we have

$$\frac{\#T_m}{\#A_{\overline{X}}^{n_0 m}} = \frac{1}{q^{\deg(\lambda(y)) + \deg(\varphi^{-1}(y))}}. \tag{6.3.9.3}$$

Now observe that if we replace our choice of $n_0$ by a multiple, the open subset $Y^\circ \subset Y$ and $\lambda$ remain the same, but we can decrease the size of $\tilde{\epsilon}$ by making such a choice of $n_0$. Since the right side of (6.3.9.2) is larger than the right side of (6.3.9.3) for $\tilde{\epsilon}$ sufficiently small this gives a contradiction. We conclude that the two maps (6.3.9.1) agree.

**Lemma 6.3.10.** *Let $k$ be a field, let $S$ be a normal, quasi-projective $k$-variety, and let $T/k$ be a proper $k$-variety. Let $f \colon |S| \to |T|$ be a continuous map of topological spaces that is a homeomorphism onto an open subset of $|T|$. Assume that there exists a dense open subset $U \subset S$ and a morphism of schemes $\tilde{f}_U \colon U \to T$ whose associated morphism of topological spaces $|\tilde{f}_U| \colon |U| \to |T|$ agrees with the restriction of $f$. Then there exists a unique morphism of schemes $\tilde{f} \colon S \to T$ whose associated morphism of topological spaces is $f$ and that restricts to $\tilde{f}_U$ on $U$.*

*Proof.* Since $S$ is normal and $T$ is proper, there exists an open subset $S^\circ \subset S$ containing $U$ and with complement of codimension $\geq 2$ such that $\tilde{f}_U$ extends to a morphism of schemes

$$\tilde{f}_{S^\circ} \colon S^\circ \to T.$$

We claim that $\tilde{f}_{S^\circ}$ induces $f|_{|S^\circ|}$ on topological spaces.

If $s \in S^\circ$ is a closed point, then by Bertini's theorem (or in the finite field case Poonen-Bertini) there exist effective irreducible divisors $D_1, \ldots, D_r \subset S^\circ$, with $D_i \cap U$ nonempty for all $i$, such that

$$\{s\} = D_1 \cap \cdots \cap D_r.$$

Since $f$ is a homeomorphism onto an open subset of $T$ we can further arrange that

$$\{f(s)\} = \overline{f(D_1)} \cap \cdots \cap \overline{f(D_r)},$$

where $\overline{f(D_i)}$ is the closure of $f(D_i)$ in $|T|$. Since

$$\tilde{f}_{S^\circ}(s) \subset \overline{f(D_1)} \cap \cdots \cap \overline{f(D_r)},$$

we conclude that $\tilde{f}_{S^\circ}(s) = f(s)$. This shows that $\tilde{f}_{S^\circ}$ agrees with $f$ on all closed points and therefore also on all points.

We are therefore reduced to the case when the complement of $U$ in $Y$ has codimension $\geq 2$. In this case, the morphism $\tilde{f}_U$ extends to a map $\tilde{f} \colon Y \to T$ by the same argument as in the proof of (5.1.4), and repeating the previous argument we see that $\tilde{f}$ induces $f$ on topological spaces.                              □

**6.3.11.** By (6.3.10) we therefore get a morphism of schemes

$$u \colon Y \to X$$

whose associated morphism of topological spaces is $\varphi^{-1}$.

For $n$ sufficiently big, the line bundle $\mathcal{O}_Y(n)$ can be represented by an effective divisor $D \subset Y$ all of whose irreducible components occur with multiplicity 1 and have nonempty intersection with $Y^\circ$. The divisor $\varphi(D)$ then represents the line bundle $\mathcal{O}_X(n)$, and we have a nonzero map

$$u^*\mathcal{O}_X(-n) = u^*\mathcal{O}_X(-\varphi^{-1}(D)) \to \mathcal{O}_Y(-D) = \mathcal{O}_Y(-n).$$

Since $\varphi$ induces an isomorphism on class groups, we conclude that this map is an isomorphism, so $u$ extends to a map of polarized schemes

$$u \colon (Y, \mathcal{O}_Y(n)) \to (X, \mathcal{O}_X(n)),$$

for all $n$ sufficiently big.

Since the cardinalities of the linear systems $\Gamma(X, \mathcal{O}_X(n))$ and $\Gamma(Y, \mathcal{O}_Y(n))$ are the same for all $n$, we conclude that $u$ induces an isomorphism of graded rings

$$A_X(n) \to A_Y(n)$$

for all $n$ sufficiently big. This implies that $u$ is an open immersion. Indeed if $\overline{X}$ (resp. $\overline{Y}$) is the closure of $X$ (resp. $Y$) in the projective embedding defined by $\Gamma(X, \mathcal{O}_X(n))$ (resp. $\Gamma(Y, \mathcal{O}_Y(n))$) then we see that $u$ induces an isomorphism between the homogeneous coordinate rings of $\overline{X}$ and $\overline{Y}$, and therefore $u$ is an open immersion inducing an isomorphism of topological spaces, whence an isomorphism.

This completes the proof of (4.1.14). $\qquad\qquad\qquad\qquad\qquad\qquad\square$

# Chapter Seven

## Topological geometry

In this chapter we prove the main reconstruction results over uncountable fields of characteristic 0. There are two reasons to single out this case. The results we have are stronger—we can prove that the theorems hold in the minimal possible dimension, namely 2—and the proofs are technically simpler. In addition, this chapter serves as a model for the later reconstruction results. Many of its ideas and results resurface in technically more complicated forms.

The main result of this chapter, proved in Section 7.6, says that for a normal projective $k$-variety $X$ of dimension at least 2 over an uncountable field, linear equivalence of divisors on $X$ is determined by $|X|$. Combining this with (4.1.14), we then have that $X$ is uniquely determined as a scheme by $|X|$ up to unique isomorphism. (Note that $k$ is uncountable if and only if $|X|$ has uncountably many points, so this is a topological assumption.)

The main technical tool for this is a study of pencils of divisors. Pencils are algebraic notions, so we start by writing down topological properties of pencils, leading to the notion of topological pencils in (7.3.1). A key question is to understand which topological pencils come from actual pencils (in which case we say that the topological pencil is algebraic). This is studied in Section 7.3. The main result is the algebraicity criterion (7.3.4), which will be used again in Chapter 8. An immediate consequence is that, over uncountable fields, every topological pencil is algebraic; see (7.3.5). This is the only place where uncountability is used, and the reason why this case is much easier.

Another notion that is important in algebraic geometry is the degree of a subvariety (with respect to an ample divisor). In Section 7.4 we prove that having a degree function is equivalent to knowing which topological pencils are algebraic. One direction, which works over any field, follows from (7.3.4). Constructing a degree function using algebraic pencils is harder, and works only in characteristic 0; see (7.4.15).

Then in Section 7.5 we prove that if $X$ is a normal, projective, geometrically irreducible variety of dimension $\geq 2$ over an infinite field and $\deg_H$ a degree function on divisors, then $|X|$ and $\deg_H$ determine linear equivalence; see (7.5.8). This leads to the proof of the main result in Section 7.6.

The method of Section 7.5 also relies heavily on pencils. Specifically, the main problem is to decide which members of a topological pencil are members of the corresponding algebraic pencil; we call these *true members*.

Instead of a complete solution, we only deal with 'well behaved' linear systems. We find sufficient conditions for

- linearity of a pencil (in (7.5.4)),
- membership in a pencil (in (7.5.5)), and
- linear equivalence of reduced divisors (in (7.5.7)).

Then, in (7.5.8), we see that linear equivalences between reduced divisors generate the full linear equivalence relation.

## 7.1   PENCILS

We collect various results from the classical theory of pencils. Most of these are (or have been) well known, but precise references are hard to find. For our purposes the most important is (7.1.17). We start with some general remarks on Chow varieties.

**Definition 7.1.1.** Let $g\colon X \to S$ be a projective morphism of pure relative dimension $n$ with a relatively ample divisor $H$. Assume for simplicity that the fibers are geometrically normal. The *Chow variety* of divisors parametrizing relative divisors of $H$-degree $d$ is an $S$-scheme $\mathrm{Chow}_d^1(X/S)$ with a universal family

$$\mathrm{Univ}_d^1(X/S) \overset{\pi}{\lhook\joinrel\longrightarrow} X \times \mathrm{Chow}_d^1(X/S)$$

$$u \downarrow \qquad\qquad\qquad\qquad (7.1.1.1)$$

$$\mathrm{Chow}_d^1(X/S).$$

It has the following properties (see [Kol96, Theorem 3.21]).

(1) $\mathrm{Chow}_d^1(X/S)$ is seminormal and projective over $S$.

(2) $u$ is projective, of pure relative dimension $n - 1$, and its fibers have $\pi^* H$-degree $d$.

(3) Assume that we have a diagram $T \leftarrow D_T \overset{c}{\hookrightarrow} X \times T$, where $T$ is a seminormal scheme, $D_T \to T$ is projective, of pure relative dimension $n-1$, and all fibers have $c^* H$-degree $d$. Then there is a unique commutative diagram

$$
\begin{array}{ccc}
D_T & \longrightarrow & \mathrm{Univ}_d^1(X/S) \\
\downarrow & & \downarrow \\
T & \longrightarrow & \mathrm{Chow}_d^1(X/S).
\end{array}
$$

The *Chow variety* of divisors is then $\mathrm{Chow}^1(X/S) := \amalg_d \mathrm{Chow}_d^1(X/S)$. It has countably many connected components, all of which are projective over $S$.

Readers interested in the projective case can skip to (7.1.7); next we discuss the proper case.

Assume now that $X/S$ is only proper. It is almost certain that $\mathrm{Chow}^1(X/S)$ exists as an algebraic space and satisfies the properties (1)–(3). However, this is not treated in the literature, but see [BM20] for the complex case. The following results allow us to go around this problem. We start with a variant of [FKL16, 3.3].

**Lemma 7.1.2.** *Let $g\colon Y \to X$ be a proper, birational morphism of normal varieties and $D$ a divisor on $Y$. Let $\{E_i : i \in I\}$ be the $g$-exceptional divisors. Then there is a divisor $E = \sum e_i E_i$ such that for every $m \geq 0$ and every $g$-numerically trivial Cartier divisor $T$ the following hold.*

(1) *$g_* \mathcal{O}_Y(m(D + E) + T)$ is reflexive.*

(2) *$g_* \mathcal{O}_Y(m(D + E) + T + \sum_i c_i E_i) = g_* \mathcal{O}_Y(m(D + E) + T)$ for every $c_i \geq 0$.*

*Proof.* We may assume that $X$ is affine. Let $\mathcal{L}$ be a reflexive sheaf on $Y$. Then $g_* \mathcal{L}$ is reflexive if and only if $\mathrm{depth}_x \, g_* \mathcal{L} \geq 2$ for every point $x \in X$ of codimension $\geq 2$. The latter holds if and only if every section of $\mathcal{L}$ over $Y \setminus g^{-1}(x)$ extends to a section of $\mathcal{L}$. In particular, $\mathrm{depth}_x \, g_* \mathcal{L} \geq 2$ except possibly at the generic points of the $g(E_i)$. We also see that (1) and (2) are equivalent.

By localization and induction we may assume that the claim holds outside a single point $x \in X$. We can then move the exceptional divisors not contained in $g^{-1}(x)$ to $D$ and work with the divisorial irreducible components $\{E_j : j \in J\}$ of $g^{-1}(x)$. We may also assume that $g$ is projective with a very ample divisor $H$. Let now $S \subset Y$ be a normal surface obtained by intersecting $\dim Y - 2$ general members of $|H|$. We may also assume that the $F_j := E_j|_S$ are irreducible. On a normal surface the intersection numbers of proper curves are well defined and the intersection matrix of the $F_j$ is negative definite by the Hodge index theorem. Thus there is a linear combination $\sum a_j F_j$ such that $\sum a_j F_j + D|_S$ is numerically trivial on every $F_j$.

We claim that $E := \sum \lceil a_j \rceil E_j$, where $\lceil a_j \rceil$ denotes the round-up of $a_j$, has the required property. That is, a rational section $\sigma$ of $\mathcal{O}_Y(m(D+E)+T)$ cannot have poles only along $\sum E_j$. The restriction of such a $\sigma$ to $S$ would give a rational section of $\mathcal{O}_S(m(D|_S + E|_S) + T|_S)$ with poles only along $\sum F_j$.

Note that $\sigma|_S$ would be a section of sheaf of the form $\mathcal{O}_S(\sum c_j F_j + T')$, where $c_j \geq 0$ and $T' = m(\sum a_j F_j + D|_S) + T|_S$ is numerically trivial on every $F_j$. Choose the $c_j$ the smallest possible. Using the Hodge index theorem again, we get that there is a $j_0$ such that $(F_{j_0} \cdot \sum c_j F_j) < 0$. Thus every section of $\mathcal{O}_S(\sum c_j F_j + T')$ vanishes along $F_{j_0}$. (This is clear if $\mathcal{O}_S(\sum c_j F_j + T')$ is invertible; a simple computation shows that the singularities work in our favor.) That is, $\sigma|_S$ is a section of $\mathcal{O}_S(\sum c_j F_j - F_{j_0} + T')$, contradicting our choice of the $c_j$.  $\square$

**Corollary 7.1.3.** *Let $X$ be a proper, geometrically normal variety over a field $k$ and $D$ a Weil divisor on $X$. Let $g\colon Y \to X$ be a proper, birational morphism. Assume that $Y$ is projective and $Y \to \mathbf{Alb}(Y) = \mathbf{Alb}(X)$ is a morphism. Fix an*

*ample divisor $H$ on $Y$. Then there is a $d(D) > 0$ such that $\deg_H g_*^{-1} D' \le d(D)$
for every effective divisor $D'$ on $X$ that is algebraically equivalent to $D$.*

*Proof.* As explained in [Ful98, Example 10.3.4], $g_*^{-1} D'$ is algebraically equivalent
to some $g_*^{-1} D + \sum m_i E_i$ where $E_i$ are the $g$-exceptional divisors. Thus $g_*^{-1} D'$ is
linearly equivalent to some $g_*^{-1} D + T + \sum m_i E_i$ where $T$ is numerically trivial.
By (7.1.2) the $m_i$ are bounded from above, independent of $D'$. $\qquad\square$

Combining (7.1.3) with (7.1.1 (3)) we get the following.

**Corollary 7.1.4.** *Let $X$ be a proper, geometrically normal variety over a field
$k$ and $D$ a Weil divisor on $X$. Then all divisors algebraically equivalent to $D$
are parametrized by a $k$-scheme of finite type.* $\qquad\square$

**Example 7.1.5.** Hironaka's example in [Har77, B.3.4.1] is a smooth, proper
3-fold over $\mathbb{C}$, containing a curve $C$ such that all curves algebraically equivalent
to $C$ are *not* parametrized by a $\mathbb{C}$-scheme of finite type.

Its exceptional divisor is a proper non-normal surface $E$. The curve $C$ is
a divisor on it and all divisors algebraically equivalent to it are again *not*
parametrized by a $\mathbb{C}$-scheme of finite type. Also, $E$ contains a nonzero, effective
divisor that is algebraically equivalent to 0.

Applying (7.1.3) to $D, 2D, \ldots$ gives the following.

**Corollary 7.1.6.** *On a proper, geometrically normal variety $X$ an effective
divisor is algebraically equivalent to 0 if and only if it is the 0 divisor.* $\qquad\square$

Next we recall the definitions and basic facts about pencils in modern language; see [Zar41] for a classical exposition.

**Definition 7.1.7.** Let $X$ be an integral $k$-variety.

(1) A *pencil of divisors*, or just a *pencil*, on $X$ is a nonconstant, rational map
$\pi \colon X \dashrightarrow C$ to an integral, nonsingular and projective curve $C$. If we want to
emphasize the distinction with the notion of t-pencil, introduced in (7.3.1)
below, we sometimes refer to a pencil of divisors as an *algebraic pencil*.

(2) The indeterminacy locus $B \subset X$ of a pencil $\pi$ is called the *base locus* of the
pencil.

(3) For a closed point $c \in C$, the closure of $\pi^{-1}(c) \subset X$ is called a *fiber* of
the pencil. We frequently denote it by $D_c$. The fibers over $k$-points are the
*members* of the pencil. The fibers of a pencil are Cartier divisors on $X \setminus B$.

(4) A pencil is called *linear* if $C \cong \mathbb{P}_k^1$, *rational* if $C$ is a smooth, geometrically rational curve, and *irrational* if the geometric genus $g(C)$ is greater
than 0. (Over imperfect fields there are pencils that are neither rational nor
irrational. They will not come up for us.)

(5) Fix an algebraic closure $k \hookrightarrow \overline{k}$ and let $\pi$ be a pencil. If $X$ is geometrically

integral, then $\pi_{\overline{k}}\colon X_{\overline{k}} \dashrightarrow C_{\overline{k}}$ is also a pencil; its members are the *geometric members* of $\pi$. In traditional terminology

$$|D|_{\mathrm{alg}} := \{D_c : c \in C(\overline{k})\}$$

is an *(algebraic) pencil of divisors* parametrized by the curve $C$. We use $|D|_{\mathrm{alg}}$ to emphasize that $C$ can be a non-rational curve.

(6) If a pencil $\pi$ factors as $X \dashrightarrow C' \dashrightarrow C$, where $\deg(C'/C) > 1$, then $\pi'\colon X \dashrightarrow C'$ is another pencil. We say that $\pi$ is *composite with* $\pi'$. In this case each fiber of $\pi$ is a union of certain fibers of $\pi'$.

If no such $C'$ exists then $\pi$ is called *non-composite*. Every pencil is composite with a unique non-composite pencil.

**Remark 7.1.8.** Let $\pi\colon X \dashrightarrow C$ be a pencil over a field $k$.

(1) If $k$ is perfect, then we can describe the members as follows. Let $\overline{c} \in C_{\overline{k}}$ be a geometric point lying over $c \in C$. As $\sigma$ runs through $\mathrm{Gal}(\overline{k}/k)$, $D_c := \cup_\sigma D_{\overline{c}}^\sigma$ is a finite union, giving a divisor defined over $k$.

(2) If $k = \overline{k}$ then the notions member, fiber, geometric member coincide. The distinction between these three notions is not systematic in the literature. Thus the phrase 'let $|D_\lambda|$ be a pencil' may mean that $D_\lambda$ runs through all members, fibers, or geometric members of a pencil. We keep the very convenient pencil notation $|D|$, but specify whether we work with members, fibers, or geometric members.

(3) If $C \cong \mathbb{P}^1$ and $D_1, D_2$ are distinct members of a pencil $\pi$, then the pencil can be identified with the linear system $|D_1, D_2|$.

(4) (Warning) The definition of linear system frequently allows fixed components; those without fixed components are called *mobile*. In this terminology, our pencils are the *mobile pencils*.

**Definition 7.1.9.** Let $X$ be a normal variety. Two divisors $D_1, D_2$ are called *linearly similar* if there are nonzero integers $m_1, m_2$ such that $m_1 D_1 \sim m_2 D_2$. We denote it by $D_1 \sim_s D_2$.

This notion will play a central role starting in Section 9.4, but it also gives a convenient way to talk about pencils, since all fibers of a rational pencil are linearly similar to each other. We say that two pencils $|D_1|, |D_2|$ are *linearly similar* if every fiber of $|D_1|$ is linearly similar to fiber of $|D_2|$. We denote this by $|D_1| \sim_s |D_2|$. If $A$ is a divisor, we define $|D_1| \sim_s A$ analogously.

The numerical versions of these notions are the following. First, two real-valued functions $f_1, f_2$ are called *similar* if $f_1 = cf_2$ for some positive constant $c \in \mathbb{R}$.

Let $X$ be a proper variety. Two $\mathbb{Q}$-Cartier divisors $D_1, D_2$ are called *numerically similar*—denoted by $D_1 \equiv_s D_2$—if the functions $C \mapsto (D_i \cdot C)$ are similar (as functions from the set of all curves on $X$ to $\mathbb{R}$). We use the same terminology for pencils of $\mathbb{Q}$-Cartier divisors.

The basic characterization of non-composite pencils is due to Bertini, but it may have been first fully proved in [vdW37].

**Theorem 7.1.10.** *Let $X$ be an integral variety over a perfect field $k$ and*

$$\pi\colon X \dashrightarrow C$$

*a pencil. The following are equivalent.*

(1) *Almost all fibers of $\pi$ are irreducible and reduced.*

(2) *Almost all geometric fibers of $\pi$ are irreducible and reduced.*

(3) *$\pi$ is not composite with any other pencil.*

(4) *$k(C)$ is algebraically closed in $k(X)$.*

(5) *$k(C)$ is algebraically closed in $k(X)$ and $k(X)$ is a separable extension of $k(C)$.*

*Proof.* The claims all follow from Stein factorization, except for two issues.

To see that $(4) \Rightarrow (5)$, assume to the contrary that $k(X)/k(C)$ is not separable. Then, by MacLane's Separating transcendence basis theorem (see [Wei62, p.18], [Lan02, Sec.VIII.4] or [Eis95, p.558]), $k(X) \otimes_{k(C)} k(C)^{1/p}$ is nonreduced. If $k$ is perfect and $\dim C = 1$ then $\deg\big[k(C)^{1/p} : k(C)\big] = p$, thus $k(X) \otimes_{k(C)} k(C)^{1/p}$ has degree $p$ over $k(X)$. So, if it is nonreduced, then

$$\mathrm{red}\big(k(X) \otimes_{k(C)} k(C)^{1/p}\big) = k(X).$$

Thus $k(C)^{1/p} \subset k(X)$, contradicting (4).

Next (5) implies that the generic fiber is irreducible and generically reduced, but we claim that it is reduced. Let $B$ denote the base locus. Assume first that $X \setminus B$ is $S_2$ and let $D$ be a fiber. Since $D \setminus B$ is a Cartier divisor, it is $S_1$, hence reduced if and only if it is generically reduced. Taking its closure does not add any embedded points. In general, $X \setminus B$ is $S_1$, but then it is $S_2$ except at finitely many points by (9.2.3). Thus the previous argument shows that all but finitely many fibers are irreducible and reduced. □

**Lemma 7.1.11.** *Let $X$ be a normal, proper variety and $\pi\colon X \dashrightarrow C$ a pencil with base locus $B$. Then $B$ is also the intersection of any two fibers of $\pi$.*

*Proof.* To see this let $\Gamma \subset X \times C$ be the closure of the graph of $\pi$ with projection $\rho\colon \Gamma \to X$. If $\pi$ is not a morphism at $x \in X$ then $\rho^{-1}(x)$ is positive dimensional, hence dominates $C$. Thus $x$ is contained in every fiber of $\pi$. □

The next classical claims help us recognize rational and linear pencils.

**Lemma 7.1.12** (Rationality test). *Let $X$ be a normal, projective variety over a perfect field $k$. Let $\pi\colon X \dashrightarrow C$ be a pencil with parameter curve $C$. If $\pi$ has a smooth basepoint then $\pi$ is rational.*

*Proof.* Let $\rho\colon \Gamma \to X$ be as in (7.1.11). By a lemma of Abhyankar, if $x$ is smooth then $\rho^{-1}(x)$ is rationally connected over $k(x)$; see, for example, [Kol96, VI.1.9]. Thus $C$ is a geometrically rational curve.  $\square$

**Lemma 7.1.13** (Linearity test I). *Let $X$ be a normal, projective variety over a perfect field $k$. Let $\pi\colon X \dashrightarrow C$ be a rational pencil with base locus $B$ and parameter curve $C$. Then $C \cong \mathbb{P}^1$ in any of the following cases.*

(1) *$X$ has a smooth $k$-point.*

(2) *There is a geometrically irreducible subvariety $W \not\subset B$ that is contained in a fiber of $\pi$.*

(3) *There is a fiber $D_c$ of $\pi$ that is smooth at some point of $B$.*

*Proof.* If $X$ has a smooth $k$-point then so does $C$ by Nishimura's lemma, and then $C \cong \mathbb{P}^1$. (See [KSC04, p.183] for a very simple proof of Nishimura's lemma.)

For (2), note that $\pi(W \setminus B) \subset C$ is geometrically irreducible, hence a $k$-point.

For (3), assume that a fiber $D_c$ is smooth at $x \in B$. As we noted in (7.1.7), $D_c$ becomes the union of $\deg\!\big(k(c)/k\big)$ geometric fibers over $\bar{k}$, and they all contain $B$. If $D_c$ is smooth at $x$ then $k(c) = k$.  $\square$

**Proposition 7.1.14.** *A proper variety has only countably many irrational pencils.*

*Proof.* We may as well assume that the base field is perfect and the variety is normal.

*First proof.* Let $\pi\colon X \dashrightarrow C$ be an irrational pencil. It gives $X \dashrightarrow C \to \mathrm{Jac}(C)$. By the universal property of the Albanese variety (11.3.9), we can factor it as $X \dashrightarrow \mathbf{Alb}(X) \twoheadrightarrow \mathrm{Jac}(C)$. Thus irrational pencils are in one-to-one correspondence with abelian quotients $\tau\colon \mathbf{Alb}(X) \twoheadrightarrow B$ such that the composite $X \dashrightarrow \mathbf{Alb}(X) \twoheadrightarrow B$ has a 1-dimensional image. Let $L_B$ be an ample line bundle on $B$. Then $\tau$ is determined by the numerical equivalence class of $\tau^* L_B$, which is an element of the countable Néron-Severi group.

*Second proof.* First assume that $X$ is smooth and projective. Then irrational pencils are basepoint-free by (7.1.12), hence by (7.1.16) each irrational pencil corresponds to a 1-dimensional connected component of the Chow variety of divisors on $X$ by (7.1.16). The Chow variety has countably many irreducible components by (7.1.1 (2)).

Now suppose $X$ is an arbitrary proper variety over $k$. Let $X' \to X$ be a dominant morphism. We see that the set of pencils on $X$ injects into the set of pencils on $X'$ (since the set of subfields of $k(X)$ injects into the set of subfields of $k(X')$). Choosing $X'$ to be a resolution or an alteration of $X$ reduces the result to the smooth projective case. □

The following lemmas help us recognize fibers of a pencil.

**Lemma 7.1.15.** *Let $X$ be a normal, proper, irreducible variety over a perfect field $k$ and $\pi\colon X \to C$ a basepoint-free pencil. Let $E_1,\ldots,E_r$ be irreducible divisors contained in fibers such that their union does not contain the support of any fiber. Let $E_0$ be an irreducible fiber. Then $\sum_{i=0}^r m_i E_i$ is algebraically equivalent to 0 if and only if it is identically 0.*

*Proof.* Using [Ful98, Example 10.3.4], we may assume that $X$ is projective. Taking general hypersurface sections reduces us to the case when $X$ is a surface, which we may even assume nonsingular.

It is now enough to show that the intersection matrix $(E_i \cdot E_j)$ for $1 \le i, j \le r$ is negative definite. This can be done one fiber at a time. After reindexing, we may assume that $E_1 + \cdots + E_m$ is a maximal connected component of $E_1 + \cdots + E_r$. Then $E_1, \cdots, E_m$ are contained in a fiber $F$ and there is an irreducible curve $E^* \subset F$ that is different from the $E_i$ but is not disjoint from them. Then $(F - E^*) \cdot E_i \le 0$ for every $i$ and $(F - E^*) \cdot E_j < 0$ for some $1 \le j \le m$. By a lemma on quadratic forms, this implies that $(E_i \cdot E_j)$ for $1 \le i, j \le m$ is negative definite; see [Kol13, 10.3.4]. □

**Lemma 7.1.16.** *Let $X$ be a normal, proper, irreducible variety over a perfect field $k$ and $|D|$ a basepoint-free pencil. Let $A \subset X$ be a connected, effective divisor such that $A \equiv_s |D|$. Then $A$ is a (rational) multiple of a fiber of $\pi$.*

*Proof.* Let $D \in |D|$ be an irreducible fiber not contained in $A$. Since $A|_D \equiv_s D|_D = 0$, we see that $A$ is disjoint from $D$ by (7.1.6), hence $A$ is contained in some fiber $D_0$ of $|D|$. There is a largest $c \in \mathbb{Q}$ such that $D_0 - cA$ is effective. We are done if $D_0 - cA = 0$. Otherwise the support of $D_0 - cA$ is a proper subset of $D_0$ and it is numerically equivalent to a multiple of a fiber. This is impossible by (7.1.15). □

The next result will allow us to reconstruct pencils from topological data.

**Proposition 7.1.17.** *Let $X$ be a proper, geometrically normal, irreducible variety over a field $k$, $\{D_i : i \in I\}$ an infinite set of irreducible Weil divisors, and $B \subsetneq X$ a closed subset. Assume that*

(1) *the $D_i$ are algebraically equivalent to each other,*

(2) *$D_i \cap D_j \subset B$ for every $i \ne j \in I$, and*

(3) *$D_i \not\subset B$ for every $i \in I$.*

*Then there is a unique non-composite pencil of divisors $\pi\colon X \dashrightarrow C$ such that all the $D_i$ are*

(4) *fibers of $\pi$ if $k$ is perfect, and*

(5) *reductions of fibers of $\pi$ in general.*

*Proof.* By (7.1.4), there is scheme of finite type parametrizing all effective divisors algebraically equivalent to the $D_i$. Denote it by $\mathrm{Chow}_X^D$.

Thus the closure of the infinite set of points $\{[D_i] : i \in I\} \subset \mathrm{Chow}_X^D$ contains a positive dimensional irreducible component $Z$. Set $J := \{i \in I : [D_i] \subset Z\}$.

There is a universal family $u : U \to Z$ with canonical map $\chi : U \to X$. For a divisor $D$ with $[D] \in Z$ write $D^u \subset U$ for the fiber of $u$ over $[D] \in Z$. Note that $\chi : D^u \to D$ is an isomorphism for general $[D] \in Z$, but $D^u$ can have some embedded points in general.

The image $\chi(U)$ contains a countable set of distinct divisors $\{D_i : i \in J\}$, hence $\chi$ is dominant. We claim that $\chi$ is generically purely inseparable.

Let $X^\circ \subset X$ be a dense, nonsingular, open set such that $\chi$ is flat over $X^\circ$ and let $U^\circ \subset U$ be its preimage.

Let $x \in D_i \cap X^\circ$ be a closed point for some $i \in J$. We can view $x$ as a point $x^u \in D_i^u$, and then $\chi(x^u) = x$. We need to show that $\mathrm{Supp}(\chi^{-1}(x)) = \{x^u\}$.

To see this, note that any other point of $\chi^{-1}(x)$ would lie on another fiber $D^u$. Since $D$ is irreducible but not equal to $D_i$, the intersection $D_i \cap D$ has codimension 2 in $X$ and is not contained in $B$. We check below that then the same holds for an open and dense set of fibers $D_z$. Since the $\{D_i : i \in J\}$ are dense in $Z$, this contradicts assumption (2).

Thus $\chi^{-1}(D_i \cap X^\circ) \subset U^\circ$ is a Cartier divisor that intersects $D^u \cap U^\circ$ properly, so $u : \chi^{-1}(D_i) \to Z$ is dominant. Thus there is a dense, open $Z^\circ \subset Z$ contained in $u(\chi^{-1}(D_i))$.

Set $J^\circ := \{j : [D_j] \in Z^\circ\} \setminus \{i\}$. Then

$$\cup_{j \in J^\circ} D_j^u \cap \chi^{-1}(D_i) \text{ is dense in } \chi^{-1}(D_i).$$

Taking its image by $\chi$ we get that

$$\cup_{j \in J^\circ} D_j \cap D_i \text{ is dense in } D_i.$$

This is impossible since, by assumption, this set is contained in $D_i \cap B$.

Thus $\chi^{-1}(x)$ is a single point for a dense set of points $x \in X$, hence $\chi$ is generically purely inseparable. In particular, $\dim U = \dim X$ and so $\dim Z = 1$. After composing with a power of the Frobenius, $u \circ \chi^{-1}$ gives a pencil $\pi_Z : X \dashrightarrow Z$ such that the $\{D_i : i \in J\}$ are fibers of $\pi$, at least set theoretically. Taking Stein factorization gives a non-composite pencil $\pi : X \dashrightarrow C$.

If $E \subset X$ is any irreducible divisor that is not contained in a fiber of $\pi$, then $\pi|_E : E \dashrightarrow C$ is dominant, hence $\cup_{j \in J^\circ} D_j \cap E$ is dense in $E$. Applying this to $D_i$ for $i \in I \setminus J$ shows that every $D_i$ is contained in a fiber of the pencil $\pi$. Since there are only finitely many reducible fibers, at least one of the $D_i$ is a fiber. Then all the $D_i$ are fibers because they are algebraically equivalent to a fiber, so if they were a proper subset of a fiber, their complement would be numerically trivial, contradicting (7.1.15).

Finally note that all but finitely many fibers are irreducible and reduced if $k$ is perfect. In general, all but finitely many fibers are irreducible.   □

The following result is proved in [BPS16], sharpening earlier versions of [Tot00, Per06].

**Theorem 7.1.18.** *Let $X$ be a normal, projective, irreducible $k$-variety and $\{D_i : i \in I\}$ pairwise disjoint divisors. Then*

(1) *either $|I| \leq \rho^{cl}(X_{\overline{k}}) - 1$,*

(2) *or all the $D_i$ are contained in members of a basepoint-free pencil.*

*Proof.* Let $D_i'$ be a geometric irreducible component of $D_i$. We apply [BPS16] to the $D_i'$. Thus we get that either (1) holds or the $D_i'$ are contained in members of a basepoint-free pencil $|M|$ defined over $\overline{k}$.

By (7.1.15), there can be at most $\rho^{cl}(X_{\overline{k}}) - 2$ of the $D_i'$ that are components of reducible members of $|M|$. Thus either (1) holds, or there is a divisor, say $D_1'$, that is an irreducible member of $|M|$. Any divisor disjoint from an irreducible member of $|M|$ is contained in a finite union of members of $|M|$. Thus all the other $D_i$ are contained in finite unions of members of $|M|$. So $|M|$ is defined over $k$ if $k$ is perfect, and $|qM|$ is defined over $k$ for some power $q$ of the characteristic in general.   □

## 7.2   FIBERS OF FINITE MORPHISMS

Let $\pi \colon X \to Y$ be a quasi-finite, dominant morphism of $k$-varieties. For any closed point $y \in Y$, the fiber $X_y$ is a finite $k(y)$-scheme. We will need to understand which finite $k(y)$-schemes occur. The extreme cases are especially important for us. These are

- $X_y$ is reduced and irreducible;
- $X_y$ is a disjoint union of copies of $k(y)$.

The first condition is the topic of Hilbert's irreducibility theorem and its generalizations. We consider such questions in (8.6.6) and the discussion immediately afterward. We focus on the second case here.

Note that such questions can be viewed as attempted generalizations of Chebotarev's density theorem, which describes the density of primes with given splitting behavior in an extension of number fields. Density does not seem to make sense in general, but the first result, proved in [Poo01], shows that there are infinitely many completely split points in any separable field extension.

**Proposition 7.2.1.** *Let $C$ be a geometrically reduced $k$-curve and $\pi \colon C \to \mathbb{P}^1$ a quasi-finite, separable morphism. Then there are infinitely many separable points*

$p_j \in \mathbb{P}^1$ such that $\pi^{-1}(p_j)$ is a reduced, disjoint union of copies of $p_j$.

*Proof.* Let $C_i$ be the irreducible components of $C$. Let $D$ be the normalization of $\mathbb{P}^1$ in a Galois closure of a composite of the $k(C_i)/k(\mathbb{P}^1)$. If $p_j$ works for $\sigma \colon D \to \mathbb{P}^1$ and $\pi$ is étale over $p_j$ then $p_j$ works for $C \to \mathbb{P}^1$.

If $k$ is infinite, then a general pencil of very ample divisors gives a separable morphism $\rho \colon D \to \mathbb{P}^1 := \mathbb{P}^1$ such that

$$(\rho, \sigma) \colon D \to D' \subset \mathbb{P}^1 \times \mathbb{P}^1 \qquad (7.2.1.1)$$

is birational onto its image. (We use the notation $\mathbb{P}^1$ to distinguish between the two factors.) Let $S \subset D$ be the union of the preimage of Sing $D'$, the ramification locus of $\sigma$ and the ramification locus of $\rho$.

Pick any $c \in \mathbb{P}^1(k) \setminus \rho(S)$. Let $p_D \in \rho^{-1}(c)$ be any closed point, $p'_D$ its image in $D'$ and $p := \sigma(p_D) \in \mathbb{P}^1$. Then

$$k(p_D) = k(p'_D) = k(c) \otimes_k k(p) = k(p).$$

Since $D/\mathbb{P}^1$ is Galois, the same holds for all points in $\sigma^{-1}(p)$.

If $k$ is finite, choose $q = p^r$ such that $D$ decomposes into $m$ irreducible components that are geometrically irreducible. Then $D$ has about $mq$ points in $\mathbb{F}_q$. All these map to $\mathbb{F}_q$ points in $\mathbb{P}^1$. We show that for most of them, their image is not defined over a proper subfield of $\mathbb{F}_q$. All proper subfields of $\mathbb{F}_q$ have at most $\sqrt{q}$ elements and the number of maximal ones equals the number of prime divisors of $r$, so there are at most $\log_2 r$ of them. Thus at most $\log_2 r \cdot \sqrt{q}$ points of $\mathbb{P}^1(\mathbb{F}_q)$ are in a smaller subfield and these have at most $\deg \pi \cdot \log_2 r \cdot \sqrt{q}$ preimages in $D$. So for $q \gg 1$, almost all $\mathbb{F}_q$ points of $D$ map to points of $\mathbb{P}^1$ whose residue field is $\mathbb{F}_q$. $\square$

The following versions of (7.2.1) are also useful.

**Claim 7.2.2.** *Let $X, Y$ be geometrically reduced $k$-schemes and $\pi \colon X \to Y$ a quasi-finite, separable morphism. Then there is a Zariski dense set of closed, separable points $p_j \in Y$ such that $\pi^{-1}(p_j)$ is a reduced, disjoint union of copies of $p_j$.*

*Proof.* We can replace $Y$ by a general curve $B \subset Y$. The curve case is reduced to (7.2.1) by composing with a quasi-finite, separable morphism $B \to \mathbb{P}^1$. $\square$

**Claim 7.2.3.** *Let $X, Y$ be geometrically reduced $k$-schemes and $\pi \colon X \to Y$ a quasi-finite, separable morphism. Then there are infinitely many irreducible divisors $D_j \subset Y$ such that $\pi^{-1}(D_j)$ is a reduced union of divisors $D_j^i$, such that each $D_j^i \to D_j$ is birational.*

*Proof.* As in (7.2.1) and in (7.2.2), we may assume that $Y = \mathbb{P}^n$ and $X \to \mathbb{P}^n$ is Galois. If $k$ is infinite, the proof works as in (7.2.1), but we replace (7.2.1.1) by

$$(\rho, \sigma) \colon X \to X' \subset \mathbb{P}^1 \times \mathbb{P}^n.$$

We can also reduce (7.2.3) to (7.2.2) by taking a coordinate projection $p\colon \mathbb{P}^n \dashrightarrow \mathbb{P}^{n-1}$ and working with the generic fibers of $Y \dashrightarrow \mathbb{P}^{n-1}$ and $X \dashrightarrow \mathbb{P}^{n-1}$.  $\square$

The first application is a topological formula for the degree of a morphism.

**Lemma 7.2.4.** *Let $g\colon Y \to X$ be a generically finite and separable morphism of $k$-varieties of pure dimension $d$. Assume that $X$ is normal. Then*

$$\deg g = \max\{\#|g^{-1}(x)| : x \in X, \#|g^{-1}(x)| < \infty\}.$$

*Proof.* We may assume that $g$ is quasi-finite. Then $x \mapsto \dim_{k(x)} \mathcal{O}_{g^{-1}(x)}$ is lower semi-continuous and equal to the degree on a dense open set $X^\circ \subset X$. We may also assume that $g$ is étale over $X^\circ$.

Note that $\#|g^{-1}(x)| \le \dim_{k(x)} \mathcal{O}_{g^{-1}(x)}$ and equality holds if and only if $g^{-1}(x)$ is a union of copies of $\{x\}$. The existence of such points is guaranteed by (7.2.2).  $\square$

As a special case, we get a simple topological formula computing the intersection number of a curve with a pencil of divisors.

**Notation 7.2.5.** Let $X$ be a normal, proper variety over a field $k$ of characteristic 0, and $|D|$ an algebraic pencil of $\mathbb{Q}$-Cartier divisors with parameter curve $C$. Let $\{D_c : c \in C(\overline{k})\}$ be its geometric members. For any 1-cycle $Z \subset X$, the intersection number $(Z \cdot D_c)$ is independent of $c$. We denote it by $(Z \cdot |D|)$.

**Corollary 7.2.6.** *Let $X$ be a normal, proper variety over a field $k$ of characteristic 0, and $|D|$ an algebraic, $\mathbb{Q}$-Cartier pencil with parameter curve $C$. Let $A \subset X$ be an irreducible curve disjoint from $\mathrm{Bs}\,|D|$. Then*

$$\big(A \cdot |D|\big) = \max\{\#|A \cap D_c| : c \text{ is a closed point of } C\}. \quad \square$$

**Remark 7.2.7.** If $\operatorname{char} k = p > 0$ then we get the formula

$$\big(A \cdot |D|\big) = p^m \cdot \max\{\#|A \cap D_c| : c \text{ is a closed point of } C\},$$

where $m$ is the degree of inseparability of $A \to C$. In our applications the latter is an unknown (and in fact unknowable) number.

This gives us a topological way to compute the prime-to-$p$ part of intersection numbers between curves and algebraic pencils. This does carry a lot of information, but it is much harder to exploit.

Another application is the following, whose proof was suggested by Michael Larsen; see (8.6.1) for the definition of locally finite fields.

**Lemma 7.2.8.** *Let $k$ be a field that is not locally finite and $T$ a nontrivial algebraic torus over $k$. Then $\operatorname{rank}_{\mathbb{Q}} T(k) = \infty$.*

*Proof.* The torus $T$ is defined over a finitely generated subfield, so we may as well assume that $k$ is either a number field or the field of functions of a geometrically integral curve $C$ over a subfield $k_0 \subset k$. Over a dense, open, regular subset $U \subset C$ we have a torus $T_U \to U$.

Assume to the contrary that $t_1, \ldots, t_s \in T(k)$ generate a maximal rank subgroup. We can view the $t_i$ as rational sections of $T_U \to U$. After further shrinking $U$ we may assume that they are all regular sections. If $t \in T(k)$ then $t^n \in \langle t_1, \ldots, t_s \rangle$ for some $n > 0$. Thus $t$ is a rational section that is also finite over $U$, hence a regular section. Thus every rational section of $T_U \to U$ is regular.

Next we show that this is not the case. The torus $T$ is isomorphic to $\mathbb{G}_m^r$ over $k^{\mathrm{sep}}$, hence there is a finite, separable field extension $K/k$ such that $T_K \cong \mathbb{G}_m^r$. (Such a $K$ is called a *splitting field* of $T$.) After further shrinking $U$ we assume that we have a finite morphism $\pi \colon V \to U$ such that $T_V \cong V \times \mathbb{G}_m^r$. Let now $p \in U$ be a point with preimages $p_1, \ldots, p_r \in V$ such that $k(p_i) \cong k(p)$ for every $i$ and $\pi$ is étale over $p$. In the geometric cases this is possible by (7.2.2), while for number fields this follows from the Chebotarev density theorem.

Then $T_V$ has a rational section $s_V$ that has a pole along $p_1$ but is regular at $p_2, \ldots, p_r$. Then $\mathrm{norm}_{K/k}(s_V)$ is a rational section of $T_U$ with a pole at $p$, a contradiction.                                                                                    □

The following more precise version was communicated to us by B. Poonen; we do not use it.

**Theorem 7.2.9.** *Let $k$ be an infinite, finitely generated field and $T$ a nontrivial algebraic torus over $k$. Then $T(k) \cong A \times \mathbb{Z}^{\omega_0}$, where $A$ is a finite abelian group and $\mathbb{Z}^{\omega_0}$ is a free abelian group.*

## 7.3   TOPOLOGICAL PENCILS

Given a pencil $\pi \colon X \dashrightarrow C$ with base locus $B$, we get a map of topological spaces $|\pi| \colon |X \setminus B| \to |C|$. All information is contained in the collection of the fibers of $|\pi|$. Its abstract properties define topological pencils. The main question is then to understand which topological pencils come from algebraic pencils.

**Definition 7.3.1.** Let $X$ be an irreducible, normal $k$-variety. A *topological pencil* or *t-pencil* is a collection of effective, reduced divisors $\{D_\lambda : \lambda \in \Lambda\}$ such that the following hold.

(1) Every closed point of $X$ is contained in some $D_\lambda$.

(2) Almost all of the $D_\lambda$ are irreducible.

(3) There is a closed subset $B \subset X$ of codimension $\geq 2$ such that

    (a)  $D_\lambda \cap D_\mu \subset B$ for all $\lambda \neq \mu \in \Lambda$, and

    (b)  $D_\lambda \cap D_\mu = B$ for almost all $\lambda \neq \mu \in \Lambda$.

(4) Each $D_\lambda \setminus B$ is connected.

Here 'almost all' means that there is a cofinite subset $\Lambda' \subset \Lambda$ such that the claims hold for all $\lambda, \mu \in \Lambda'$. $B$ is called the *base locus*. We call a t-pencil *ample* if almost all members are ample $\mathbb{Q}$-Cartier divisors.

If $\dim X = 1$ then there is a unique t-pencil, and its members are the closed points of $X$. Thus the notion is interesting only for $\dim X \geq 2$.

The main examples are the following.

**Example 7.3.2.** Let $X$ be a normal, irreducible $k$-variety and $|D|$ an algebraic pencil given by $\pi \colon X \dashrightarrow C$ with parameter curve $C$ and base locus $B$.

Given a closed point $c \in C$, let $\{D_{c,j} : j = 1 \ldots, r_c\}$ be the closures of the connected components of $\operatorname{red} D_c \setminus B$. The set of all $D_{c,j}$ forms a t-pencil with base locus $B$. We denote it by $|D|^t$ and write its members as $\{D_\lambda : \lambda \in \Lambda\}$. These are the *algebraic t-pencils*.

If $|D|$ is composite with $|D'|$, then $|D|^t = |D'|^t$.

Note that $|D|^t$ is ample if and only if the fibers of $|D|$ are ample $\mathbb{Q}$-Cartier divisors.

While this makes sense over any field, we can use t-pencils effectively only if the divisors $D_\lambda$ are also geometrically reduced. Thus $k$ should be perfect. As in (7.1.7), we can then describe the members as follows.

Let $\bar{c} \in C_{\bar{k}}$ be a geometric point lying over $c \in C$ and $D_{\bar{c},j}$ a connected component of $\operatorname{red} D_{\bar{c}} \setminus B$. As $\sigma$ runs through $\operatorname{Gal}(\bar{k}/k)$, $D_{c,j} := \cup_\sigma D_{\bar{c},j}^\sigma$ is a finite union, giving a divisor defined over $k$. Note also that for $\sigma_1, \sigma_2 \in \operatorname{Gal}(\bar{k}/k)$, the divisors $D_{\bar{c},j}^{\sigma_1}$ and $D_{\bar{c},j}^{\sigma_2}$ are either identical or their intersection is contained in $B$. This shows that a member of $|D|^t$ is either contained in a member of $|D|$, or every irreducible component of it is geometrically reducible.

**Proposition 7.3.3** (Algebraicity criterion I). *Let $X$ be a normal, irreducible $k$-variety and $\{D_\lambda : \lambda \in \Lambda\}$ a t-pencil. Assume that there is an algebraic pencil $|A|$ given by $\pi \colon X \dashrightarrow C$ such that infinitely many fibers of $\pi$ are members of $\{D_\lambda : \lambda \in \Lambda\}$. Then $\{D_\lambda : \lambda \in \Lambda\} = |A|^t$.*

*Proof.* We claim that every $D_\lambda$ is contained in a fiber of $\pi$; the rest is then clear.

Let $\{A_i = D_{\mu_i} : i \in I\}$ be the fibers of $\pi$ that are members of $\{D_\lambda : \lambda \in \Lambda\}$. If a given $D_\lambda$ is not contained in a fiber of $\pi$, then $\pi \colon D_\lambda \setminus B \to C$ is dominant. Thus $\cup_i A_i \cap D_\lambda$ is dense in $D_\lambda$. But

$$\cup_i A_i \cap D_\lambda = \cup_i D_{\mu_i} \cap D_\lambda \subset B,$$

which is a contradiction.                                                    $\square$

**Proposition 7.3.4** (Algebraicity criterion II). *Let $X$ be a geometrically normal, proper, irreducible variety over an infinite field. A t-pencil $\{D_\lambda : \lambda \in \Lambda\}$ is* *algebraic if and only if there is an infinite subset $\Lambda^* \subset \Lambda$ such that the $\{D_\lambda : \lambda \in \Lambda^*\}$ are algebraically (or numerically) equivalent to each other.*

*Proof.* Let $p: X \dashrightarrow C$ be an algebraic pencil. Let $\tau: C \to \mathbb{P}^1$ be a nonconstant morphism of degree $d$. For every $p \in \mathbb{P}^1(k)$ the fiber $\tau^{-1}(p)$ is a union of closed points of degrees $\leq d$. Thus there is a $d' \leq d$ such that $C$ has infinitely many closed points $c_i \in C$ of degree $d'$. The corresponding fibers are algebraically equivalent to each other.

Conversely, assume that the $\{D_\lambda : \lambda \in \Lambda^*\}$ are algebraically equivalent to each other. By (7.1.17) there is an algebraic pencil $|A|$ given by $\pi: X \dashrightarrow C$ such that infinitely many of the fibers of $\pi$ are members of $\{D_\lambda : \lambda \in \Lambda\}$. Thus $\{D_\lambda : \lambda \in \Lambda\} = |A|^t$ by (7.3.3) $\qquad\square$

The correspondence between t-pencils and algebraic pencils works best over uncountable fields.

**Corollary 7.3.5.** *Let $X$ be a geometrically normal, proper variety over an uncountable field $k$. Then every t-pencil on $X$ is algebraic.*

*Proof.* Let $\{D_\lambda : \lambda \in \Lambda\}$ be a t-pencil with base locus $B$. Pick any $D_{\lambda_0}$, a point $x \in D_{\lambda_0} \setminus B$, and an irreducible curve $x \in C$ that is not contained in $D_{\lambda_0}$. Then $C$ is not contained in any other $D_\lambda$, so the $C \cap D_\lambda$ are all finite. They cover all closed points of $C$, thus $|\Lambda|$ is uncountable.

By (7.1.4), the Chow variety of divisors has only countably many irreducible components. Thus one of them contains uncountably many of the $D_\lambda$. So $\{D_\lambda : \lambda \in \Lambda\}$ is algebraic by (7.3.4). $\qquad\square$

As the following example shows, over a countable field t-pencils need not come from algebraic pencils (this more or less must be the case by (10.3.1)).

**Example 7.3.6** (t-pencils over countable fields). Let $X$ be a normal, projective variety of dimension $\geq 2$ over an infinite field $K$ and $L$ a very ample line bundle on $X$.

Pick any $s_1 \in H^0(X, L^{m_1})$ and $s_2 \in H^0(X, L^{m_2})$. Assume that we already have $s_i \in H^0(X, L^{m_i})$ for $i = 1, \ldots, r$ such that $\mathrm{Supp}(s_i = s_j = 0)$ is independent of $1 \leq i < j \leq r$. Set $M = \prod_i m_i$,

$$S_{r+1} := \left( \prod_i s_i^{M/m_i} \right) \left( \sum_i s_i^{-M/m_i} \right), \text{ and } T_{r+1} := \prod_i s_i.$$

Choose $s_{r+1} = S_{r+1} + g \cdot T_{r+1}$ for a general $g \in H^0(X, L^{n_r})$, where $n_r = M(r-1) - \sum_i m_i$. Then $(s_{r+1} = 0)$ is irreducible and $\mathrm{Supp}(s_i = s_j = 0)$ is independent of $1 \leq i < j \leq r+1$.

If $K$ is countable then we can order the points of $X$ as $x_1, x_2, \ldots$ and we can choose the $s_i$ such that $\prod_{i=1}^r s_i$ vanishes on $x_1, \ldots, x_r$ for every $r$. Then the resulting $D_i := (s_i = 0)$ is a t-pencil that does not correspond to any algebraic pencil.

## 7.4  DEGREE FUNCTIONS AND ALGEBRAIC PENCILS

**Definition 7.4.1.** Let $X$ be an irreducible, normal, proper variety. A *degree function* on $d$ cycles is a real-valued function that is

(1) linear, namely $\deg(Z_1 + Z_2) = \deg(Z_1) + \deg(Z_2)$, and

(2) respects algebraic equivalence, namely $\deg(Z) = 0$ if $Z$ is algebraically equivalent to 0.

The main example is of course the following. Let $A$ be a $\mathbb{Q}$-Cartier divisor on $X$. Then $\deg_A(Z) := (Z \cdot A^d)$ is a degree function.

The most important cases are when $A$ is ample. With this in mind, a degree function deg is called *ample* if it is

(3) positive, that is, $\deg(Z) > 0$ if $Z$ is effective and nonzero, and

(4) bounded, that is, there is an ample divisor $H$ and positive constants $c_i$ such that $c_1 \deg_H(Z) \le \deg(Z) \le c_2 \deg_H(Z)$ for every effective cycle $Z$.

If $X$ has an ample degree function deg, then (7.3.4) gives a characterization of algebraic t-pencils.

A key step in the proof of the main theorem is to construct $\deg_H$ out of topological data. However, we are able to get only $c \cdot \deg_H$ for some *completely unknown* positive constant $c \in \mathbb{Q}$.

For arbitrary algebraic pencils, we turn (7.2.5) and (7.2.6) into a definition.

**Definition 7.4.2.** Let $X$ be a normal, projective variety over a field $k$ of characteristic 0 and $|D|$ an algebraic pencil with corresponding topological pencil $|D|^t$. Let $C \subset X$ be a curve disjoint from $\mathrm{Bs}\,|D|^t$. Set

$$\mathfrak{d}_{|D|}(C) := \max\{\#|C \cap D_\lambda| : D_\lambda \in |D|^t\}.$$

We can use this to decide when two algebraic, $\mathbb{Q}$-Cartier pencils are numerically similar, after recalling some of the basic results of [Kle66] in slightly modified forms.

**Definition 7.4.3.** Let $X$ be a normal, projective variety. Two 1-cycles $C_1, C_2$ are *numerically equivalent* if $(C_1 \cdot D) = (C_2 \cdot D)$ for every Cartier divisor $D$. Let $H$ be an ample divisor. Then $mH + D$ is ample for $m \gg 1$ and $D = (mH + D) - mH$. Thus $C_1 \equiv C_2$ if and only if $(C_1 \cdot H) = (C_2 \cdot H)$ for every ample Cartier divisor $H$.

**Definition 7.4.4.** Two $\mathbb{Q}$-Cartier divisors $D_1, D_2$ are called *numerically equivalent* if $(C \cdot D_1) = (C \cdot D_2)$ for every curve $C \subset X$.

**Lemma 7.4.5.** *Let $D$ be a $\mathbb{Q}$-Cartier divisor and $H$ an ample divisor. Then $D$ is numerically trivial if and only if $(H^{n-1} \cdot D) = (H^{n-2} \cdot D^2) = 0$, where*

$n = \dim X$.

*Sketch of proof.* We need to show that $(C \cdot D) = 0$ for every curve $C \subset X$. Now $C$ is contained in an $H$-complete intersection surface $S$. After normalizing we are reduced to the $n = 2$ case, which is the Hodge index theorem.   $\square$

**Lemma 7.4.6.** *Let $D_1, D_2$ be $\mathbb{Q}$-Cartier divisors. Then $D_1 \equiv D_2$ if and only if $(C \cdot D_1) = (C \cdot D_2)$ for every irreducible, ample-sci curve $C$. As in (11.9.1), ample-sci means set-theoretic complete intersection of ample divisors.*

*Sketch of proof.* Set $D = D_1 - D_2$. We can unify the conditions

$$(H^{n-1} \cdot D) = (H^{n-2} \cdot D^2) = 0$$

as

$$\left(H^{n-2} \cdot (mH + D) \cdot D\right) = 0 \ \text{ for all } \ m \gg 1.$$

Since $mH + D$ is ample for $m \gg 1$, we can think of $H^{n-2} \cdot (mH + D)$ as an ample-sci curve. It is actually enough to work with self-intersections of the form $(mH + D)^{n-1}$ since

$$H^{n-2} \cdot D \in \left\langle (H^i \cdot D^{n-1-i}) : i = 0, \ldots, n-1 \right\rangle = \left\langle (mH + D)^{n-1} : m \gg 1 \right\rangle. \ \square$$

Note finally that ample-sci curves can be moved away from any codimension 2 subset, hence we get the following variant.

**Lemma 7.4.7.** *Let $D$ be a $\mathbb{Q}$-Cartier divisor on $X$ and $B \subset X$ a subset of codimension $\geq 2$. Then $D \equiv 0$ if and only if $(C \cdot D) = 0$ for every proper curve $C \subset X \setminus B$.*   $\square$

Combining (7.2.6) and (7.4.7), we get the following.

**Corollary 7.4.8.** *Let $X$ be a normal, projective variety over a field $k$ of characteristic $0$. Let $|D^1|$ and $|D^2|$ be $\mathbb{Q}$-Cartier, algebraic pencils. Then $|D^1| \equiv_s |D^2|$ if and only if $\partial_{|D^1|} \equiv_s \partial_{|D^2|}$ as functions on the set of proper curves contained in $X \setminus (\mathrm{Bs}\,|D^1| \cup \mathrm{Bs}\,|D^2|)$.*   $\square$

This is a good topological description of numerical similarity, but the $\mathbb{Q}$-Cartier condition is not topological. The following definitions aim to fix this.

**Definition 7.4.9.** Let $X$ be a normal variety and $\{Z_i \subset X : i \in I\}$ be irreducible, closed subvarieties. We say that a t-pencil $|D|^t = \{D_\lambda : \lambda \in \Lambda\}$ is in *general position* for $\{Z_i : i \in I\}$ if the following hold.

(1) None of the $Z_i$ is contained in a member of $|D|^t$.

(2) $Z_i \cap \mathrm{Bs}\,|D|^t$ has codimension $\geq 2$ in $Z_i$ for every $i$.

We also say that these $Z_i$ are in general position for $|D|^t$.

**Definition 7.4.10.** Let $X$ be a normal, projective variety over a field $k$ of characteristic 0. A set $\{|D^i| : i \in I\}$ of algebraic pencils is called *compatible* if $\mathfrak{d}_{|D_i|} \equiv_s \mathfrak{d}_{|D_j|}$ (on the set of curves where both are defined) for every $i, j \in I$.

The set $\{|D^i| : i \in I\}$ is called *complete* if for every finite set of closed subvarieties $Z_j \subset X$ there is an $i \in I$ such that $|D^i|$ is in general position for all the $Z_j$.

The set $\{|D^i| : i \in I\}$ is called *ample* if for every pair of closed points $p, q \in X$, there is a $|D^i|$ such that $p \in \operatorname{Bs}|D^i|$ but $q \notin \operatorname{Bs}|D^i|$.

**Example 7.4.11.** Let $H$ be an ample divisor on $X$. The set of all $\mathbb{Q}$-Cartier pencils of divisors linearly similar to $H$ is ample, complete, and compatible.

An important observation is that the converse also holds.

**Proposition 7.4.12.** *Let $X$ be a normal, projective variety over a field $k$ of characteristic 0 and $\{|D^i| : i \in I\}$ a complete and compatible set of pencils. Then there is a Cartier divisor $H$ such that $\deg_H \equiv_s \mathfrak{d}_{|D^i|}$ for every $i \in I$.*

*If $\{|D^i| : i \in I\}$ is ample then $H$ is ample.*

*Proof.* Let $\{Z_j \subset X : j \in J\}$ be the closures of the non-Cartier centers as in (11.3.19). Let $I' \subset I$ be the subset indexing those pencils that are in general position for $\{Z_j \subset X : j \in J\}$. Then $\{|D^i| : i \in I'\}$ is still complete and compatible. These $|D^i|$ are $\mathbb{Q}$-Cartier and numerically similar by (7.4.8). Thus there is a Cartier divisor $H$ whose degree function matches all their degree functions, and then by the definition of compatibility, matches the degree functions of $D^i$ for all $i \in I$.

If $\{|D^i| : i \in I\}$ is ample, we check the Nakai-Moishezon criterion. Let $Z \subset X$ be an irreducible subvariety. Pick closed points $p, q \in Z$. By assumption there is a $|D^i|$ such that $p \in \operatorname{Bs}|D^i|$ but $q \notin \operatorname{Bs}|D^i|$. Let $D^i_c$ be a general fiber. Then $D^i_c \equiv_s H$ and $(Z \cap D^i_c) \subsetneq Z$ is nonempty. By induction we get that $(Z \cdot H^{\dim Z}) > 0$. $\qquad\square$

So far we have constructed a degree function on curves. Now we extend it to all algebraic cycles.

**Definition 7.4.13.** Let $S$ be a set and $g \colon S \to \mathbb{R}$ a function. The *generic minimum* of $g$, denoted by $\operatorname{genmin}(g) \in \mathbb{R} \cup \{\infty\}$, is the supremum of $c \in \mathbb{R}$ such that $\{s : g(s) < c\}$ is finite.

**Lemma 7.4.14.** *Let $X$ be a normal, projective variety over a field $k$ of characteristic 0 and $Z \subset X$ an irreducible subvariety of dimension $r \geq 2$. Let $|D|$ be an ample, algebraic pencil given by $\pi \colon X \dashrightarrow C$. Assume that $|D|$ is in general position for $Z$. Let $d \in \mathbb{N}$ be the smallest such that $C$ has infinitely many points of degree $d$. Then*

$$d \cdot (Z \cdot |D|^r) = \operatorname{genmin}\{((Z \cap D_\lambda) \cdot |D|^{r-1}) : \lambda \in \Lambda\}.$$

*Proof.* With finitely many exceptions, $Z \cap D_\lambda$ is reduced and

$$\left(Z \cdot D_\lambda \cdot |D|^{r-1}\right) = \left((Z \cap D_\lambda) \cdot |D|^{r-1}\right). \qquad \square$$

We can summarize these results as follows.

**Theorem 7.4.15.** *Let $X$ be a normal, projective variety of dimension $\geq 2$ over a field $k$ of characteristic $0$. Let $\mathbf{P} := \{|D^i| : i \in I\}$ be an ample, complete, compatible set of algebraic pencils. Then $\mathbf{P}$ determines a similarity class of degree functions $\deg_{\mathbf{P}}$ on all algebraic $r$-cycles on $X$ for every $1 \leq r \leq \dim X$. Furthermore, $\deg_{\mathbf{P}} \equiv_s \deg_H$ for some ample divisor $H$.*

*Proof.* First, (7.4.12) defines $\deg_{\mathbf{P}}$ on curves. Next assume that we already extended $\deg_{\mathbf{P}}$ to cycles of dimension $\leq r - 1$.

Fix now an $r$-dimensional, irreducible cycle $Z_0$. Pick $\{D_\lambda^i : \lambda \in \Lambda\}$ in general position for $Z_0$. If $Z$ is in general position for $|D^i|^t$ then, by (7.4.14),

$$\frac{\operatorname{genmin}\{\deg_{\mathbf{P}}(Z \cap H_\lambda) : \lambda \in \Lambda\}}{\operatorname{genmin}\{\deg_{\mathbf{P}}(Z_0 \cap H_\lambda) : \lambda \in \Lambda\}} = \frac{(Z \cdot H^r)}{(Z_0 \cdot H^r)}.$$

Thus the formula

$$\deg_{\mathbf{P}}(Z) := \frac{\operatorname{genmin}\{\deg_{\mathbf{P}}(Z \cap H_\lambda) : \lambda \in \Lambda\}}{\operatorname{genmin}\{\deg_{\mathbf{P}}(Z_0 \cap H_\lambda) : \lambda \in \Lambda\}}$$

defines a degree function similar to $\deg_H$ on those $r$-cycles that are in general position for $|D^i|$. It is normalized by the condition $\deg_{\mathbf{P}}(Z_0) = 1$.

For any $Z$ we can choose $|D^i|$ in general position for $Z$ and $Z_0$, so the definition works for any $Z$. $\qquad \square$

**Remark 7.4.16.** To be precise, we proved that if $Z_1, Z_2$ are nonzero, effective $r$-cycles then

$$\frac{\deg_{\mathbf{P}}(Z_1)}{\deg_{\mathbf{P}}(Z_2)} = \frac{(Z_1 \cdot H^r)}{(Z_2 \cdot H^r)}.$$

This implies that $\deg_{\mathbf{P}} = c_r \cdot \deg_H$, where the constant depends on the dimension. There does not seem to be any way of assigning a specific value to the $c_r$.

## 7.5  DEGREE FUNCTIONS AND LINEAR EQUIVALENCE

**Definition 7.5.1.** Let $X$ be a normal variety, $|D|$ an algebraic, non-composite pencil given by $\pi \colon X \dashrightarrow C$, and $|D|^t = \{D_\lambda : \lambda \in \Lambda\}$ the corresponding t-pencil.

We say that $D_\mu$ is a *true member* of $|D|^t$ if there is a $c \in C(k)$ such that $D_c$ is generically reduced and $D_\mu = \operatorname{red} D_c$.

If $|D|$ is a non-composite, linear pencil with parameter curve $C \cong \mathbb{P}^1$, then $D_c$ is a true member of $|D|^t$ for all but finitely many $c \in \mathbb{P}^1(k)$, but in general there may not be any true members.

Assume that $X$ is projective and let deg be an ample degree function on divisors. We say that $D_\mu$ is a *generically minimal member* of $|D|^t$ if

$$\deg D_\mu = \operatorname{genmin}\{\deg D_\lambda : \lambda \in \Lambda\}.$$

The corresponding index set is denoted by $\Lambda^{\mathrm{gmin}} \subset \Lambda$. (This notion a priori depends on deg, but this will not be important for us.)

If $|D|$ is non-composite and linear, then a true member is also generically minimal. Understanding the converse will be a key question for us.

The following example shows that a linear pencil can have generically minimal members that are not true members.

**Example 7.5.2.** Start with the pencil on $\mathbb{P}^2$ given in affine coordinates as

$$\left|(u^2 + v^2 + u)(v^2 + u), (u^2 + v^2 + v)(u^2 + v)\right|.$$

Its general member is a quartic with a node at the origin, but it has two members that split into conics.

Next make a change of variables $u = x + iy, v = x - iy$. The resulting pencil $|D|$ is still defined over $\mathbb{Q}$ but now it has a conjugate pair of reducible members. Thus we obtain that

$$\begin{aligned}
&\left((x+iy)^2 + (x-iy)^2 + (x-iy)\right)\left((x+iy)^2 + (x-iy)^2 + (x+iy)\right) \\
&= (2x^2 + 2y^2 + x - iy)(2x^2 + 2y^2 + x + iy) \\
&= (2x^2 + 2y^2 + x)^2 + y^2, \text{ and}
\end{aligned}$$

$$\begin{aligned}
&\left((x-iy)^2 + (x+iy)\right)\left((x+iy)^2 + (x-iy)\right) \\
&= \left(x^2 - y^2 + x + i(y - 2xy)\right)\left(x^2 - y^2 + x + i(y - 2xy)\right) \\
&= (x^2 - y^2 + x)^2 + (y - 2xy)^2
\end{aligned}$$

both give degree 4 false members of $|D|^t$.

**Lemma 7.5.3.** *Let $X$ be a normal, projective variety over a perfect field $k$ and* deg *an ample degree function on divisors. Let $|D|$ be a rational, algebraic pencil with parameter curve $C$ and corresponding topological pencil $|D|^t = \{D_\lambda : \lambda \in \Lambda\}$. Then there is a cofinite subset $\Lambda^* \subset \Lambda^{\mathrm{gmin}}$ such that the $\{D_\lambda : \lambda \in \Lambda^*\}$ are linearly equivalent to each other.*

*Proof.* If $|D|$ is linear, then we can take $\Lambda^*$ correspond to irreducible members of $|D|$. Otherwise we can take $\Lambda^*$ correspond to irreducible members of $|D|$ after a quadratic extension of $k$. □

**Lemma 7.5.4** (Sub-membership test). *Let $X$ be a normal, projective variety over a perfect field $k$ and* deg *an ample degree function on divisors. Let $|D|$ be a non-composite, algebraic pencil with parameter curve $C$ and corresponding* *topological pencil* $|D|^t$. *Assume that some* $D_\mu \in |D|^t$ *is generically smooth along a geometrically irreducible subvariety* $W \subset D_\mu$. *Then*

(1) $D_\mu$ *is contained in a member of* $|D|$,

(2) $C$ *has a $k$-point, and*

(3) *if $|D|$ is rational then it is linear.*

*Proof.* As we noted in (7.3.2), a member $D_\lambda$ of $|D|^t$ is either contained in a member of $|D|$, or every irreducible component of it is geometrically reducible. In the latter case if a geometric irreducible component of $D_\lambda$ contains $W$ then so do its conjugates, hence $D_\lambda$ is singular along $W$. The rest follows from (7.1.13).  □

**Lemma 7.5.5** (True membership test). *Let $X$ be a normal, projective variety over a perfect field $k$ and* deg *an ample degree function on divisors. Let $|D|^t = \{D_\lambda : \lambda \in \Lambda\}$ be a non-composite, rational, algebraic $t$-pencil.*
  *Assume that $D_\mu$ is a generically minimal member that is generically smooth along a geometrically irreducible subvariety $W \subset D_\mu$. Then $|D|^t$ is linear and $D_\mu$ is a true member.*

*Proof.* By (7.5.4), $|D|$ is linear and there is a $c \in C(k)$ such that $D_\mu \subset D_c$. Since $|D|$ is linear, all but finitely many of its members are generically minimal. If $D_\mu \subset D_c$ then $\deg_L D_\mu \leq \deg_L D_c = \deg_L |D|$ and equality holds if and only if $D_\mu = D_c$, that is, if and only if $D_\mu$ is a true member of $|D|^t$.  □

**Lemma 7.5.6.** *Let $X$ be a normal variety and $A_1, A_2$ reduced divisors without common irreducible components. Let $|D|^t$ be a $t$-pencil in general position for the irreducible components of $A_1 + A_2$. Then there is a cofinite subset $\Lambda^* \subset \Lambda^{\mathrm{mem}}$ such that, for every $\lambda_1 \neq \lambda_2 \in \Lambda^*$, the subsets*

$$(A_i \cup D_{\lambda_i}) \setminus \big((A_1 \cup D_{\lambda_1}) \cap (A_2 \cup D_{\lambda_2})\big) \tag{7.5.6.1}$$

*are connected.*

*Proof.* Let $A' \subset A_1$ be an irreducible component. All the $A' \cap D_\lambda$ are distinct divisors, and only finitely many of them have an irreducible component that is contained in $A_1 \cap A_2$. For all the others, $A' \cap D_\lambda$ contains an irreducible divisor $A'_\lambda \subset A'$ that is not contained in $A_2$ or in any $D_\mu$ for $\mu \neq \lambda$. This $A'_\lambda$ connects $A'$ to $D_{\lambda_1}$ inside (7.5.6.1). We can apply this to each irreducible component of the $A_i$.
  Thus, with finitely many exceptions, $D_\lambda$ is irreducible and each irreducible component of $A_1$ is connected to it inside (7.5.6.1). Thus (7.5.6.1) is connected.
  □

**Theorem 7.5.7.** *Let $X$ be a normal, projective, geometrically irreducible variety of dimension $\geq 2$ over an infinite, perfect field. Assume that we have an ample degree function* deg *on divisors. Let $A_1, A_2$ be reduced divisors without* *common irreducible components. Then $A_1, A_2$ are linearly equivalent if and only if there is a closed subset $W \subset X$ of codimension $\geq 2$ such that the following holds:*

*Let $|D|^t = \{D_\lambda : \lambda \in \Lambda\}$ be any algebraic t-pencil in general position for the irreducible components of $A_1 + A_2$ and such that $\mathrm{Bs}\,|D|^t \not\subset W$. Then there is a cofinite subset $\Lambda^* \subset \Lambda^{\mathrm{gmin}}$ such that, for every $\lambda_1 \neq \lambda_2 \in \Lambda^*$, the divisors $A_1 + D_{\lambda_1}$ and $A_2 + D_{\lambda_2}$ are generically minimal members of an algebraic t-pencil $|G|^t$ (which depends on the $A_i, D_{\lambda_i}$).*

*Proof.* Assume that $A_1 \sim A_2$ and choose $W \supset \mathrm{Sing}\,X$. Then $|D|^t$ has a smooth basepoint, hence it is rational by (7.1.12). Then, by (7.5.3) we can choose $\Lambda^*$ such that the corresponding members are linearly equivalent. Thus $D_{\lambda_1} \sim D_{\lambda_2}$ and then $A_1 + D_{\lambda_1} \sim A_2 + D_{\lambda_2}$. Thus they span a linear t-pencil $|G|^t$. By (7.5.6) $A_1 + D_{\lambda_1}$ and $A_2 + D_{\lambda_2}$ are generically minimal members of $|G|^t$.

Conversely, we may choose $|D|^t$ such that it is linear, ample, has a smooth basepoint, and all but finitely many of its generically minimal members are geometrically irreducible. Since

$$\mathrm{Bs}\,|D|^t = D_{\lambda_1} \cap D_{\lambda_2} \subset \mathrm{Bs}\,|G|^t,$$

we see that $|G|^t$ also has a smooth basepoint, hence it is rational by (7.1.13). Next note that $A_i + D_{\lambda_i}$ is generically smooth along the geometrically irreducible subvariety $D_{\lambda_i}$, hence $|G|^t$ is linear by (7.5.4), and the $A_i \cup D_{\lambda_i}$ are members by (7.5.6) and true members by (7.5.5).

Thus $A_1 + D_{\lambda_1} \sim A_2 + D_{\lambda_2}$, and also $A_1 \sim A_2$.    $\square$

**Corollary 7.5.8.** *Let $X$ be a normal, projective, geometrically irreducible variety of dimension $\geq 2$ over an infinite, perfect field and $\deg$ an ample degree function on divisors. Then $|X|$ and $\deg$ determine linear equivalence.*

*Proof.* As we noted earlier, (7.3.4) gives a characterization of algebraic t-pencils using $\deg$. Then, by (7.5.7), $|X|$ and $\deg$ determine linear equivalence of reduced divisors. By (7.5.9) this gives linear equivalence for all divisors.    $\square$

**Lemma 7.5.9.** *Let $X$ be a proper, normal variety over an infinite field. Let $T(X) \subset \mathrm{WDiv}(X)$ denote the subgroup generated by all $A_1 - A_2$ such that $A_1 \sim A_2$ and the $A_i$ are reduced. Then $T(X)$ is the subgroup of all divisors linearly equivalent to 0.*

*Proof.* Let $\pi\colon X' \to X$ be a proper, birational morphism such that $X'$ is projective. Let $H'$ be an ample divisor on $X'$ and set $H = \pi_*(H')$. In general, $H$ is not ample, but if $A$ is an divisor on $X$ then $|A + mH|$ gives a birational map and $\mathrm{Bs}\,|A + mH|$ has codimension $\geq 2$ for $m \gg 1$.

Suppose given an effective divisor $\sum_{i=1}^{n} a_i A_i$ on $X$. For large enough $m$ and a general member $H_m \in |mH|$, there is an integral divisor $A_0$ that is distinct from $A_1, \ldots, A_n$ and such that $A_1 + H_m - A_0 \in T(X)$. Thus

$$H_m + \sum_{i=1}^{n} a_i A_i - \left(A_0 + (a_1 - 1)A_1 + \sum_{i=2}^{n} a_i A_i\right) \in T(X).$$

By induction on $\sum a_i$, we see that for all sufficiently large $m$, for any $d > \sum a_i$ and general members $H_m^{(1)}, \ldots, H_m^{(d)}$, there is an integral divisor $A_\infty$ such that

$$H_m^{(1)} + \cdots + H_m^{(d)} + \sum u_i A_i - A_\infty \in T(X).$$

Given linearly equivalent effective divisors $A = \sum_{i=1}^{n} a_i A_i$ and $B = \sum_{j=1}^{m} b_j B_j$, choose $d > \max\{\sum a_i, \sum b_j\}$. By the above argument, we get $A_\infty$ and $B_\infty$ as above. We can thus arrange that $A_\infty - B_\infty \in T(X)$, hence $A - B \in T(X)$ as claimed. $\qquad\qquad\square$

## 7.6   UNCOUNTABLE FIELDS

The main result of this chapter is the following.

**Theorem 7.6.1.** *Let $X$ be a normal, projective variety of dimension at least 2 over an uncountable field $k$ of characteristic 0. Then linear equivalence of divisors is determined by $|X|$.*

*Proof.* Since $k$ is uncountable, every t-pencil on $X$ is algebraic by (7.3.5). Thus we know which t-pencils are algebraic. Then $X$ has a degree function by (7.4.15), and we conclude by (7.5.8). $\qquad\qquad\square$

Note that $k$ is uncountable if and only if $|X|$ has uncountably many points, so this is a topological assumption.

Applying (4.1.14) yields the following.

**Corollary 7.6.2.** *Let $X$ be a normal, projective variety of dimension at least 2 over an uncountable field of characteristic 0. Then the scheme structure of $X$ is uniquely determined by its Zariski topological space $|X|$.* $\qquad\qquad\square$

# Chapter Eight

## The set-theoretic complete intersection property

Let $X$ be a normal, projective variety over some field $K$ and $C \subset X$ an irreducible curve. In Sections 8.2 and 8.3 we study which finite subsets of $C$ are obtained as the intersection of $C$ with a divisor, a condition that depends only on the topology of the pair $|C| \subset |X|$.

Somewhat surprisingly, at the most basic level the answer is governed by the base field $K$. More precisely, it is determined by the qualitative behavior of the groups of $K$-points of abelian varieties over $K$. There are three classes of fields $K$ for which the 'size' of $A(K)$ is about the same for every abelian variety.

- *Finite fields*: For these, $A(K)$ is finite. More generally, if $K$ is *locally finite* (8.6.1) then $A(K)$ is a torsion group.
- *Number fields*: For these, $A(K)$ is finitely generated by the Mordell-Weil theorem. More generally, the same holds for fields that are finitely generated over a prime field [LN59].
- *Geometric fields*: For these, $A(K)$ has infinite rank. This holds, for example, if $K$ is algebraically closed, except for $\overline{\mathbb{F}}_p$. These are called anti-Mordell-Weil fields [IL19]; see (8.6.5) for the main examples.

Roughly speaking, our results show how to read off closely related properties of $K$ from the topology of $|X|$ in the first two cases and to recognize rational curves on $X$ in the third case.

## 8.1  SUMMARY OF RESULTS

The conclusions are most complete for locally finite fields.

**Theorem 8.1.1.** *Let $X$ be an irreducible, quasi-projective variety of dimension $\geq 2$ over a perfect field $K$. The following are equivalent.*

(1) *For every irreducible curve $C \subset X$ and every finite, closed subset $P \subset C$, there is a divisor $D \subset X$ such that $C \cap D = P$ (as sets).*

(2) *$K$ is locally finite.*

(3) *$A(K)$ is a torsion group for every algebraic group $A$ over $K$.*

We have a more complicated characterization of the Mordell-Weil case.

**Theorem 8.1.2.** *Let $X$ be an irreducible, quasi-projective variety of dimension $\geq 2$ over a field $K$. The following are equivalent.*

(1) *For every irreducible curve $C \subset X$, there is a finite, closed subset $\Sigma \subset C$ such that for every finite, closed subset $P \subset C$, there is a divisor $D \subset X$ such that $P \subset C \cap D \subset P \cup \Sigma$ (as sets).*

(2) $A(K)$ has finite $\mathbb{Q}$-rank for every abelian variety $A$ over $K$.

In the geometric cases our considerations yield a Zariski-topological characterization of rational curves. The complete statement in (8.3.7) needs several definitions, so here we state it somewhat informally.

**Proposition 8.1.3.** *Let $X$ be an irreducible, quasi-projective variety of dimension $n \geq 2$ over an anti-Mordell-Weil field and $C \subset X$ an irreducible, geometrically reduced curve. One can decide using only the topology of the pair $|C| \subset |X|$ whether $C$ is rational or not.*

The last result is especially useful if $X$ contains many rational curves, for example, for $X = \mathbb{P}^n$. However, in Sections 8.4–8.5 we get better results by observing that, from the topological point of view,

$$(\text{hyperplane}) \cup (\text{line}) \subset \mathbb{P}^n$$

is a very unusual configuration. This leads to the following simpler special case of Main Theorem (1.2.1).

**Theorem 8.1.4.** *Let $L$ be a field of characteristic 0 and $K$ an arbitrary field. Let $Y_L$ be a normal, projective, geometrically irreducible variety of dimension $n \geq 2$ over $L$ such that $|Y_L|$ is homeomorphic to $|\mathbb{P}^n_K|$. Then $K \cong L$ and $Y_L \cong \mathbb{P}^n_L$.*

## 8.2    SET-THEORETIC COMPLETE INTERSECTION PROPERTY

**Definition 8.2.1.** Let $X$ be a variety and $Z \subset X$ a closed, irreducible subset. We say that $Z$ has the *set-theoretic complete intersection property*—or that $Z$ is *scip*—if the following holds.

(1) Let $D_Z \subset Z$ be a closed subset of pure codimension 1. Then there is an effective divisor $D_X \subset X$ such that $\mathrm{Supp}(D_X \cap Z) = D_Z$.

(See (8.4.1) for $Z$ reducible.)

In some cases only 'nice' subvarieties $D_Z \subset Z$ are set-theoretic complete intersections. It is usually hard to formulate this in general, but the next variant allows us to ignore finitely many 'bad' points of $Z$.

We say that $Z$ is *generically scip* if there is a finite (not necessarily closed) subset $\Sigma_Z \subset Z$ such that the following holds.

(2) Let $D_Z \subset Z$ be a closed subset of pure codimension 1 that is disjoint from $\Sigma_Z$. Then, for every finite (not necessarily closed) subset $\Sigma_X \subset X \setminus D_Z$, there is an effective divisor $D_X \subset X$ that is disjoint from $\Sigma_X$ and such that $\mathrm{Supp}(D_X \cap Z) = D_Z$.

As a simple example, the quadric cone $Q \subset \mathbb{P}^4$ is not scip (over $\mathbb{C}$) but it is generically scip with $\Sigma_Z = \{\text{vertex}\}$ and $\Sigma_X$ arbitrary.

The introduction of $\Sigma_Z$ means that we do not have to worry about some very singular points on $Z$. This is especially clear on curves, where we may assume that $\Sigma_Z$ contains all singular points. The introduction of $\Sigma_X$ makes finding $D_X$ harder. However, if $X$ is normal and $\Sigma_X$ contains all non-Cartier centers of $X$ (11.3.19), then $D_X$ is a Cartier divisor. Thus, we can usually work with the Picard group of $X$ (for which there are solid references), rather than the class group (for which modern references seem to be lacking). We will mostly work with generically scip, but the following is an open question.

**Question 8.2.2.** Does scip imply generically scip?

It is clear that these notions depend only on the topological pair $|Z| \subset |X|$.

At the beginning we study the case when $Z$ is an irreducible curve, but later we need to understand many cases when $Z$ is reducible, and not even pure dimensional.

We check in (8.2.11) that being generically scip is invariant under purely inseparable morphisms and purely inseparable base field extensions. Thus, in order to save considerable trouble with nonreduced group schemes, we usually work over perfect base fields.

**Lemma 8.2.3.** *Let $X$ be a normal, quasi-projective variety over a perfect field $k$ and $C \subset X$ an integral, generically scip curve. Then*

$$\mathrm{coker}\big[\mathrm{Pic}(X) \to \mathrm{Pic}(C)\big] \text{ is a torsion group.}$$

*If $X$ is in addition factorial, then the same conclusion holds under the assumption that $C \subset X$ is integral and scip.*

*Proof.* We may assume that $\Sigma_Z \supset \mathrm{Sing}\, C$ and $\Sigma_Z \cup \Sigma_X$ contains all non-Cartier centers of $X$ (11.3.19). Let $p \in C \setminus \Sigma_Z$ be a point. By assumption there is an effective, Cartier divisor $D_p$ such that $\mathrm{Supp}(D_p \cap C) = \{p\}$. We do not know the intersection multiplicity at $p$, so we can only say that $\mathcal{O}_X(D_p)|_C \cong \mathcal{O}_C(m[p])$ for some $m > 0$. (Here we use that $p$ is a regular point.) That is, $\mathcal{O}_C(p)$ is torsion in $\mathrm{coker}\big[\mathrm{Pic}(X) \to \mathrm{Pic}(C)\big]$. Since the $\mathcal{O}_C(p)$ generate $\mathrm{Pic}(C)$, we are done.

For the case that $X$ is factorial and $C$ is scip, because $X$ is factorial all Weil divisors are Cartier, and we can apply the same argument to any $p \in C \setminus \mathrm{Sing}\, C$. $\square$

The rest of Sections 8.2 and 8.3 is essentially devoted to trying to understand the converse of (8.2.3). Let us see first that the direct converse does not hold.

**Example 8.2.4.** Let $C \subset \mathbb{P}^2$ be a smooth cubic over a number field. By the Mordell-Weil theorem $\mathrm{Pic}(C)$ is finitely generated. Choose points $\{p_i \in C : i \in I\}$ such that $\mathcal{O}_C(1)$ and the $\mathcal{O}_C(p_i)$ form a basis of $\mathrm{Pic}(C) \otimes \mathbb{Q}$.

Let $X$ be obtained by blowing up the points $p_i \in C \subset \mathbb{P}^2$. Let $E_i \subset X$ be the exceptional curves and let $C_X \subset X$ denote the birational transform of $C$. Note that $\mathrm{Pic}(X)$ is spanned by the $E_i$ and the pullback of $\mathcal{O}_{\mathbb{P}^2}(1)$. Thus, $\mathrm{Pic}(X) \to \mathrm{Pic}(C_X)$ is an injection with torsion cokernel.

**Claim 8.2.5.** $C_X \subset X$ *is not generically scip.*

*Proof.* Choose $n_i > 0$ and let $p \in C_X$ be a closed point such that $n[p] \sim \sum n_i[p_i]$ where $n = \sum n_i$. Assume that $\{p\} = \mathrm{Supp}(C_X \cap D)$ for some effective divisor $D \subset X$. Then

$$\mathcal{O}_X(nD)|_{C_X} \cong \mathcal{O}_{C_X}(nm[p]) \cong \mathcal{O}_{C_X}\left(\sum mn_i[p_i]\right)$$

for some $m > 0$. Since $\mathrm{Pic}(X) \to \mathrm{Pic}(C_X)$ is an injection, this implies that $D \sim \sum mn_i[E_i]$. But then $D = \sum mn_i[E_i]$ and so $C_X \cap D = \{p_i : i \in I\}$. $\square$

The following is a partial converse of (8.2.3).

**Lemma 8.2.6.** *Let $X$ be a projective variety over a field $k$ and $C \subset X$ a reduced, irreducible curve. Assume that*

$$\mathrm{coker}\left[\mathrm{Pic}^\circ(X) \to \mathrm{Pic}^\circ(C)\right] \text{ is a torsion group.}$$

*Then $C$ is scip and generically scip.*

*Proof.* Let $L$ be an ample line bundle such that $H^1(X, L^m \otimes T \otimes I_C) = 0$ for every $m \geq 1$ and every $T \in \mathrm{Pic}^\circ(X)$, where $I_C \subset \mathcal{O}_X$ denotes the ideal sheaf of $C$. Then $H^0(X, L^m \otimes T) \to H^0(C, L^m \otimes T|_C)$ is surjective for every $m \geq 1$. Set $d = \deg_C L$. For a point $p \in C$ let $P$ be a Cartier divisor on $C$ whose support is $p$ and set $r = \deg P$. (We can take $P = p$ if $p$ is a regular point.) Then $L_C^r(-dP) \in \mathrm{Pic}^\circ(C)$, thus there is an $m \geq 1$ and $T \in \mathrm{Pic}^\circ(X)$ such that

$$L_C^{mr}(-mdP) \cong T^{-1}|_C.$$

This gives a section $s_C \in H^0(C, L^{mr} \otimes T|_C)$ whose divisor is $mdP$. It lifts to a section $s \in H^0(X, L^m \otimes T)$ and $D := Z(s)$ works, verifying that $C$ is scip.

To verify that $C$ is generically scip, we must choose a section non-vanishing on a finite set $\Sigma_X$. We may assume that $\Sigma_X$ consists of closed points, choose $m$ large enough that we can lift a section of $L^m \otimes T$ from $C \cup \Sigma_X$ to $X$, and extend $s_C$ to a section on $C \cup \Sigma_X$ non-vanishing on $\Sigma_X$, and then lift to $X$, to achieve this. $\square$

The next example shows that $C$ can be scip even if

$$\operatorname{coker}\left[\operatorname{Pic}^{\circ}(X) \to \operatorname{Pic}^{\circ}(C)\right]$$

is non-torsion.

**Example 8.2.7.** Again let $C \subset \mathbb{P}^2$ be a smooth cubic over a number field. Choose a finite subset $L_i \in \operatorname{Pic}^{\circ}(C)$, closed under inverse, that generates a full rank subgroup.

Choosing general sections in each $L_i^{-1}(3)$, their zero sets $P_i \subset C$ are irreducible and distinct. Let $S_i \to \mathbb{P}^2$ denote the blowup of $P_i$ and $C_i \subset S_i$ the birational transform of $C$. Then $\mathcal{O}_{S_i}(C_i)|_{C_i} \cong L_i$ and $\mathcal{O}_{S_i}(C_i)$ is *nef*, that is, it has non-negative degree on every curve.

Finally, consider the diagonal embedding $C \subset \prod_i C_i \subset \prod_i S_i =: X$. We claim that $C \subset X$ is scip.

The key point is that $\{T|_C : T \in \operatorname{Pic}(X), T \text{ is nef and } \deg T|_C = 0\}$ contains a full rank subgroup of $\operatorname{Pic}^{\circ}(C)$. By Fujita's vanishing theorem [Laz04, 1.4.35], there is an ample line bundle $L$ such that $H^1(X, L^m \otimes T \otimes I_C) = 0$ for every $m \geq 1$ and every nef $T$. The rest of the argument in the proof of (8.2.6) works.

**8.2.8** (Cokernel of $\operatorname{Pic}(X) \to \operatorname{Pic}(Y)$). (See (11.3.2) for definitions and notation involving the Picard group.)

If $X$ is a proper variety then $\operatorname{Pic}(X)$ is an extension of $\operatorname{NS}(X)$ by $\operatorname{Pic}^{\circ}(X)$. While $\operatorname{NS}(X)$ is always a finitely generated abelian group, $\operatorname{Pic}^{\circ}(X)$ can be trivial or very large, depending on the ground field and $X$. However, $\mathbf{Pic}^{\circ}(X)$ is an algebraic group and $\mathbf{Pic}^{\circ}(X)(k)/\operatorname{Pic}^{\circ}(X)$ is torsion. Thus, if $p \colon Y \to X$ is a morphism, we aim to understand $p^* \colon \operatorname{Pic}(X) \to \operatorname{Pic}(Y)$, in terms of

$$p^* \colon \mathbf{Pic}^{\circ}(X) \to \mathbf{Pic}^{\circ}(Y) \text{ and } p^* \colon \operatorname{NS}(X) \to \operatorname{NS}(Y). \qquad (8.2.8.1)$$

We have better theoretical control of these maps since the first is a map of abelian varieties and the second a map of finitely generated abelian groups.

**Proposition 8.2.9.** *Let* $p \colon Y \to X$ *be a morphism of proper $k$-varieties. Then*

$$\operatorname{rank}_{\mathbb{Q}} \operatorname{coker}\left[\operatorname{Pic}(X) \to \operatorname{Pic}(Y)\right] \geq$$
$$\geq \operatorname{rank}_{\mathbb{Q}} \operatorname{coker}\left[\mathbf{Pic}^{\circ}(X) \to \mathbf{Pic}^{\circ}(Y)\right](k) - \operatorname{rank}_{\mathbb{Q}} \operatorname{NS}(X).$$

*Proof.* Because $0 \to \operatorname{Pic}^{\circ}(X) \to \operatorname{Pic}(X) \to \operatorname{NS}(X) \to 0$ is exact, and the same exact sequence exists for $Y$, by the snake lemma there is an exact sequence

$$\ker\left[\operatorname{NS}(X) \to \operatorname{NS}(Y)\right] \to \operatorname{coker}\left[\operatorname{Pic}^{\circ}(X) \to \operatorname{Pic}^{\circ}(Y)\right] \to \operatorname{coker}\left[\operatorname{Pic}(X) \to \operatorname{Pic}(Y)\right],$$

so

$$\operatorname{rank}_{\mathbb{Q}} \operatorname{coker}\left[\operatorname{Pic}(X) \to \operatorname{Pic}(Y)\right] \geq$$
$$\operatorname{rank}_{\mathbb{Q}} \operatorname{coker}\left[\operatorname{Pic}^{\circ}(X) \to \operatorname{Pic}^{\circ}(Y))\right] - \operatorname{rank}_{\mathbb{Q}} \ker\left[\operatorname{NS}(X) \to \operatorname{NS}(Y)\right].$$

The result follows on noting that

$$\mathrm{rank}_{\mathbb{Q}}\ker\big[\mathrm{NS}(X)\to\mathrm{NS}(Y)\big]\le\mathrm{rank}_{\mathbb{Q}}\mathrm{NS}(X)$$

and

$$\mathrm{rank}_{\mathbb{Q}}\mathrm{coker}\big[\mathrm{Pic}^\circ(X)\to\mathrm{Pic}^\circ(Y)\big]-\mathrm{rank}_{\mathbb{Q}}\mathrm{coker}\big[\mathbf{Pic}^\circ(X)\to\mathbf{Pic}^\circ(Y)\big](k),$$

since the maps $\mathrm{Pic}^\circ(Z)\otimes\mathbb{Q}\to\mathbf{Pic}^\circ(Z)(k)\otimes\mathbb{Q}$ are isomorphisms for proper $k$-varieties and $A\mapsto A(k)\otimes\mathbb{Q}$ is an exact functor of commutative group varieties (11.3.26). □

The first application is a characterization of locally finite fields.

**Theorem 8.2.10.** *Let $X$ be an irreducible, quasi-projective variety of dimension $\ge 2$ over a perfect field $k$. The following are equivalent.*

(1) *$k$ is locally finite (8.6.1).*

(2) *Every irreducible curve $C\subset X$ is scip.*

(3) *Every irreducible curve $C\subset X$ is generically scip.*

*Proof.* Assume first that $k$ is locally finite and let $\overline{X}\supset X$ be a compactification. Let $C\subset X$ be an irreducible curve with closure $\overline{C}$. If $k$ is locally finite then $\mathrm{Pic}^\circ(\overline{C})(k)$ is torsion by (11.3.23 (1)), hence $\overline{C}$ is scip and generically scip by (8.2.6), and so is $C\subset X$. It remains to prove that $(2)\Rightarrow(1)$ and $(3)\Rightarrow(1)$. We handle the case of (3) first.

Note that if (3) holds for $X$ then it holds for every open subset of it; we may thus assume that $X$ is normal (or even smooth). Let $\overline{X}\supset X$ be a normal compactification such that $\mathbf{Cl}^\circ(\overline{X})\cong\mathbf{Pic}^\circ(\overline{X})$ as in (11.3.5).

If $k$ is not locally finite then let $\overline{C}\subset\overline{X}^{\mathrm{ns}}$ be an irreducible curve with a single node that is in $X$. Note that $\mathbf{Pic}^\circ(\overline{X})$ is an abelian variety (11.3.2) and $\mathbf{Pic}^\circ(\overline{C})$ contains a $k$-torus (11.3.24). Thus, $\mathrm{coker}\big[\mathbf{Pic}^\circ(\overline{X})\to\mathbf{Pic}^\circ(\overline{C})\big]$ contains a $k$-torus, hence its $\mathbb{Q}$-rank is infinite (11.3.23). Thus,

$$\mathrm{rank}_{\mathbb{Q}}\mathrm{coker}\big[\mathrm{Pic}(\overline{X})\to\mathrm{Pic}(\overline{C})\big]=\infty$$

by (8.2.9) and so $\overline{C}$ is not generically scip by (8.2.3).

If $\overline{C}\backslash X$ consist of $m_\infty$ points, then the kernel of the restriction map $\mathrm{Pic}(\overline{C})\to\mathrm{Pic}(C)$ has rank $\le m_\infty$, thus we still have $\mathrm{rank}_{\mathbb{Q}}\mathrm{coker}\big[\mathrm{Pic}(X)\to\mathrm{Pic}(C)\big]=\infty$, hence $C$ is not generically scip by (8.2.3).

To handle (2), we again may pass to an open subset, and so we may assume that $X$ is smooth, hence factorial. This allows us to apply (8.2.3) to deduce that $C$ is not scip. □

We will use this in (10.3.2) below to strengthen the results of [WK81].

**Lemma 8.2.11.** *Let* $p\colon X' \to X$ *be a morphism between normal, projective varieties. Let* $C' \subset X'$ *be an irreducible curve. Set* $C := p(C')$ *and assume that* $k(C')/k(C)$ *is purely inseparable.*

*Then* $C$ *generically scip* $\Rightarrow C'$ *generically scip.*

*Proof.* Using (11.3.20) there is a finite subset $\Sigma_X \subset X$ that contains all non-Cartier centers and such that $C' \setminus p^{-1}(\Sigma_X) \to C \setminus \Sigma_X$ is a bijection.

Pick any $q' \in C' \setminus p^{-1}(\Sigma_X)$ and set $q = p(q')$. There is a divisor $D(q)$ such that $C \cap D(q) = \{q\}$. Then $D(q)$ is Cartier, hence its pullback gives a divisor $D(q')$ such that $C' \cap D(q') = \{q'\}$. $\qquad\square$

## 8.3 MORDELL-WEIL FIELDS

The Mordell-Weil theorem says that if $A$ is an abelian variety over a number field $k$ then $A(k)$ is a finitely generated group; fields with this property are Mordell-Weil fields; see (8.6.3). Our results are not sensitive to torsion in $A(k)$, this is why we need the concept of $\mathbb{Q}$-Mordell-Weil fields where $\mathrm{rank}_{\mathbb{Q}} A(k)$ is always finite; see (8.6.3).

$\mathbb{Q}$-Mordell-Weil fields have a nice characterization involving complete intersections on curves.

**Definition 8.3.1.** Let $X$ be a variety and $C \subset X$ a curve.

(1) $C$ is *scip with defect* $\Sigma \subset C$ if, for every closed, finite subset $P \subset C$, there is an effective divisor $D = D(C, P) \subset X$ such that $P \subset \mathrm{Supp}(D \cap C) \subset P \cup \Sigma$.

(2) $C$ is *scip with finite defect* if it is scip with defect $\Sigma$ for some finite subset $\Sigma \subset C$.

It is clear that these depend only on the topological pair $|C| \subset |X|$.

Note that being scip with finite defect is invariant under birational maps that are isomorphisms at the generic point of $C$; we just need to add to $\Sigma$ all the indeterminacy points that lie on $C$.

Let $X, Y$ be irreducible varieties and $\pi\colon Y \to X$ a dominant, finite morphism. Let $C_X \subset X$ be a curve with reduced preimage $C_Y \subset Y$. If $C_Y$ is scip with finite defect $\Sigma_Y$ then $C_X$ is scip with finite defect $\Sigma_X := \pi(\Sigma_Y)$.

Thus, most questions about these notions can be reduced to normal, projective varieties.

**Lemma 8.3.2.** *Let* $X$ *be a normal, projective variety over a perfect field* $k$ *and* $C \subset X$ *an irreducible curve with normalization* $\pi\colon \overline{C} \to C$. *Then* $C$ *is scip with finite defect if and only if*

$$\mathrm{rank}_{\mathbb{Q}} \mathrm{coker}\left[\mathbf{Pic}^{\circ}(X) \to \mathbf{Jac}(\overline{C})\right](k) < \infty. \qquad (8.3.2.1)$$

*Proof.* Note first that the difference between $\mathrm{rank}_{\mathbb{Q}} \mathrm{coker}\big[\mathrm{Pic}(X) \to \mathrm{Pic}(\overline{C})\big]$ and $\mathrm{rank}_{\mathbb{Q}} \mathrm{coker}\big[\mathbf{Pic}^{\circ}(X) \to \mathbf{Jac}(\overline{C})\big](k)$ is at most $\mathrm{rank}_{\mathbb{Q}} \mathrm{NS}(X)$.

Assume that $C$ is scip with finite defect. By (11.3.21) there is a finite set $\Sigma_X \subset X$ such that $C \setminus \Sigma_X$ is smooth and every Weil divisor on $X$ not containing $\Sigma_X$ is Cartier along $C \setminus \Sigma_X$. Let the defect set be $\Sigma \subset C \setminus \Sigma_X$ and $\overline{\Sigma} \subset \overline{C}$ its preimage. Let $\Gamma \subset \mathrm{Pic}(\overline{C})$ be the subgroup generated by all $\overline{q} \in \overline{\Sigma}$.

Pick any closed point $p \in C \setminus \Sigma_X$. By assumption, we have an effective Cartier divisor $D_p$ such that $p \in \mathrm{Supp}(D_p \cap C) \subset \{p\} \cup \Sigma$. This shows that, for some $m > 0$,

$$m[p] \in \big\langle \Gamma, \mathrm{Im}\big[\mathrm{Pic}(X) \to \mathrm{Pic}(\overline{C})\big]\big\rangle.$$

Since these $\{[p] : p \in C \setminus \Sigma\}$ generate $\mathrm{Pic}(\overline{C})$, we get that

$$\mathrm{rank}_{\mathbb{Q}} \mathrm{coker}\big[\mathrm{Pic}(X) \to \mathrm{Pic}(\overline{C})\big] \leq \mathrm{rank}_{\mathbb{Q}} \Gamma.$$

Conversely, assume that (8.3.2.1) holds. Using embedded resolution of curves, we may assume that $C$ is smooth, so $\overline{C} = C$. Then there is a finite subset $\{F_i : i \in I\} \subset \mathrm{Jac}(C)$, closed under inverses, that generates $\mathrm{coker}\big[\mathrm{Pic}^{\circ}(X) \to \mathrm{Jac}(C)\big]$ modulo torsion. Fix a point $p_0 \in C \setminus \Sigma$ and an ample line bundle $L$ on $X$ such that $\deg_C L = c \deg p_0$ for some $c > 0$ and $L$, restricted to $c$, has a section with zero set $D$ containing $p_0$. Set $d_0 = \deg p_0$ and choose $r_1$ such that $r_1[p_0] + F_i \sim G_i$, where the $G_i$ are effective and disjoint from $\Sigma_X$.

Now pick any $p \in C \setminus \Sigma_X$ and set $d = \deg p$. Then

$$L\big(-c[p_0] + d[p_0] - d_0[p]\big)|_C \in \mathrm{Jac}(C).$$

So, by assumption, there are non-negative $m_i$ and $T \in \mathrm{Pic}^{\circ}(X)$ such that

$$L\big(-c[p_0] + d[p_0] - d_0[p]\big)|_C \sim_{\mathbb{Q}} \mathcal{O}_C\big(\textstyle\sum_i m_i F_i\big) \otimes T^{-1}.$$

Set $e := d - c + r_1 \sum m_i$. We can rewrite this as

$$(L \otimes T)|_C \sim_{\mathbb{Q}} \mathcal{O}_C\big(d_0[p] - e[p_0] + \textstyle\sum_i m_i G_i\big).$$

So, for sufficiently divisible natural numbers $r > 0$,

$$(L^r \otimes T^r)|_C \cong \mathcal{O}_C\big(r d_0[p] - r e[p_0] + r\textstyle\sum_i m_i G_i\big).$$

If $e \leq 0$, then for large enough $r$, the constant 1 section on the right side lifts to a section of $L^r \otimes T^r$. If $e > 0$, then

$$\big(L^{r(1+e)} \otimes T^r\big)|_C \cong \mathcal{O}_C\big(r d_0[p] + re(D - [p_0]) + \textstyle\sum_i m_i G_i\big).$$

As before, for large enough $r$, the constant 1 section of the right side lifts to a section of $L^{r(1+e)} \otimes T^r$. Combining the two cases, $C$ is scip with defect $\Sigma = \mathrm{Supp}(p_0 + D + \sum G_i)$.  $\square$

**Lemma 8.3.3** (Curves with large Jacobians). *Let $X$ be a geometrically normal, projective variety of dimension $\geq 2$ and $A$ an abelian variety over $k$. Then there is an irreducible, projective curve $C \subset X^{\mathrm{ns}}$ such that there is a $\mathbb{Q}$-injection (that is, with finite kernel)*

$$A \hookrightarrow_{\mathbb{Q}} \mathrm{coker}\big[\mathbf{Pic}^\circ(X) \to \mathbf{Jac}(\overline{C})\big].$$

*Proof.* Let $\overline{C} \subset A^\vee \times X$ be a curve that is a general, irreducible, complete intersection of ample divisors; we use [Poo08] in case $k$ is finite. Let $C \subset X$ be the image of the second coordinate projection $\pi \colon \overline{C} \to C$. Then $C \subset X^{\mathrm{ns}}$ and $\overline{C}$ is the normalization of $C$. (In fact, $\overline{C} \cong C$ if $\dim X \geq 3$ and $k$ is infinite.) By (11.3.13) the natural map

$$A \times \mathbf{Pic}^\circ(X) = \mathbf{Pic}^\circ(A^\vee) \times \mathbf{Pic}^\circ(X) \to \mathbf{Jac}(\overline{C})$$

is $\mathbb{Q}$-injective, giving a $\mathbb{Q}$-injection $A \hookrightarrow_{\mathbb{Q}} \mathrm{coker}\big[\mathbf{Pic}^\circ(X) \to \mathbf{Jac}(\overline{C})\big]$.       □

**Theorem 8.3.4.** *Let $X$ be an irreducible, quasi-projective variety of dimension $\geq 2$ over a field $k$. The following are equivalent.*

(1) *$k$ is $\mathbb{Q}$-Mordell-Weil.*

(2) *Every irreducible curve $C \subset X$ is scip with finite defect.*

*Proof.* By (8.6.4 (3)) we may assume that $k$ is perfect. As we observed in (8.3.1), it is enough to prove (1) $\Rightarrow$ (2) for normal, projective varieties. If $k$ is $\mathbb{Q}$-Mordell-Weil then (2) holds for these by (8.3.2).

Conversely, if every irreducible curve $C \subset X$ is scip with finite defect then the same holds for the smooth locus $X^{\mathrm{ns}} \subset X$. Let $X' \supset X^{\mathrm{ns}}$ be a normal compactification. If $C' \subset X'$ has nonempty intersection with $X^{\mathrm{ns}}$ then $C'$ is also scip with finite defect.

Now let $A$ be an abelian variety over $k$. By (8.3.3) there is an irreducible, projective curve $C' \subset X'$ and a $\mathbb{Q}$-injection

$$A \hookrightarrow_{\mathbb{Q}} \mathrm{coker}\big[\mathbf{Pic}^\circ(X') \to \mathbf{Jac}(\overline{C}')\big].$$

By (8.3.2) $\mathrm{coker}\big[\mathbf{Pic}^\circ(X') \to \mathbf{Jac}(\overline{C}')\big](k)$ has finite $\mathbb{Q}$-rank, and so does $A(k)$. Thus $k$ is $\mathbb{Q}$-Mordell-Weil.       □

By (8.3.2), being scip with finite defect depends on the interaction of $\mathbf{Pic}^\circ(X)$ and $\mathbf{Jac}(\overline{C})$. The following definition is designed to get rid of the influence of $\mathbf{Pic}^\circ(X)$.

**Definition 8.3.5.** Let $X$ be a variety, $C \subset X$ an irreducible curve. We say that $C$ is *absolutely scip with finite defect* if the following holds.

(1) Let $C' \neq C$ be any irreducible curve. Then there are finite subsets $\Sigma \subset C$ and $C \cap C' \subset \Sigma' \subset C'$ such that for every finite subset $P \subset C$ there is an effective divisor $D \subset X$ such that $P \subset \mathrm{Supp}\big(D \cap (C \cup C')\big) \subset P \cup \Sigma \cup \Sigma'$.

Note that $P$ is a subset of $C$ only. This has the following effect.

Let $\Gamma' \subset \mathrm{Jac}(\overline{C}')$ be the subgroup generated by the preimages of $\Sigma'$. Let $\Gamma'_X \subset \mathrm{Pic}(X)$ be the preimage of $\Gamma'$ under $\mathrm{Pic}(X) \to \mathrm{Jac}(\overline{C}')$. Then the class of $D$ has to be in $\Gamma'_X$.

If $C'$ is general ample curve then the kernel of $\mathrm{Pic}(X) \to \mathrm{Jac}(\overline{C}')$ is torsion, thus $\Gamma'_X$ is a finitely generated group.

Now when we run the proof of (8.3.2) for $C \subset X$, instead of the whole $\mathrm{Pic}(X)$, we have only $\Gamma'_X$ to choose $D$ from. The condition (8.3.2.1) now becomes

$$\mathrm{rank}_{\mathbb{Q}} \, \mathrm{coker}\big[\Gamma'_X \to \mathrm{Pic}(\overline{C})\big] < \infty.$$

Since $\Gamma'_X$ is finitely generated, this holds if and only if $\mathrm{rank}_{\mathbb{Q}} \, \mathrm{Pic}(\overline{C}) < \infty$, and we get the following.

**Proposition 8.3.6.** *Let $X$ be a normal, projective variety of dimension $\geq 2$ over a perfect field $k$ and $C \subset X$ an irreducible curve. Then $C$ is absolutely scip with finite defect if and only if $\mathrm{Pic}(\overline{C})$ has finite $\mathbb{Q}$-rank.* $\square$

This is especially useful over fields where the opposite of the Mordell-Weil theorem happens; these are the *anti-Mordell-Weil* fields (8.6.5). For varieties over such fields we can recognize rational curves using only their set-theoretic intersection properties.

Putting together (8.3.6) with (8.6.5) gives the topological characterization of rational curves.

**Corollary 8.3.7.** *Let $k$ be a perfect, anti-Mordell-Weil field, $X$ an irreducible, quasi-projective $k$-variety of dimension $\geq 2$ and $C \subset X$ an irreducible curve. Then $C$ is absolutely scip with finite defect if and only if every irreducible component of $C_{\overline{k}}$ is rational.* $\square$

## 8.4 REDUCIBLE SCIP SUBSETS

**Definition 8.4.1.** Let $X$ be a variety and $Z \subset X$ a closed subset. We say that $Z$ is *scip* if the following holds.

(1) Let $D_Z \subset Z$ be a closed subset of pure codimension 1 that has nonempty intersection with every irreducible component of $Z$. Then there is an effective divisor $D_X \subset X$ such that $\mathrm{Supp}(D_X \cap Z) = D_Z$.

We say that $Z$ is *generically scip* if the following holds.

(2) There is a (not necessarily closed) finite subset $\Sigma_Z \subset Z$ such that if $D_Z$ in

(1) is disjoint from $\Sigma_Z$, then, for every (not necessarily closed) finite subset $\Sigma_X \subset X \setminus D_Z$, we can find $D_X \subset X$ as in (1) that is also disjoint from $\Sigma_X$.

It is clear that these depend only on the topological pair $|Z| \subset |X|$. Also, if (8.4.1 (2)) holds for some $\Sigma_Z$ then it also holds for every larger $\Sigma'_Z \supset \Sigma_Z$. We usually just take the union $\Sigma := \Sigma_Z \cup \Sigma_X$ large enough.

If $Z$ is scip (resp. generically scip) then any union of its irreducible components is also scip (resp. generically scip).

**Example 8.4.2.** In $\mathbb{P}^n$ with coordinates $x_0, \ldots, x_n$, set $L_1 = (x_1 = \cdots = x_i = 0)$ and $L_2 = (x_{i+1} = \cdots = x_n = 0)$. We claim that $L_1 \cup L_2$ is generically scip. First we discuss the case $\Sigma_X$ empty. Given divisors $D_{Z_i} \subset L_i$ not containing $L_1 \cap L_2 = (1{:}0{:}\cdots{:}0)$, they can be given as zero sets of polynomials

$$D_{Z_1} = Z\big(g_1(x_0, x_{i+1}, \ldots, x_n)\big) \text{ and } D_{Z_2} = Z\big(g_2(x_0, x_1, \ldots, x_i)\big).$$

We may assume that $g_i(1, 0, \ldots, 0) = 1$. Then

$$D_X := Z\big(g_1^{\deg g_2} + g_2^{\deg g_1} - x_0^{\deg g_1 \deg g_2}\big)$$

satisfies $\operatorname{Supp}\big(D_X \cap (L_1 \cup L_2)\big) = D_{Z_1} \cup D_{Z_2}$.

We can modify this construction to allow for nonempty sets $\Sigma_X$. We may assume that $\Sigma_X$ consists of closed points by choosing a specialization of each point of $\Sigma_X$. Then we can choose $g_1, g_2$ so that $\deg g_1 \deg g_2$ is high enough that $H^1(\mathcal{O}_{\mathbb{P}^n}(\deg g_1 \deg g_2) \otimes I_{L_1 \cup L_2 \cup \Sigma_X}$ is zero. It follows that every section of $\mathcal{O}(\deg g_1 \deg g_2)$ on $L_1 \cup L_2 \cup \Sigma_X$ extends to a section of $\mathcal{O}(\deg g_1 \deg g_2)$ on $\mathbb{P}^n$. Then we choose a section that agrees with $g_1^{\deg g_2} + g_2^{\deg g_1} - x_0^{\deg g_1 \deg g_2}$ on $L_1, L_2$ and does not vanish at any point of $\Sigma_X$.

We prove a partial converse in (8.4.6).

Next we prove a general result about reducible scip subschemes.

**Notation 8.4.3.** For a $k$-scheme $Y$ we use $k[Y] := H^0(Y, \mathcal{O}_Y)$ to denote the ring of regular functions. If $Y$ is normal and proper then $k[Y] = k$ if and only if $Y$ is geometrically integral.

If $Y$ is reduced then $k(Y)$ denotes the ring of rational functions. $Y$ is irreducible if and only if $k(Y)$ is a field. If $Y_i$ are the irreducible components of $Y$ then $k(Y) \cong \oplus_i k(Y_i)$.

**Proposition 8.4.4.** *Let $X$ be a normal, projective $k$-variety such that $\rho(X) = 1$. Let $Y, W \subset X$ be reduced, irreducible subvarieties such that $Y \cap W$ is 0-dimensional. Assume that $k$ is not locally finite. Then $Y \cup W$ is generically scip (8.4.1) if and only if the following hold.*

(1) *$Y$ and $W$ are generically scip,*

(2) *$Y \cap W$ is irreducible,*

(3) *either $k[\mathrm{red}(Y \cap W)]/k[W]$ or $k[\mathrm{red}(Y \cap W)]/k[Y]$ is purely inseparable, and*

(4) *if char $k = 0$ then $Y \cap W$ is reduced.*

*Proof.* Assume first that $Y \cup W$ is generically scip. Choose any $\Sigma$ that contains $\Sigma(Y \cup W)$ (9.2.2) and the non-Cartier centers of $X$ (11.3.19). Let $L$ be an ample line bundle on $X$ such that $H^0(X, L) \to H^0(Y \cap W, L_{Y \cap W})$ is surjective. Choose sections $s_Y, s_W \in H^0(X, L)$ that are nowhere zero on $\Sigma$. Write $Z(s_Y|_Y) = \sum_i a_i A_i$ and $Z(s_W|_W) = \sum_j b_j B_j$. By assumption, for every $i, j$ there is a divisor $D_{ij} \subset X$ such that $D_{ij}|_{Z \cup W} = c_{ij} A_i + d_{ij} B_j$ for some $c_{ij}, d_{ij} > 0$. The $D_{ij}$ are Cartier since they are disjoint from $\Sigma$.

For each $j$ a suitable positive linear combination of the $D_{ij}$ gives a divisor $D_j$ such that $D_j|_Y$ is a multiple of $Z(s_Y|_Y)$ and $D_j|_W$ is a multiple of $B_j$. Then we can take a suitable positive linear combination $D$ of the $D_j$ such that $D|_Y$ is a multiple of $Z(s_Y|_Z)$ and $D|_W$ is a multiple of $W(s_W|_W)$.

Since $\rho(X) = 1$, after passing to a suitable power we may assume that

$$D = Z(s) \text{ for some } s \in H^0(X, L^m) \text{ and } m > 0. \tag{8.4.4.1}$$

As we check in (9.2.4) below, this implies that

$$\begin{aligned} s|_Y &= u_Y s_Y^m|_Y & \text{for some} \quad u_Y \in k[Y]^\times \text{ and} \\ s|_W &= u_W s_W^m|_W & \text{for some} \quad u_W \in k[W]^\times, \end{aligned} \tag{8.4.4.2}$$

and hence

$$\begin{aligned} (s_Y/s_W)^m|_{Y \cap W} &= u_W|_{Y \cap W} \cdot u_Y^{-1}|_{Y \cap W} \\ &\in \mathrm{Im}\big[k[W]^\times \times k[Y]^\times \to k[Y \cap W]^\times\big]. \end{aligned} \tag{8.4.4.3}$$

At the beginning we can choose $s_Y/s_W$ to be an arbitrary element of $k[Y \cap W]^\times$, hence we conclude that

$$k[Y \cap W]^\times / k[W]^\times \times k[Y]^\times \text{ is a torsion group.} \tag{8.4.4.4}$$

Now (11.3.32) shows that either $k[\mathrm{red}(Y \cap W)]/k[W]$ or $k[\mathrm{red}(Y \cap W)]/k[Y]$ is purely inseparable, proving (3). Since $Y, W$ are irreducible, $k[Y]$ and $k[W]$ are finite field extensions of $k$. Thus, $k[\mathrm{red}(Y \cap W)]$ is a finite field extension of $k$, hence $Y \cap W$ is irreducible. Finally, (4) follows from (11.3.30).

Conversely, assume that (1)–(4) hold. Let $A_Y$ and $B_W$ be effective divisors on $Y$ and $W$ that are disjoint from $\Sigma$, obtained as restrictions of Cartier divisors from $X$. Since $\rho(X) = 1$, there is a power $L^m$ and sections $\sigma_Y \in H^0(Y, L^m|_Y)$ and $\sigma_W \in H^0(W, L^m|_W)$ defining $A_Y$ and $B_W$.

Suitable powers of $\sigma_Y^r$ and $\sigma_W^r$ can be glued to a section of $\sigma_{Y \cup W} \in H^0(Y \cup W, L^{mr}|_{Y \cup W})$ if and only if

$$\sigma_Y \sigma_W^{-1} \in k[Y \cap W]^\times / k[W]^\times \times k[Y]^\times \text{ is torsion.} \tag{8.4.4.5}$$

This is guaranteed by (3) and (4) using (11.3.32).

Once $A_Y \cup B_W$ is defined as the zero set of a section $\sigma_{Y \cup W}$, we can lift (a possibly higher power of) it to a section $\sigma_X \in H^0(X, L^N)$ that is non-vanishing on $\Sigma_X$, and $D_X := \operatorname{Supp} Z(\sigma_X)$ shows that $Y \cup W$ is generically scip. ☐

**Corollary 8.4.5.** *Let $X$ be a smooth, projective $k$-variety such that $\rho(X) = 1$ and $\operatorname{char} k = 0$. Let $C \subset X$ be a geometrically connected curve and $D \subset X$ a divisor. If $C \cup D$ is generically scip then $(C \cdot D) = 1$.* ☐

The following strong converse to (8.4.2) also illustrates the big difference between fields of characteristic 0, fields of positive characteristic, and subfields of $\overline{\mathbb{F}}_p$.

**Corollary 8.4.6.** *Let $k$ be a field, $D \subset \mathbb{P}^n_k$ an irreducible divisor and $C \not\subset D$ an irreducible, ample-sci (11.2.1) curve. Then $C \cup D$ is scip (resp. generically scip) if and only if one of the following holds.*

(1) $\operatorname{char} k = 0$, *$C$ is a line and $D$ is a hyperplane.*

(2) $\operatorname{char} k > 0$, $\operatorname{Supp}(C \cap D)$ *is a single $k^{\mathrm{ins}}$-point, and $C$, $D$ are both scip (resp. generically scip).*

(3) *$k$ is locally finite and $D$ is scip (resp. generically scip).*

*Proof.* Assume first that $\operatorname{char} k = 0$. Since $C$ is ample-sci, it is geometrically connected, hence, by (8.4.5), $(C \cdot D) = 1$, so $\deg C = 1$ and $\deg D = 1$. If $\operatorname{char} k > 0$ but $k$ is not locally finite, then $\operatorname{Supp}(C \cap D)$ is a single $k^{\mathrm{ins}}$-point by (8.4.4). If $C \cup D$ is scip (resp. generically scip) then $C$ and $D$ are both scip (resp. generically scip). This shows that the conditions of (2–3) are necessary. Their sufficiency also follows from (8.4.4). ☐

Note that if $C$ is smooth and rational, then it is scip. If $n \geq 4$ and $D$ is smooth then it is scip. Let $D \subset \mathbb{P}^n$ be a smooth hypersurface and $n \geq 3$. If $\deg D \leq n$ then there are lots of smooth rational curves that meet $D$ in 1 point only. If $\deg D \geq n + 2$ then there should be few such curves, but there are examples of arbitrary large degree.

Looking at the above proofs shows that there should be even fewer generically scip reducible subsets if $\rho(X) > 1$, but for now we have the following slightly weaker result.

**Proposition 8.4.7.** *Let $X$ be a normal, projective $k$-variety where $k$ is not locally finite. Let $Y, W \subset X$ be reduced, irreducible subvarieties such that $Y \cap W$ is 0-dimensional. Assume that $Y \cup W$ is generically scip (8.4.1). Then*

(1) $Y \cap W$ *is irreducible,*

(2) *either $k[\operatorname{red}(Y \cap W)]/k[W]$ or $k[\operatorname{red}(Y \cap W)]/k[Y]$ is purely inseparable, and*

(3) *if* char $k = 0$ *then*

$$\dim_k \ker\big[k[Y \cap W] \to k[\mathrm{red}(Y \cap W)]\big] \leq \frac{\rho(X) - 1}{\deg[k : \mathbb{Q}]}.$$

*Proof.* Choose $\Sigma$ to contain $\Sigma(Y \cup W)$ and the non-Cartier centers of $X$ (11.3.19). As in (11.3.6), let $\mathrm{WDiv}(X, \Sigma)$ denote the group of Weil divisors whose support is disjoint from $\Sigma$. These are all Cartier by our choice of $\Sigma$. We get restriction maps $r_Y \colon \mathrm{WDiv}(X, \Sigma) \to \mathrm{WDiv}(Y, \Sigma)$ and $r_W \colon \mathrm{WDiv}(X, \Sigma) \to \mathrm{WDiv}(W, \Sigma)$. These descend to maps of the Picard groups $\bar{r}_Y \colon \mathrm{Pic}(X) \to \mathrm{Pic}(Y)$ and $\bar{r}_W \colon \mathrm{Pic}(X) \to \mathrm{Pic}(W)$, which do not depend on $\Sigma$. Set $K_Y(X) := \ker \bar{r}_Y$, $K_W(X) := \ker \bar{r}_W$, and $K_{YW}(X) = K_Y(X) \cap K_W(X)$.

As in (11.3.6), the kernels of $\bar{r}_Y$ and $\bar{r}_W$ define closed subgroups $\mathbf{K}_Y(X) \subset \mathbf{Pic}(X)$ and $\mathbf{K}_W(X) \subset \mathbf{Pic}(X)$. Their intersection is denoted by $\mathbf{K}_{YW}(X)$.

Let $B$ be a divisor in $\mathrm{WDiv}(X, \Sigma)$ whose class $[B]$ lies in $K_{YW}(X)$. Then $B|_Y = Z(s_Y)$, where $s_Y$ is unique up to $k[Y]^\times$, and $B|_W = Z(s_W)$, where $s_W$ is unique up to $k[W]^\times$; here we use that $\Sigma \supset \Sigma(Y \cup W)$ and (9.2.1). Restricting both to $Y \cap W$ we get

$$s_Y|_{Y \cap W} \cdot s_W^{-1}|_{Y \cap W} \in k[Y \cap W]^\times / \big(k[W]^\times \times k[Y]^\times\big),$$

which defines a homomorphism

$$K_{YW}(X) \to k[Y \cap W]^\times / \big(k[W]^\times \times k[Y]^\times\big).$$

As in (11.3.27), we get a homomorphism of algebraic groups

$$\partial_{YW} \colon \mathbf{K}_{YW}(X) \to \big(\mathscr{R}_k^{Y \cap W} \mathbb{G}_m\big) / \big(\mathscr{R}_k^W \mathbb{G}_m \times \mathscr{R}_k^Y \mathbb{G}_m\big).$$

Note that $\mathbf{K}_{YW}(X, \Sigma) \cap \mathbf{Pic}^\circ(X)$ is an abelian variety (11.3.9.7), hence a positive dimensional subgroup of it has no morphisms to a linear algebraic group. Set

$$\mathrm{NS}_{YW}(X) := K_{YW}(X) / \big(K_{YW}(X) \cap \mathbf{Pic}^\circ(X)(k)\big).$$

Thus, $\partial_{YW}$ factors through

$$\mathrm{NS}_{YW}(X) \to \big(\mathscr{R}_k^{Y \cap W} \mathbb{G}_m\big) / \big(\mathscr{R}_k^W \mathbb{G}_m \times \mathscr{R}_k^Y \mathbb{G}_m\big).$$

Let $\Gamma_{YW}(X)$ denote its image.

By the above, $\Gamma_{YW}(X)$ is the image of a subgroup of $\ker\big[\mathrm{NS}(X) \to \mathrm{NS}(Y)\big]$ (modulo torsion) and $\mathrm{NS}(X) \to \mathrm{NS}(Y)$ has rank at least 1. All we need from this is that

$$\mathrm{rank}_{\mathbb{Q}} \Gamma_{YW}(X) \leq \rho(X) - 1.$$

Now we start to follow the proof of (8.4.4). The departure from it happens at (8.4.4.5), where now $\sigma$ is not a section of $L^m$, but of some $L^m(B)$ for some

$B \in K_{YW}(X)$. Thus, we conclude that

$$(s/t)^m|_{Y \cap W} \;=\; u_W|_{Y \cap W} \cdot u_Y^{-1}|_{Y \cap W} \cdot \gamma \\ \in\; \mathrm{Im}\big[k[W]^\times \times k[Y]^\times \to k[Y \cap W]^\times\big],$$ (8.4.7.1)

for some $\gamma \in \Gamma_{YW}(X)$. We can arrange $s/t$ to be an arbitrary element of $k[Y \cap W]^\times$, hence we conclude that

$$k[Y \cap W]^\times / k[W]^\times \times k[Y]^\times \times \Gamma_{YW}(X) \text{ is a torsion group.}$$ (8.4.7.2)

Now we use (11.3.32) to get that $Y \cap W$ is irreducible and (11.3.30) implies (2). Finally, we get that

$$\mathrm{rank}_{\mathbb{Q}} \ker\big[k[Y \cap W] \to k[\mathrm{red}(Y \cap W)]\big] \le \mathrm{rank}_{\mathbb{Q}} \Gamma_{YW}(X) \le \rho(X) - 1.$$

Thus, (3) follows from (11.3.23 (3)). □

In characteristic 0 we can reformulate the bound of (8.4.7 (3)) as follows.

**Corollary 8.4.8.** *Let $X$ be a normal, projective variety over a field $k$ of characteristic 0. Let $Z, W \subset X$ be reduced, irreducible subvarieties such that $Z \cap W$ is 0-dimensional. Assume that $Z \cup W$ is generically scip (8.4.1). Then $Z \cap W$ is irreducible and*

$$\dim_k k[Z \cap W] \le \max\{\dim_k k[Z], \dim_k k[W]\} + \frac{\rho(X) - 1}{\deg[k : \mathbb{Q}]}. \quad \square$$

**Example 8.4.9.** Combining the ideas of (8.4.2) with (8.4.4) we get a method to recognize $k$-points. The assumptions are restrictive, but this gives the first indication that one can get detailed scheme-theoretic information from the topology. However, scip and generically scip turn out to be too restrictive in general; searching for a more flexible variant leads to the notion of linkage in Section 9.6.

**Claim 8.4.10.** *Let $X$ be a smooth, projective $k$-variety of dimension $\ge 7$ such that $\rho(X) = 1$. Assume that $k$ is perfect and not locally finite. Then $p \in X$ is a $k$-point if and only if there are 3-dimensional, set-theoretic complete intersections $Z, W \subset X$ such that*

(1) $\mathrm{Supp}(Z \cap W) = \{p\}$ *and*

(2) $Z \cup W$ *is generically scip.*

*Proof.* Assume that $p \in X$ is a $k$-point and let $Z, W \subset X$ be 3-dimensional, smooth, complete intersections such that $Z \cap W = \{p\}$ as schemes. Lefschetz theorem tells us that if $D_Z \subset Z$ is any divisor then (some multiple of) it is a complete intersection. Arguing as in (8.4.2) we see that $Z \cup W$ is generically scip.

Conversely, $k[Z] = k[W] = k$ since $Z, W$ are set-theoretic complete intersections (11.2.2), thus (8.4.4) says that $p \in X(k)$. □

Note that the bound $\dim X \geq 7$ can be improved to $\dim X \geq 5$ if the Noether-Lefschetz theorem applies over $k$; see (10.6.6) for such cases.

## 8.5   PROJECTIVE SPACES

We study the scip property for the union of a curve and of a divisor. As we observed in Section 8.4, this happens very rarely, and it leads to the following stronger version of (8.1.4)

**Theorem 8.5.1.** *Let $L$ be a field of characteristic $0$ and $K$ an arbitrary field. Let $Y_L$ be a normal, projective, geometrically irreducible $L$-variety of dimension $n \geq 2$ and $\Phi : |\mathbb{P}^n_K| \sim |Y_L|$ a homeomorphism. Then*

(1) $Y_L \cong \mathbb{P}^n_L$,

(2) $K \cong L$, and

(3) $\Phi$ *is the composite of a field isomorphism $\varphi : K \cong L$ and of an automorphism of $\mathbb{P}^n_K$.*

We will not build on (8.5.1) in later sections—instead, the proof serves as a first example of the kinds of arguments we will make afterward. Thus, the proof of (8.5.1) uses a result from Section 9.4 without circularity.

We start with an easy to prove but interesting special case of (8.5.1).

**8.5.2** (Proof of (8.5.1) when $Y_L \cong \mathbb{P}^n_L$). Let $H \subset \mathbb{P}^n_K$ be a hyperplane and $\ell \subset \mathbb{P}^n_K$ a line not contained in $H$. Then $\ell \cup H$ is scip by (8.4.1 (2)), hence so is $\Phi(\ell) \cup \Phi(H) \subset \mathbb{P}^n_L$. So $\Phi(H) \subset \mathbb{P}^n_L$ is a hyperplane by (8.4.6 (1)). By taking intersections, we see that $\Phi$ gives an isomorphism of the projective geometries $\mathrm{Proj}^n_K$ and $\mathrm{Proj}^n_L$. By (1.1.2) this is induced by a field isomorphism $\varphi : K \cong L$.

Composing $\Phi$ with the natural isomorphism induced by $\varphi^{-1}$, we get a homeomorphism $\Psi \colon |\mathbb{P}^n_K| \to |\mathbb{P}^n_K|$ that is the identity on $K$-points. It remains to show that it is the identity on all points. Let $C \subset \mathbb{P}^n_K$ be a $K$-rational curve. It has infinitely many $K$-points and these are fixed by $\Psi$. Thus, $C \cap \Psi(C)$ is infinite, hence $C = \Psi(C)$. However, we do not yet know that $\Psi|_C$ is the identity.

Let $p \in \mathbb{P}^n_K$ be a closed point. Assume that there are $K$-rational curves $C_\lambda \subset \mathbb{P}^n_K$ such that $\{p\} = \cap C_\lambda$. Then $\{\Psi(p)\} = \cap \Psi(C_\lambda) = \cap C_\lambda = \{p\}$, as needed.

It remains to construct such curves $C_\lambda$. For this we can work in an affine chart $p \in \mathbb{A}^n_K \subset \mathbb{P}^n_K$ with coordinates $x_i$. Note that $K(p)/K$ is a finite, separable extension, hence can be generated by a single element $z_p \in K(p)$. We can thus write $x_i(p) = h_i(z_p)$ for some $h_i \in K[t]$.

Let $g(t) \in K[t]$ be the minimal polynomial of $z_p$; we can then identify $z_p$ with a root of $g$ in $\overline{K}$.

For $i \in \{1, \ldots, n\}$ and $a \in K$ let $C_{i,a}$ be the image of

$$\tau_{i,a} : t \mapsto \big(h_1(t), h_2(t), \ldots, h_n(t)\big) + ag(t)e_i,$$

where $e_i$ is the $i$th standard basis vector.

The $C_{i,a}$ are $K$-rational curves, hence stabilized by $\Phi$.

**Claim 8.5.3.** $\cap_{i,a} C_{i,a} = \{p\}$.

*Proof.* First, the $\tau_{i,a}$ all map $z_p \in \mathbb{A}^1(\overline{K})$ to $p$, so $p \in \cap_{i,a} C_{i,a}$. To see the converse, assume that $p \neq q \in \cap_{i,a} C_{i,a}$. After permuting the coordinates we may assume that $p_n \neq q_n$. If $q = \tau_{1,a}(z')$ then $h_n(z') = q_n$. The equation $h_n(*) = q_n$ has finitely many solutions $z'_j$, and they are all different from $z_p$. Then, for all but finitely many $a \in K$, $h_1(z'_i) + ag(z'_i) \neq q_i$ for every $i$. Thus, $q \notin C_{1,a}$. $\qquad\square$

In positive characteristic the proof above and (8.4.4) give the following.

**Claim 8.5.4.** *Let $K, L$ be perfect fields that are not locally finite, $n \geq 2$, and $\Phi : |\mathbb{P}^n_K| \sim |\mathbb{P}^n_L|$ a homeomorphism. Then $\Phi$ induces a bijection of sets $\mathbb{P}^n(K) \leftrightarrow \mathbb{P}^n(L)$ (but we do not know that the linear structure is preserved).*

*Proof.* Let $H \subset \mathbb{P}^n_K$ be a hyperplane and $\ell \subset \mathbb{P}^n_K$ a line not contained in $H$. Then $\ell \cup H$ is scip by (8.4.1 (2)), hence so is $\Phi(\ell) \cup \Phi(H) \subset \mathbb{P}^n_L$. So $\Phi(H \cap \ell) = \Phi(H) \cap \Phi(\ell)$ is an $L^{\mathrm{ins}}$-point by (8.4.6 (2)). Since $L$ is perfect, $L^{\mathrm{ins}} = L$. Thus, every $K$-point is sent to an $L$-point. Applying the same argument to $\Phi^{-1}$, we see that every $L$-point is sent to a $K$-point, and so $\Phi$ is a bijection. $\qquad\square$

The following lemma, which essentially says that pencils determine higher-dimensional linear systems, is longer to state than to prove.

**Lemma 8.5.5.** *Let $Y$ be a normal, projective variety over a field $L$. Let $K$ be an infinite field and $e_0, \ldots, e_n \in K\mathbb{P}^n$ independent points. Assume that we have a map*

$$\Phi : K\mathbb{P}^n \to (\text{effective Weil divisors on } Y)$$

*with the following property.*

(1) *For $r = 1, \ldots, n$ there are Zariski open subsets $\emptyset \neq U_{r-1} \subset \langle e_0, \ldots, e_{r-1} \rangle$ such that, for every $p \in U_{r-1}$, the divisors $\{\Phi(q) : q \in \langle p, e_r \rangle\}$ are $L$-members of a linear pencil on $Y$.*

*Then there is a Zariski open subset $W \subset K\mathbb{P}^n$ such that the divisors*

$$\{\Phi(q) : q \in W\}$$

*are $L$-members of a linear system of dimension $\leq n$ on $Y$.* $\qquad\square$

**8.5.6** (Proof of (8.5.1) in general). First note that $K$ is not locally finite by (8.2.10).

For any $H \in |H|^{\mathrm{set}}$, its image $\Phi(H)$ is ample by (9.4.15). Thus, $\Phi(H)$ is geometrically connected, and so are the $\Phi$-images of the lines since they are set-theoretic complete intersections of divisors $\Phi(H_i)$ (11.2.2).

Set $Z := \Phi^{-1}(\mathrm{Sing}(Y_L))$ and pick any $K$-point $x_0 \notin Z$. Let $\ell$ be a line through $x_0$ and $H$ a plane through $x_0$ but not containing $\ell$. Then $\ell \cup H$ is scip, hence so is $\Phi(\ell) \cup \Phi(H)$. Since $\Phi(\ell)$ and $\Phi(H)$ are both geometrically connected, (8.4.7 (2)) shows that $y_0 := \Phi(x_0)$ is an $L$-point and (8.4.8) gives that

$$\dim_L L\big[\Phi(\ell) \cap \Phi(H)\big] \leq 1 + \frac{\rho(Y) - 1}{\deg[L : \mathbb{Q}]}. \tag{8.5.6.1}$$

Since $y_0$ is a smooth point of $Y$, $\big(\Phi(\ell) \cdot \Phi(H)\big) = \dim_L L\big[\Phi(\ell) \cap \Phi(H)\big]$. So the $\Phi(H)$ have bounded intersection number with a curve that is an intersection of ample divisors, hence they form a bounded family.

Let $|D| \subset |H|$ be a pencil of hyperplanes whose base locus contains $x_0$, and $|D|^{\mathrm{t}} = \{D_\lambda : \lambda \in \Lambda\}$ the corresponding t-pencil (7.3.1). Thus, $\{\Phi(D_\lambda) : \lambda \in \Lambda\}$ is a t-pencil on $Y_L$.

There are infinitely many hyperplanes among the $\{D_\lambda\}$ and, as we noted, their $\Phi$-images form a bounded family of divisors. Thus, $\{\Phi(D_\lambda) : \lambda \in \Lambda\}$ is algebraic (7.3.4). Since $y_0$ is a smooth $L$-point, $\{\Phi(D_\lambda) : \lambda \in \Lambda\}$ is linear (7.5.4) and the images of the $K$-hyperplanes are true members (7.5.5). Thus, by (8.5.5), the $\{\Phi(H) : H \in |H|^{\mathrm{set}}\}$ span an $n$-dimensional linear system $|H|^Y$, which is basepoint-free since already the $\{\Phi(H) : H \in |H|^{\mathrm{set}}\}$ have no point in common. So $|H|^Y$ gives a morphism $g \colon Y \to \mathbb{P}^n_L$. Since any hyperplane $H$ has nonempty intersection with every curve, the same holds for $\Phi(H)$, so $g \colon Y \to \mathbb{P}^n_L$ is finite and $|H|^Y$ is ample.

We can also obtain $y_0$ as $\Phi(H_1) \cap \cdots \cap \Phi(H_n)$, or as a fiber of $g \colon Y \to \mathbb{P}^n_L$. Since char $L = 0$, general fibers of $g$ are reduced. Thus, $g \colon Y \to \mathbb{P}^n_L$ is finite and of degree 1, hence an isomorphism. The rest now follows from (8.5.2).  $\qquad\square$

## 8.6   APPENDIX: SPECIAL FIELDS

We discuss various classes of fields that were used earlier.

**Lemma 8.6.1.** *For a field $k$ the following are equivalent.*

(1) *Every finitely generated subfield of $k$ is finite.*

(2) *$k$ is an algebraic extension of a finite field.*

(3) *$k$ is isomorphic to a subfield of $\overline{\mathbb{F}}_p$ for some $p > 0$.*

(4) *$G(k)$ is a torsion group for every algebraic group $G$ over $k$.*

(5) *$A(k)$ is a torsion group for every abelian variety $A$ over $k$.*

(6) *There is a $C > 0$ such that* $\operatorname{rank}_{\mathbb{Q}} A(k) \leq C$ *for every abelian variety $A$ over $k$.*

*Proof.* The only non-obvious claim is that (5) and (6) are equivalent to the others. If $k$ is not locally finite then it contains either $\mathbb{Q}$ or $\mathbb{F}_p(t)$. In both cases, there is an abelian variety $A$ over $k$ with arbitrarily large $\operatorname{rank}_{\mathbb{Q}} A(k)$. For example, if $E$ is an elliptic curve of rank $\geq 1$ then $E^m$ has rank $\geq m$. □

**Definition 8.6.2.** If one of the equivalent conditions in 8.6.1 hold, then we say that $k$ is *locally finite.*[*]

**Definition 8.6.3.** A field $k$ is *Mordell-Weil* (resp. *$\mathbb{Q}$-Mordell-Weil*) if for every abelian variety $A$ over $k$, the group of its $k$-points $A(k)$ is finitely generated (resp. has finite $\mathbb{Q}$-rank).

**Remark 8.6.4.** (1) By [LN59], every finitely generated field is Mordell-Weil.

(2) Weil restriction (cf. [BLR90, Sec.7.6]) shows that these properties are invariant under finite field extensions. Since every abelian variety is a quotient of a Jacobian, it is equivalent to asking that $\operatorname{Jac}(C)$ have these properties for every smooth projective curve $C$ over $k$ (11.3.26).

(3) It is nor clear how much the class of $\mathbb{Q}$-Mordell-Weil fields differs from the class of Mordell-Weil fields. If $\operatorname{char} k = p > 0$ and $a \in A(k^{1/p})$ then $a^p \in A(k)$. Thus, $k$ is $\mathbb{Q}$-Mordell-Weil if and only if its inseparable closure $k^{\mathrm{ins}}$ is $\mathbb{Q}$-Mordell-Weil.

Note also that $\overline{\mathbb{F}}_p$ is $\mathbb{Q}$-Mordell-Weil but not Mordell-Weil.

(4) By [Fal94] if $k$ is $\mathbb{Q}$-Mordell-Weil and $\operatorname{char} k = 0$, then every curve of genus $\geq 2$ has only finitely many $k$-points.

**Definition 8.6.5.** Following [IL19], a field $k$ is called *anti-Mordell-Weil* if

(1) the $\mathbb{Q}$-rank of $A(k)$ is infinite for every positive dimensional abelian variety.

In particular, an anti-Mordell-Weil field $k$ is not locally finite. If the latter holds then the $\mathbb{Q}$-rank of $T(k)$ is infinite for every $k$-torus $T$ (11.3.23 (4)), hence (8.6.5 (1)) can be restated as:

(2) The $\mathbb{Q}$-rank of $A(k)$ is infinite for every positive dimensional semi-abelian variety $A$.

If $\operatorname{char} k = 0$, then $k$ is not a finite extension of $\mathbb{Q}$ by the Mordell-Weil theorem, hence the $\mathbb{Q}$-rank of $U(k)$ is infinite for every unipotent group (11.3.23 (3)). Thus, (8.6.5 (1)) and (8.6.5 (2)) are further equivalent to:

---

[*]The terminology is not standard in English.

(3) The $\mathbb{Q}$-rank of $A(k)$ is infinite for every positive dimensional commutative algebraic group $A$.

*Warning.* Note that if $\mathrm{char}\, k = p > 0$ then $U(k)$ is $p$-power torsion for every unipotent group; this creates a crucial difference between 0 and positive characteristics for us.

The following are examples of anti-Mordell Weil fields:

(4) algebraically closed fields, save for $\overline{\mathbf{F}}_p$,

(5) $\mathbb{R}$ and all real closed fields,

(6) $\mathbb{Q}_p$, more generally quotient fields of Henselian, local domains,

(7) large fields [Kob06, FP10, FP19] that are not locally finite, where a field $k$ is *large* (also called ample, fertile or anti-Mordell) if $C(k)$ is either empty or infinite for every smooth curve $C$. (The last case implies the earlier ones.)

**Definition 8.6.6.** A field $k$ is *Hilbertian* if for every irreducible polynomial $f(x, y) \in k[x, y]$ such that $\partial f / \partial y \neq 0$, there are infinitely many $c \in k$ such that $f(x, c) \in k[x]$ is irreducible. (We follow [FJ08, Chap. 12] with adding the separability condition.)

Equivalently, for every smooth, irreducible curve $C$ and every basepoint-free linear system $|M|$ that defines a separable map $C \to \mathbb{P}^1$, there are infinitely many irreducible members $M_c \in |M|$. This also implies that, for every irreducible variety $X$ over $k$, and every mobile linear system $|M|$ that defines a separable map $X \to \mathbb{P}^N$, there is a dense set $\Lambda \subset |M|(k)$ such that $M_\lambda \in |M|$ is irreducible for $\lambda \in \Lambda$.

Hilbert proved that number fields are Hilbertian. In fact, every finitely generated, infinite field is Hilbertian. A finite, separable extension of a Hilbertian field is Hilbertian, and so is any purely inseparable extension. See [Lan62, Chap. VIII] or [FJ08, Chaps. 12–13] for these and many other facts.

**Theorem 8.6.7.** *Let $k$ be a $\mathbb{Q}$-Mordell-Weil field of characteristic 0. Then $k$ is Hilbertian, hence the Bertini-Hilbert dimension (9.5.5) of $k$ equals 1.*

*Proof.* The hardest part is a theorem of [Fal94] which implies that a smooth, projective curve of genus $\geq 2$ has only finitely many $k$-points. The rest follows from (8.6.8). $\qquad\square$

**Theorem 8.6.8.** *Let $k$ be an infinite field. Assume that every irreducible curve of geometric genus $\geq 2$ has only finitely many $k$-points. Then $k$ is Hilbertian.*

*Proof.* Let $C$ be an irreducible, projective curve and $\pi \colon C \to \mathbb{P}^1$ a finite, separable morphism of degree $d \geq 2$. Pick $p \in \mathbb{P}^1(k)$. If $\pi^{-1}(p)$ is reducible, then one of its irreducible components has degree $\leq d/2$. Let $\pi^{(r)} \colon C^{(r)} \to \mathbb{P}^1$ denote the $r$-fold symmetric fiber product of $\pi \colon C \to \mathbb{P}^1$ with itself, that is, the quotient of the $r$-fold fiber product $C \times_{\mathbb{P}^1} \cdots \times_{\mathbb{P}^1} C$ by the symmetric group $S_r$ permuting the factors. Then the set of reducible $k$-fibers equals

$$\mathrm{RedFib}(\pi):= \cup_{r \leq d/2} \pi^{(r)}\big(C^{(r)}(k)\big) \subset \mathbb{P}^1(k).$$

Let $B \subset C^{(r)}$ be an irreducible component. Taking its preimage in $C \times_{\mathbb{P}^1} \cdots \times_{\mathbb{P}^1} C$ and projecting to the first component gives a subvariety $C' \subset C$ whose degree over $\mathbb{P}^1$ is $\leq r \cdot \deg(B/\mathbb{P}^1)$. Since $C' = C$, we see that $\deg(B/\mathbb{P}^1) \geq 2$.

Thus, the complement of $\mathrm{RedFib}(\pi)$ is infinite by (8.6.9). $\qquad\square$

**Proposition 8.6.9.** *Let $k$ be an infinite field such that every irreducible curve of geometric genus $\geq 2$ has only finitely many $k$-points. Let $\{C_i : i \in I\}$ be finitely many irreducible, projective curves and $\pi_i \colon C_i \to \mathbb{P}^1$ finite, separable morphisms of degree $d_i \geq 2$. Then*

$$\mathbb{P}^1(k) \setminus \cup_{i \in I} \pi_i\big(C_i(k)\big) \text{ is infinite.} \qquad (8.6.9.1)$$

*Proof.* We may discard any curve $C_i$ that has only finitely many $k$-points. We may thus assume that the $C_i$ are geometrically integral. By (8.6.12) there are automorphisms $\sigma_j$ of $\mathbb{P}^1$ such that the branch loci of $\sigma_j \circ \pi_i \colon C_i \to \mathbb{P}^1$ are all disjoint. For $j = 1, 2, 3$ and $i_1, i_2, i_3 \in I$ let

$$\pi_{i_1,i_2,i_3} \colon C_{i_1,i_2,i_3} \to \mathbb{P}^1$$

denote the triple fiber product of the $\sigma_j \circ \pi_{i_j} \colon C_{i_j} \to \mathbb{P}^1$. We see in (8.6.11) that each $C_{i_1,i_2,i_3}$ is a geometrically integral curve of geometric genus $\geq 2$. It has thus finitely many $k$-points. Therefore,

$$\cup_{i_1,i_2,i_3 \in I} \pi_{i_1,i_2,i_3}\big(C_{i_1,i_2,i_3}(k)\big) \qquad (8.6.9.2)$$

is a finite subset of $\mathbb{P}^1(k)$. Its complement is the union of the three translates of (8.6.9.1). $\qquad\square$

The claims about the curves $C_{i_1,i_2,i_3}$ are geometric, hence we can check them over $\bar{k}$. We start with double fiber products.

**Lemma 8.6.10.** *Let $C_1, C_2$ be irreducible, smooth, projective curves over $\bar{k}$ and $\pi_i \colon C_i \to \mathbb{P}^1$ finite, separable morphisms of degree $d_i \geq 2$ whose branch loci are disjoint. Then $C_1 \times_{\mathbb{P}^1} C_2$ is irreducible, smooth and*

$$g\big(C_1 \times_{\mathbb{P}^1} C_2\big) = d_1 g_2 + d_2 g_1 + (d_1 - 1)(d_2 - 1).$$

*Proof.* Over any point $p \in \mathbb{P}^1$, one of the $\pi_i$ is étale, hence the fiber product is smooth. $C_1 \times_{\mathbb{P}^1} C_2 \subset C_1 \times C_2$ is the pullback of the diagonal of $\mathbb{P}^1 \times \mathbb{P}^1$. It is thus ample, hence connected and so irreducible. Hurwitz's formula now gives the genus. $\qquad\square$

We see that $g\big(C_1 \times_{\mathbb{P}^1} C_2\big) \geq 2$, unless $g_1 = g_2 = 0$ and $d_1 = d_2 = 2$. Taking triple products gives the following.

**Corollary 8.6.11.** *Let $C_1, C_2, C_3$ be irreducible, smooth, projective curves over $\overline{k}$ and $\pi_i \colon C_i \to \mathbb{P}^1$ finite, separable morphisms of degree $d_i \geq 2$ whose branch loci are pairwise disjoint. Then $C_1 \times_{\mathbb{P}^1} C_2 \times_{\mathbb{P}^1} C_3$ is irreducible, smooth and of genus $\geq 2$.*                                                                        □

**8.6.12** (Cosets covering a group). Let $G$ be a group, $H_1, \ldots, H_n$ (left or right) cosets of subgroups such that $G = \cup_i H_i$. [Neu54] proves that then the index of one of the $H_i$ in $G$ is $\leq n$.

Next let $G$ be a group acting on a set $S$ and $S_1, S_2 \subset S$ finite subsets. Then $\{g \in G : g(S_1) \cap S_2 \neq \emptyset\}$ is a union of $|S_1| \times |S_2|$ cosets of point stabilizers. If $G$ acts without finite orbits, then these stabilizers have infinite index. Thus, we conclude that there is a $g \in G$ such that $g(S_1) \cap S_2 = \emptyset$.

In our proofs the Hilbertian condition is mostly used through the following consequence.

**Lemma 8.6.13.** *Let $C$ be an irreducible, geometrically reduced, projective curve over a Hilbertian field $k$. Let $\Sigma \subset C$ be a finite subset and $Z \subset C$ a finite subscheme. Let $L$ be a line bundle on $C$ such that $\deg L \geq \deg Z + \deg \omega_C + 3$. Then every $s_Z \in H^0(Z, L|_Z)^{\times}$ can be lifted to $s_C \in H^0(C, L)$ such that $(s_C = 0)$ is irreducible, reduced and disjoint from $\Sigma$.*

*Proof.* The condition $s|_Z = c \cdot s_Z$ for some $c \in k$ determines a linear subsystem $|L, s_Z| \subset |L|$. The degree condition guarantees that it is basepoint-free and separable. Hence, it has infinitely many irreducible members. Almost all of them are disjoint from $\Sigma$.                                                                        □

It turns out that versions of (8.6.13) hold for some non-Hilbertian fields and in our proofs a weakening of it is sufficient. We discuss this in Section 9.9.

# Chapter Nine

## Linkage

In light of (4.1.14), the problem of reconstructing a variety from its topological space is at its core equivalent to the problem of reconstructing linear equivalence on divisors, which we denote by $\sim$.

After some preparatory work, in Section 9.4 we introduce a variant, called *linear similarity of divisors,* denoted by $\sim_s$. Two divisors are linearly similar if a multiple of one is linearly equivalent to a multiple of the other. This notion is closer to the topological space $|X|$, as multiplicity is not a topological notion. As it turns out, linear similarity is much easier to reconstruct than linear equivalence. With mild restrictions on $X$, we show how to recognize

- irreducible, ample $\mathbb{Q}$-Cartier divisors (9.4.15),
- linear similarity of irreducible, ample $\mathbb{Q}$-Cartier divisors (9.4.17), and
- irreducible $\mathbb{Q}$-Cartier divisors (9.4.20).

The key technical tool in this chapter is the notion of *linkage*, which addresses the following problem: given $Z$ and $W$ in $X$ such that $Z \cap W$ has dimension 0 (for example, a divisor and a curve not contained in it), when do pairs of divisors in $Z$ and $W$ come from a divisor on $X$, up to multiples?

This turns out to give surprisingly strong information on many aspects of the algebraic geometry of $X$, including residue fields of points, transversality of intersections, and numerical equivalence of divisors. In this way, we recognize the following objects/properties step by step.

- The set of $k$-points (9.6.8).
- Isomorphism of residue fields of closed points (9.6.10).
- Transversality of 0-dimensional intersections of subvarieties (9.7.5).
- Two irreducible curves having the same degree (9.8.5).
- Two irreducible divisors having the same degree (9.8.6).
- Numerical equivalence of ample divisors (9.8.15).

We can now use (7.3.4) to recognize algebraic pencils of divisors. Using (7.4.15), we can construct an ample degree function on divisors, which by (7.5.8) determines linear equivalence.

In (9.8.18) we obtain the main theorem of this section. For a normal, projective, geometrically irreducible variety, $|X|$ and $\sim_s$ together determine $\sim$, provided the characteristic is 0, and either $\dim X \geq 4$, or $\dim X \geq 3$ and $k$ is finitely generated over $\mathbb{Q}$.

## 9.1   LINKAGE OF DIVISORS

**Definition 9.1.1.** Let $X$ be a normal, projective, irreducible $k$-variety and $\mathscr{L}$ an ample line bundle on $X$. Let $Z_1, Z_2 \subset X$ be closed, irreducible subvarieties such that $\dim(Z_1 \cap Z_2) = 0$.

We say that $s_i \in H^0(X, \mathscr{L}^{m_i})$ with zero sets $H_i := (s_i = 0)$ are $\mathscr{L}$-*linked* on $Z_1 \cup Z_2$ if there is an $s \in H^0(X, \mathscr{L}^r)$ with zero set $H := (s = 0)$ such that

$$\operatorname{Supp}(H_1 \cap Z_1) \cup \operatorname{Supp}(H_2 \cap Z_2) = \operatorname{Supp}\big((Z_1 \cup Z_2) \cap H\big).$$

Note that if the Picard number of $X$ is 1, then this is clearly a question involving only the topology of $|X|$. In fact, by (9.4.18), this is almost always a question about $|X|$.

The following is an obvious sufficient condition.

**Lemma 9.1.2.** *If* $s_1^{r_1}|_{Z_1 \cap Z_2} = c \cdot s_2^{r_2}|_{Z_1 \cap Z_2}$ *for some nonzero* $c \in k^\times$ *and* $r_1, r_2 \in \mathbb{N}$*, then the* $H_i := (s_i = 0)$ *are* $\mathscr{L}$*-linked on* $Z_1 \cup Z_2$. $\square$

Note that $s_i|_{Z_1 \cap Z_2}$ can be an arbitrary element of $H^0(Z_1 \cap Z_2, \mathscr{L}|_{Z_1 \cap Z_2}) \cong H^0(Z_1 \cap Z_2, \mathcal{O}_{Z_1 \cap Z_2})$ for $\mathscr{L}$ sufficiently ample. Thus we obtain the following.

**Lemma 9.1.3** (Detecting $k$-points). *If the sufficient condition (9.1.2) is necessary and any two* $H_1, H_2$ *are* $\mathscr{L}$*-linked, then* $H^0(Z_1 \cap Z_2, \mathcal{O}_{Z_1 \cap Z_2}) \cong k$. *That is,* $Z_1 \cap Z_2$ *is a reduced $k$-point of $X$.* $\square$

This gives us a topological way to identify $k$-points and also check whether an intersection is transverse or not.

**9.1.4** (When is condition (9.1.2) necessary?). We know that $\operatorname{Supp}(s_i|_{Z_i} = 0) = \operatorname{Supp}(s|_{Z_i} = 0)$. Thus if

(1) the $Z_i$ are geometrically normal and irreducible, and

(2) the $\operatorname{Supp}(s_i|_{Z_i} = 0)$ are irreducible, then

$$s_i^{m_i}|_{Z_i} = c_i \cdot s^{n_i}|_{Z_i} \tag{9.1.4.1}$$

for some nonzero $c_i \in k^\times$ and $m_i, n_i \in \mathbb{N}$. Thus we conclude that there is a constant $c \in k^\times$ such that

$$s_1^{m_1 n_2}|_{Z_1 \cap Z_2} = c \cdot s_2^{m_2 n_1}|_{Z_1 \cap Z_2}, \tag{9.1.4.2}$$

as needed.

It remains to deal with the conditions (9.1.4 (1)) and (9.1.4 (2)). It turns out that normality can be avoided; this is worked out in Section 9.2. Instead of geometric irreducibility, the key condition is geometric connectedness, which is guaranteed if the $Z_i$ are set-theoretic complete intersections of ample divisors. Thus the troublesome condition is (9.1.4 (2)).

**9.1.5** (Irreducibility of $H \cap Z$). Let $Z \subset X$ be an irreducible subvariety and $H \subset X$ a general ample divisor. When is $H \cap Z$ irreducible?

Bertini's theorem says that this holds if $\dim Z \geq 2$. Since we also need $\dim(Z_1 \cap Z_2) = 0$, we must have $\dim X \geq 4$. For the best results we also need $Z_1 \cap Z_2$ to be a single point, which usually can be arranged only if $\dim X \geq 5$. This is the case when the methods work best.

There are two ways to lower the dimension. First, it turns out that we get almost everything if one of the $Z_i$ has dimension $\geq 2$, the other can be a curve. Thus we get all results if $\dim X \geq 4$, as in (1.2.1 (1)).

If $k$ is finitely generated over $\mathbb{Q}$, then a theorem of Hilbert guarantees that $H \cap Z$ is irreducible for most ample divisors $H$ even if $\dim Z = 1$. This property defines Hilbertian fields (8.6.6), and, for such fields, we can work with varieties of dimension $\geq 3$; leading to (1.2.1 (2)).

In order to give unified treatments, we introduce the Bertini-Hilbert dimension of fields in (9.5.5).

## 9.2   PREPARATIONS: SECTIONS AND THEIR ZERO SETS

In this section we discuss foundational results about sections and their zero sets that are needed in our study of linkage.

**9.2.1.** Let $k$ be a field and let $X$ be a normal, geometrically integral, proper $k$-variety. For a line bundle $\mathscr{L}$ on $X$ and sections $s_1, s_2 \in H^0(X, \mathscr{L})$ with corresponding divisors $Z(s_i)$ $(i = 1, 2)$ we have $Z(s_1) = Z(s_2)$ if and only if $s_1 = s_2 \cdot c$ for some $c \in k^\times$. We would like to relax the normality assumption on $X$ as much as possible while retaining the conclusion of this statement.

Let $Y$ be a reduced noetherian scheme, $\mathscr{L}$ a line bundle on $Y$ and $s \in H^0(Y, \mathscr{L})$ a section that does not vanish at any generic point of $Y$. It has *scheme-theoretic zeros* $Z(s)$ and *divisor-theoretic zeros*; the latter is the Weil divisor $\sum_\eta \mathrm{length}_{k(\eta)}(\mathscr{L}/s\mathcal{O}_Y)[\eta]$, where the summation is over all codimension 1 points of $Y$. The scheme-theoretic zeros determine the divisor-theoretic zeros, but the converse does not always hold.

We consider two genericity conditions.

(1) Every generic point of $\mathrm{Supp}(Z(s))$ is a regular codimension 1 point of $Y$.

(2) $Y$ is $S_2$ along $\mathrm{Supp}(Z(s))$.

Since $Y$ is reduced and $s$ does not vanish at any generic point of $Y$, the scheme $Z(s)$ is a Cartier divisor over a dense open subset of $Y$, and if (9.2.1(1)) holds then over a dense open of $Y$ the associated Weil divisor is Cartier and agrees with the Cartier divisor given by the restriction of $Z(s)$.

If (9.2.1(2)) holds and $s$ does not vanish at any generic point of $Y$, then the zero set $Z(s)$ has no embedded points. Indeed these assumptions imply that the map $s\colon \mathcal{O}_Y \to \mathcal{L}$ is injective which together with [Sta22, Tag 031Q] yields that $Z(s)$ is $S_1$ and therefore has no embedded points [Sta22, Tag 031Q]. It follows that in this case the scheme $Z(s)$ is determined by the associated Weil divisor. In this case we do not distinguish between the two divisors, referring to them each simply as the *zero set*, denoted by $Z(s)$.

**Notation 9.2.2.** For a reduced, noetherian scheme $Y$, let $\Sigma(Y) \subset Y$ denote the set of points $y \in Y$ such that $\mathcal{O}_{Y,y}$ is either of dimension 0, or of dimension 1 but not regular, or dimension $\geq 2$ but depth $\leq 1$.

**Lemma 9.2.3.** *Let $Y$ be a reduced, noetherian, excellent scheme. Then*

(1) $\Sigma(Y)$ *is finite, and*

(2) *a section $s$ of a line bundle $\mathcal{L}$ on $Y$ satisfies conditions (9.2.1(1)) and (9.2.1(2)) if and only if $Z(s)$ is disjoint from $\Sigma(Y)$.*

*Proof.* The ring $\mathcal{O}_{Y,y}$ has dimension 0 if and only if $y$ is a generic point. Since $Y$ is excellent, its regular locus $Y^{\mathrm{reg}} \subset Y$ is open and dense. The points such that $\mathcal{O}_{Y,y}$ is of dimension 1 but not regular are among the generic points of $Y \setminus Y^{\mathrm{reg}}$.

All the points of $Y^{\mathrm{reg}}$ are $S_2$. To prove the finiteness on non-$S_2$ points in $Y \setminus Y^{\mathrm{reg}}$, we may assume that $Y$ is affine. Let $g$ be a global section of $\mathcal{O}_Y$ that vanishes along $Y \setminus Y^{\mathrm{reg}}$ but does not vanish at any generic point of $Y$. The non-$S_2$ points in $Y \setminus Y^{\mathrm{reg}}$ are the non-$S_1$ points of the subscheme $(g = 0)$, that is, its associated points. Since $Y$ is noetherian, all three of these give finitely many points in $\Sigma(Y)$.

Statement (2) is immediate from the definition of $\Sigma(Y)$ and the above discussion. $\qquad\square$

**Lemma 9.2.4.** *Let $Y$ be a reduced, noetherian scheme, $\mathcal{L}$ a line bundle on $Y$, and $s_1, s_2 \in H^0(Y, \mathcal{L})$ sections that do not vanish at any point of $\Sigma(Y)$. Then $Z(s_1) = Z(s_2)$ if and only if $s_1 = s_2 \cdot u$ for some uniquely determined $u \in H^0(Y, \mathcal{O}_Y)^\times$.*

*Proof.* We may assume that $Y$ is affine given by a ring $A$ and that $\mathcal{L}$ is trivial. We then view $s_1$ and $s_2$ as elements of $A$.

Let $\bar{s}_2 \in A/(s_1)$ denote the image of $s_2$. The assumption $Z(s_1) = Z(s_2)$ implies that $\bar{s}_2$ vanishes at all generic points of $\mathrm{Spec}(A/(s_1))$. Since $A/(s_1)$ is $S_1$, it has no embedded points, thus $s_2 \in (s_1) \subset A$. Similarly, $s_1 \in (s_2)$, and therefore $(s_1) = (s_2)$ (equality of ideals). The lemma follows. $\qquad\square$

**Example 9.2.5.** Let $Y \subset \mathbb{P}_k^4$ be the union of $\{\underline{x} | x_1 = x_2 = 0\}$ and of $\{\underline{x} | x_3 = x_4 = 0\}$. Note that $(1:0:0:0:0)$ is a non-$S_2$ point and $H^0(Y, \mathcal{O}_Y) = k$. Consider $s(a, c) = ax_1 + cx_3 \in H^0(Y, \mathcal{O}_Y(1))$, and observe that its divisor $Z(s(a, c))$ is independent of $a, c \in k^\times$. However, $s(a, c) = s(a', c') \cdot u$ for some $H^0(Y, \mathcal{O}_Y)^\times$ if and only if $a/c = a'/c'$.

**Notation 9.2.6.** Let $Y$ be a reduced scheme, $B \subsetneq Y$ a closed subset and $\mathscr{L}$ a line bundle on $Y$. For $m > 0$ set

$$\Gamma^B(Y, \mathscr{L}, m) := \{s \in H^0(Y, \mathscr{L}^m) : \operatorname{Supp}(Z(s)) = B\}, \tag{9.2.6.1}$$

$$\Gamma^{\subset B}(Y, \mathscr{L}, m) := \{s \in H^0(Y, \mathscr{L}^m) : \operatorname{Supp}(Z(s)) \subset B\}, \tag{9.2.6.2}$$

$$\Gamma^B(Y, \mathscr{L}) := \amalg_m \Gamma^B(Y, \mathscr{L}^m) \text{ and } \Gamma^{\subset B}(Y, \mathscr{L}) := \amalg_m \Gamma^{\subset B}(Y, \mathscr{L}, m). \tag{9.2.6.3}$$

These are all unions of $k[Y]^\times$-orbits (8.4.3). Note that $\Gamma^B(Y, \mathscr{L}, m)$ is a set (it is not closed under addition in $\Gamma(Y, \mathscr{L})$) and $\Gamma^B(Y, \mathscr{L})$ and $\Gamma^{\subset B}(Y, \mathscr{L})$ are monoids.

In view of (9.4.1) and (9.4.17), $\Gamma^B(Y, \mathscr{L})$, is a very natural object to consider.

**Lemma 9.2.7.** *Let $Y$ be a reduced, projective scheme over a field, $B \subset Y$ a closed subset that is disjoint from $\Sigma(Y)$, and $\mathscr{L}$ a line bundle on $Y$.*

(1) *If $B$ is irreducible then $\Gamma^B(Y, \mathscr{L}, m)$ consists of at most one $k[Y]^\times$-orbit.*

(2) $\Gamma^{\subset B}(Y, \mathscr{L}, m)/k[Y]^\times$ *is finite.*

(3) $\Gamma^{\subset B}(Y, \mathscr{L})/k[Y]^\times$ *is a submonoid of a finitely generated free monoid.*

*Proof.* The first claim follows from (9.2.4). Indeed, if $\Gamma^B(Y, \mathscr{L}, m)$ is nonempty then $B$ is the support of an effective Weil divisor and since $B$ is irreducible it follows that $B$ is the closure of a codimension 1 point $\eta \in Y$. If $s_1, s_2 \in \Gamma^B(Y, \mathscr{L}, m)$ are two sections then the associated Weil divisors are equal to $n_i[\eta]$ for positive integers $n_i$ ($i = 1, 2$). The two integers $n_1$ and $n_2$ must be equal. Indeed, if not we obtain that a positive multiple of $[\eta]$ is trivial, which is impossible since $Y$ is a reduced projective scheme over a field. It follows that $Z(s_1) = Z(s_2)$.

To see the other two statements, let $B_i \subset B$ be the irreducible, divisorial components. Taking the length along each $B_i$ defines a map of monoids

$$\Gamma^{\subset B}(Y, \mathscr{L}) \to \oplus_i \mathbb{N},$$

which by (9.2.4) induces an inclusion

$$\Gamma^{\subset B}(Y, \mathscr{L})/k[Y]^\times \subset \oplus_i \mathbb{N},$$

proving (3). If $s \in \Gamma^{\subset B}(Y, \mathscr{L}, m)$ and $Z(s) = \sum_i m_i B_i$ then, computing the degrees (with respect to some ample divisor) gives that $\sum_i m_i \deg B_i = \deg \mathscr{L}$, hence $m_i \leq \deg \mathscr{L}$ for every $i$ and (2) follows. $\square$

**9.2.8.** Next we look at the evaluation of a section of a line bundle $\mathscr{L}$ at a point or at a 0-dimensional subscheme $V$. The twist is that we cannot distinguish two sections if their zero sets have the same support, and we also cannot distinguish various powers of $\mathscr{L}$ from each other. Thus for us the outcome of evaluation is not a single element of $H^0(V, \mathscr{L}|_V)$, but a submonoid of $\oplus_m H^0(V, \mathscr{L}^m|_V)$. Our aim is then to understand when this submonoid is small. As a further twist, we need to study this question not on the original scheme $X$, but on many of its subschemes $W \subset X$.

**Notation 9.2.9.** Let $X$ be a normal, projective, irreducible $k$-scheme, $\mathcal{L}$ a line bundle and $D$ an effective divisor on $X$. Let $W \subset X$ be a closed, integral subscheme and $V \subset W \setminus D$ a 0-dimensional subscheme. Define

$$\mathcal{R}_V^W(D, \mathcal{L}, m) := \mathrm{im}\left[ L^{W \cap D}(W, \mathcal{L}|_W, m) \to H^0(V, \mathcal{L}^m|_V) \right]$$
$$\mathcal{R}_V^W(D, \mathcal{L}) := \amalg_{m \geq 0} \mathcal{R}_V^W(D, \mathcal{L}, m).$$

Note that $\mathcal{R}_V^W(D, \mathcal{L})$ is a monoid that is closed under multiplication by $k[W]^\times = H^0(W, \mathcal{O}_W)^\times$ and, if $D \cap \Sigma(W) = \emptyset$, then $\mathcal{R}_V^W(D, \mathcal{L})/k[W]^\times$ is a submonoid of a finitely generated free monoid by (9.2.7 (3)).

The following elementary observations turn out to be crucial.

**Proposition 9.2.10.** *Using the notation and assumptions of (9.2.9), assume also that $D := Z(s)$ for some $s \in H^0(X, \mathcal{L}^r)$, and $D \cap W$ is irreducible and disjoint from $\Sigma(W)$. Then*

(1) $\mathcal{R}_V^W(D, \mathcal{L}) \cong \langle s, k[W]^\times \rangle^{\mathrm{Sat}}$, *the saturation of the submonoid of $\mathcal{R}_V^W(D, \mathcal{L})$ generated by $s$ and $k[W]^\times$.*

(2) *If $k[W] = k$ then the saturation of $\mathcal{R}_V^W(D, \mathcal{L})$ in $\amalg_{m \geq 0} H^0(V, \mathcal{L}^m|_V)$ depends only on $D, \mathcal{L}$ and $V$ (but not on $W$).*

*Proof.* The first assertion follows from (9.2.7 (1)). Indeed, $s|_W$ is the unique section of $\mathcal{L}^r|_W$ (up to $k[W]^\times$) that defines $\mathrm{Supp}(D \cap W)$. Therefore

$$\mathcal{R}_V^W(D, \mathcal{L}, rm) = s^m|_W \cdot k[W]^\times|_V = s^m|_V \cdot k[W]^\times|_V. \qquad (9.2.10.1)$$

For other values, $\mathcal{R}_V^W(D, \mathcal{L}, m')$ is either empty or consists of a single $k[W]^\times$-orbit.

For the second statement note that if $k[W] = k$ then $k[W]^\times = k^\times$ and the $k^\times$-action on $H^0(V, \mathcal{L}^m|_V)$ is independent of $W$. $\qquad \square$

**Remark 9.2.11.** If $k[W] \neq k$, we get uniqueness if $k[W]/k$ is Galois and $V$ is irreducible, or if $k[W] = k[V_{\mathrm{red}}]$ and it is separable over $k$, but neither of these conditions is easy to guarantee.

## 9.3  NÉRON'S THEOREM AND CONSEQUENCES

**Definition 9.3.1.** Let $X$ be an irreducible variety. Following [Ser89], a subset $T \subset X(k)$ is called *thin* if there is a generically finite morphism $\pi \colon Y \to X$ such that $T \subset \pi(Y(k))$ and there is no rational section $\sigma \colon X \dashrightarrow Y$.

**Remark 9.3.2.** This notion is most interesting for finitely generated, infinite fields. For such fields, $\mathbb{A}^1(k) \subset \mathbb{A}^1(k)$ is not thin; this is essentially due to Hilbert.

**Example 9.3.3.** A rather typical example to keep in mind is the following. The map $\mathbb{A}^1 \to \mathbb{A}^1$ given by $x \to x^2$ shows that the set of all squares is a thin subset of $\mathbb{A}^1(k)$.

We also need a version of this for arbitrary fields $K$:

**Definition 9.3.4.** A subset $T \subset X(K)$ is *field-locally thin* if for every finitely generated subfield $k \subset K$, the intersection $T \cap X(k)$ is thin.

**Theorem 9.3.5.** *Let $k$ be a finitely generated, infinite field. Let $U \subset \mathbb{P}^1_k$ be an open subset and $\pi \colon T^U \to U$ a smooth, projective morphism of relative dimension 1. Then there is a dense set $N(T^U) \subset U(k)$, such that the restriction map*

$$\operatorname{Pic}(T^U) \to \operatorname{Pic}(T_u) \text{ is injective for all } u \in N(T^U).$$

*Moreover, $N(T^U)$ contains the complement of a thin set.*

*Proof.* This is [Nér52b, Thm. 6].  □

**Remark 9.3.6.** A stronger version is proved in [Sil83, Thm. C], though it applies only to number fields and finite extensions of $\mathbb{F}_p(t)$.

**Corollary 9.3.7.** *Let $K$ be a field that is not locally finite. Let $S$ be a normal, projective surface over $K$ and $|C| = \{C_u : u \in \mathbb{P}^1\}$ a linear pencil of curves with finitely many basepoints $\{p_1, \ldots, p_r\}$. Assume that a general $C_u$ is smooth and $S$ is smooth along it. Let $\{B_j : j \in J\}$ be the irreducible components of the reducible members of $|C|$, plus one of the irreducible members. Let $m_{ij}$ be the intersection multiplicity of $B_j$ with a general $C_u$ at $p_i$; this is independent of $u$.*

*Then there is a dense set $N(S, |C|) \subset \mathbb{P}^1(K)$ such that, for $u \in N(S, |C|)$, all the $\mathbb{Q}$-linear relations among*

$$[p_1(u)], \ldots, [p_r(u)] \in \frac{\operatorname{Pic}(C_u)}{\operatorname{Im}[\operatorname{Cl}(S) \to \operatorname{Pic}(C_u)]}$$

*are generated by $\sum_i m_{ij}[p_i(u)] = 0$ for all $j \in J$.*

*Moreover, $N(S, |C|)$ contains the complement of a field-locally thin set.*

*Proof.* Note that the point $p_i$ is contained in every $C_u$; the notation $[p_i(u)]$ indicates that we take its class in $\operatorname{Pic}(C_u)$, which depends on $u$.

The restriction of $B_j$ to $C_u$ is $\sum_i m_{ij}[p_i(u)]$, so we do need to have the equations $\sum_i m_{ij}[p_i(u)] = 0$. The interesting part is to show that there are no other relations.

Let $T$ be the normalization of the closure of the graph of $|C| \colon S \dashrightarrow \mathbb{P}^1$. The projection $\pi_1 \colon T \to S$ is birational, with exceptional curves $E_i \subset T$ sitting over

$p_i$. Let $B_j^T \subset T$ denote the birational transform of $B_j$. Note that

$$B_j^T \sim \pi_1^* B_j - \textstyle\sum_i m_{ij} E_i.$$

The second projection $\pi_2 \colon T \to \mathbb{P}^1$ is generically smooth and the irreducible components of its reducible fibers are exactly the $B_j^T$.

Let $U \subset \mathbb{P}^1$ be the largest open set over which $\pi_2$ is smooth. By restriction we get $T^U \to U$. The Picard group of $T^U$ is then

$$\mathrm{Pic}(T^U) = \mathrm{Cl}(T)/\langle B_j^T : j \in J \rangle.$$

Choose a finitely generated subfield $k \subset K$ such that $S, |C|$, the $p_i$ and the $B_j$ are defined over $k$.

Note that $\mathrm{Cl}(T_k) = \pi_1^* \, \mathrm{Cl}(S_k) + \sum_i [E_i]$, and killing $\pi_1^* \, \mathrm{Cl}(S_k)$ gives an isomorphism

$$\mathrm{Cl}(T_k^U)/\pi_1^* \, \mathrm{Cl}(S_k) \cong \langle E_i : i \in I \rangle / \langle \textstyle\sum_i m_{ij}[p_i] : j \in J \rangle.$$

Thus all the linear relations among $[E_1], \dots, [E_r] \in \mathrm{Cl}(T_k^U)/\pi_1^* \, \mathrm{Cl}(S_k)$ are generated by $\sum_i m_{ij}[E_i] = 0$ for all $j \in J$.

We now apply (9.3.5) to get $N(T_k^U) \subset \mathbb{P}^1(k)$ such that, for $u \in N(T_k^U)$, all the linear relations among

$$[p_1(u)], \dots, [p_r(u)] \in \mathrm{Pic}(C_u)/\mathrm{Im}[\mathrm{Cl}(S_k) \to \mathrm{Pic}(C_u)]$$

are generated by $\sum_i m_{ij}[p_i(u)] = 0$ for all $j \in J$.

This is not exactly what we want since $\mathrm{Cl}(S_K)$ may be much bigger than $\mathrm{Cl}(S_k)$. However, by (11.3.26), if some $\mathscr{L}_k$ is in $\mathrm{Im}[\mathrm{Cl}(S_K) \to \mathrm{Pic}(C_u)]$, then $\mathscr{L}_k^r$ is in $\mathrm{Im}[\mathrm{Cl}(S_k) \to \mathrm{Pic}(C_u)]$ for some $r > 0$. (With a little more work one can prove the Corollary for $\mathbb{Z}$-linear relations, but this is not important for us.)  $\square$

## 9.4   LINEAR SIMILARITY

**Definition 9.4.1.** Let $X$ be a normal, integral, separated scheme. As in (7.1.9), two Weil $\mathbb{Z}$-divisors $D_1, D_2$ are *linearly similar*, denoted $D_1 \sim_{\mathrm{s}} D_2$, if there are nonzero integers $m_1, m_2$ such that $m_1 D_1 \sim m_2 D_2$.

The set of all effective divisors linearly similar to a fixed divisor $D$ is naturally an infinite union of linear systems, we denote it by $|\mathbb{Q}D|^{\mathrm{set}}$.

Let $|\mathbb{Q}D|^{\mathrm{irr}} \subset |\mathbb{Q}D|^{\mathrm{set}}$ be the subset of irreducible (but not necessarily reduced) divisors.

**Remark 9.4.2.** We will use this notion mostly when the $D_i$ are effective and $X$ is a normal scheme over a field $k$. If $X$ is proper, or, more generally, when $H^0(X, \mathcal{O}_X)$ is a finite $k$-algebra, then $m_1 D_1 \sim m_2 D_2$ implies that $m_1, m_2$ have the same sign. We always choose $m_1, m_2 > 0$.

If $\mathrm{rank}_\mathbb{Q} \, \mathrm{Cl}(X) \leq 1$ then any two effective divisors are linearly similar. Thus this notion is nontrivial only if $\mathrm{rank}_\mathbb{Q} \, \mathrm{Cl}(X) > 1$.

Some of the linear systems $|D'|^{\mathrm{set}} \subset |\mathbb{Q}D|^{\mathrm{set}}$ may be small and behave exceptionally so we introduce the following definition:

**Definition 9.4.3.** A subset $W \subset |\mathbb{Q}D|^{\mathrm{set}}$ is called *stably dense* if $W \cap |mD|^{\mathrm{set}}$ is Zariski dense in $|mD|^{\mathrm{set}}$ for $m \gg 1$.

**Remark 9.4.4.** Note that $|\mathbb{Q}D|^{\mathrm{irr}}$ need not be dense in $|\mathbb{Q}D|^{\mathrm{var}}$, but, if $k$ is infinite, $D$ is ample and $\dim X \geq 2$ (more generally, if $D$ has Kodaira dimension $\geq 2$ and there are no fixed components) then, by (11.1.15), $|\mathbb{Q}D|^{\mathrm{irr}}$ is stably dense in $|\mathbb{Q}D|^{\mathrm{set}}$.

**Remark 9.4.5.** If $\dim X = 1$ then $|\mathbb{Q}D|^{\mathrm{irr}}$ is frequently empty. This presents a serious technical difficulty in our treatment. However, if $\deg D > 0$ then $|\mathbb{Q}D|^{\mathrm{irr}}$ is stably dense in $|\mathbb{Q}D|^{\mathrm{set}}$ provided the Bertini-Hilbert dimension, defined in (9.5.5) below, is equal to 1. This will be shown in (9.5.4).

**9.4.6.** For a submonoid $M \subset \mathrm{Cl}(X)$ of effective classes one can consider a generalization of the Cox ring construction. Namely, choose for each $m \in M$ an effective divisor $D_m$ on $X$ with class $m$ and set

$$\mathrm{Cox}(X, M) := \oplus_{m \in M} \, H^0(X, \mathcal{O}_X(D_m)).$$

This is a graded $k$-vector space, well defined up to non-canonical isomorphism. We would like to endow $\mathrm{Cox}(X, M)$ with the structure of a ring compatible with the monoid structure on $M$. Let $X_{\mathrm{reg}} \subset X$ be the regular locus, which has complement of codimension $\geq 2$ since $X$ is assumed normal. To obtain the ring structure on $\mathrm{Cox}(X, M)$ it suffices to construct isomorphisms

$$\mathcal{O}_X(D_m)|_{X_{\mathrm{reg}}} \otimes \mathcal{O}_X(D_{m'})|_{X_{\mathrm{reg}}} \simeq \mathcal{O}_X(D_{m+m'})|_{X_{\mathrm{reg}}} \qquad (9.4.6.1)$$

for all $m, m' \in M$. Furthermore, these isomorphisms should satisfy a suitable associativity condition for triples of elements. This can conveniently be summarized as follows. Let $\mathscr{P}ic(X_{\mathrm{reg}})$ be the groupoid of invertible sheaves on $X_{\mathrm{reg}}$ and let $\mathscr{P}ic(X_{\mathrm{reg}}, M^{\mathrm{gp}})$ be the preimage of $M^{\mathrm{gp}} \subset \mathrm{Cl}(X_{\mathrm{reg}})$ under the natural map

$$\mathscr{P}ic(X_{\mathrm{reg}}) \to \mathrm{Cl}(X_{\mathrm{reg}}).$$

The data of compatible choices of isomorphisms (9.4.6.1) is then equivalent to the data of a section of the map of Picard categories

$$\mathscr{P}ic(X_{\mathrm{reg}}, M^{\mathrm{gp}}) \to M^{\mathrm{gp}}. \qquad (9.4.6.2)$$

By [AGV73, XVIII, 1.4.15], the obstruction to finding such a section lies in

$$\mathrm{Ext}^2(M^{\mathrm{gp}}, H^0(X, \mathcal{O}_X^\times)),$$

which vanishes since $\mathbf{Z}$ has projective dimension 1. The choice of section of (9.4.6.2) is not unique in general. In fact, by loc. cit. the group of sections, up to equivalence defined by compatible automorphisms of line bundles, is given by $\mathrm{Ext}^1(M^{\mathrm{gp}}, H^0(X, \mathcal{O}_X^\times))$, which may be nonzero if $M^{\mathrm{gp}}$ has torsion and $H^0(X, \mathcal{O}_X^\times))$ is not divisible. It follows that in general $\mathrm{Cox}(X, M)$ admits a ring structure as expected but the resulting ring is not uniquely defined in general.

**9.4.7.** We will be interested in the Cox rings for $M$, where $M$ is the union of 0 and the image of $|\mathbf{Q}D|$ in $\mathrm{Cl}(X)$, in which case we write $\mathrm{Cox}(X, |\mathbf{Q}D|)$ for $\mathrm{Cox}(X, M)$.

**9.4.8** (Restriction and linear similarity). Let $X$ be a normal variety, $Z \subset X$ a subvariety and $D_1, D_2$, effective divisors on $X$. If $D_1 \sim_{\mathrm{s}} D_2$ then (aside from some problems that appear for non-Cartier divisors), $D_1|_Z \sim_{\mathrm{s}} D_2|_Z$. For us the main interest will be the converse: if $D_1|_Z \sim_{\mathrm{s}} D_2|_Z$, when can we conclude that $D_1 \sim_{\mathrm{s}} D_2$?

Let $D$ be an irreducible divisor. We say that a subvariety $Z \subset X$ *detects linear similarity to $D$* if for any effective divisor $D'$ such that $\mathrm{Supp}(D \cap Z) = \mathrm{Supp}(D' \cap Z)$ we have $D' \sim_{\mathrm{s}} D$. It is not always easy to see when this happens, but the following is quite useful.

**Criterion 9.4.9.** Assume that $Z \cap \mathrm{Sing}\, X$ has codimension $\geq 2$ in $Z$, the kernel of $\mathrm{Cl}(X) \to \mathrm{Pic}(Z \setminus \mathrm{Sing}\, X)$ is torsion, $D$ is disjoint from $\Sigma(Z)$ (see definition 9.2.2), and $D \cap Z$ is irreducible. Then $Z$ detects linear similarity to $D$.

*Proof.* If $Z \cap \mathrm{Sing}\, X$ has codimension $\geq 2$ in $Z$ then we have a restriction map from rank 1 reflexive sheaves on $X$ (that are locally free along $\Sigma(Z)$) to rank 1 reflexive sheaves on $Z$ (that are locally free along $\Sigma(Z)$), and such a rank 1 reflexive sheaf on $Z$ is determined by the divisors of its sections by 9.2.4.     □

**Lemma 9.4.10.** *Let $X$ be a geometrically normal, projective variety over an infinite field $k$ and $D_1, \ldots, D_r$ irreducible divisors on $X$. Then linear similarity to the $D_i$ is detected by general, ample, complete intersections of dimension $\geq 2$.*

*Proof.* Let $Z \subset X$ be a general, ample, complete intersection surface. Then $\mathrm{Cl}(X) \to \mathrm{Cl}(Z)$ is an injection by (11.2.4) and $Z \cap D_i$ is irreducible and reduced for every $i$ by Bertini's theorem (11.1.15).     □

**Lemma 9.4.11.** *Let $X$ be a normal, projective variety and $C \subset X^{\mathrm{ns}}$ a smooth, projective curve. Assume that the kernel of $\mathrm{Cl}(X) \to \mathrm{Pic}(C)$ is torsion and the following holds.*

$(\star)$ *Let* $D \cap C = \{p_1, \ldots, p_r\}$. *Then the points* $p_1, \ldots, p_{r-1}$ *are linearly indepen- dent over* $\text{Im}[\text{Cl}(X) \to \text{Pic}(C)]$. *More precisely,*

$$\text{rank}_{\mathbb{Q}}(\langle p_1, \ldots, p_r \rangle \cap \text{Im}[\text{Cl}(X) \to \text{Pic}(C)]) = 1.$$

*Then* $C$ *detects linear similarity to* $D$.

*Proof.* Let $D'$ be another effective divisor such that $\text{Supp}(D' \cap C) = \text{Supp}(D \cap C)$. Note that both $D, D'$ are Cartier along $C$. Thus $D|_C = \sum d_i [p_i]$ and $D'|_C = \sum d'_i [p_i]$. By condition $(\star)$

$$m'_1 \sum d'_i [p_i] = m_1 \sum d_i [p_i]$$

for some $m'_1, m_1 > 0$, hence $m'_1 D' - m_1 D$ is in the kernel of $\text{Cl}(X) \to \text{Pic}(C)$. Thus $m_2(m'_1 D' - m_1 D) \sim 0$ for some $m_2 > 0$. $\qquad\square$

**Theorem 9.4.12.** *Let $k$ be a field that is not locally finite. Let $X$ be a ge- ometrically normal, projective variety of dimension $n \geq 2$ over $k$, $\{D_i : i = 1, \ldots, r\}$ irreducible Weil divisors and $H_1, \ldots, H_{n-1}$ ample divisors. Then, for $m_1 \gg m_2 \gg \cdots \gg m_{n-1} \gg 1$, there is a dense subset*

$$U \subset |m_1 H_1|^{\text{set}} \times \cdots \times |m_{n-1} H_{n-1}|^{\text{set}}$$

*such that, for $u \in U$, the corresponding complete intersection curve $C_u$ detects linear similarity to each $D_i$.*

*Proof.* By (9.4.10) there is a Zariski open

$$U_2 \subset |m_2 H_2|^{\text{var}} \times \cdots \times |m_{n-1} H_{n-1}|^{\text{var}}$$

such that $\text{Cl}(X) \to \text{Cl}(H_2 \cap \cdots \cap H_{n-1})$ is an injection for $(H_2, \ldots, H_{n-1}) \in U_2$ and the $D_i \cap H_2 \cap \cdots \cap H_{n-1}$ are irreducible. This reduces us to the case $n = 2$.

Thus from now on we have a normal, projective surface $X$ over $k$, $\{D_i : i = 1, \ldots, r\}$ irreducible Weil divisors on $X$, and an ample divisor $H$ on $X$.

Now choose a pencil $|C| \subset |mH|$ such that

(1) $D_0 + \cdots + D_r \in |C|$ for some irreducible curve $D_0$,

(2) all other members of $|C|$ are irreducible,

(3) the general member of $|C|$ is smooth, and

(4) $X$ and $D_0 + \cdots + D_r$ are smooth at the base locus of $|C|$.

Applying (9.3.7) to it we get a dense subset of $|C|^{\text{set}}$ where the requirements hold. $\qquad\square$

**Remark 9.4.13.** Most likely one can choose $U$ such that it contains the com- plement of a field-locally thin set.

First we recognize linear similarity of ample prime divisors.

**Notation 9.4.14** (Linear similarity of ample divisors). Let $X$ be a normal variety and $\mathrm{PDiv}(X)$ the set of prime divisors on $X$. We define a relation on $\mathrm{PDiv}(X) \times \mathrm{PDiv}(X)$ by declaring that $D_1 \sim_{\mathrm{sa}} D_2$ if and only if $D_1, D_2$ are $\mathbb{Q}$-Cartier and ample, and $D_1 \sim_{\mathrm{s}} D_2$.

Note that if $\mathrm{rank}_{\mathbb{Q}} \mathrm{Cl}(X) \leq 1$ then $D_1 \sim_{\mathrm{sa}} D_2$ for any two ample, prime divisors on $X$. In these cases the relation $\sim_{\mathrm{sa}}$ carries no extra information.

**Lemma 9.4.15.** *Let $X$ be a geometrically normal, projective variety over a field $k$. Assume that either*

(1) *$k$ is not locally finite and $\dim X \geq 2$, or*

(2) *$k$ is infinite and $\dim X \geq 3$.*

*Then an irreducible divisor $H$ is $\mathbb{Q}$-Cartier and ample if and only if for every divisor $D \subset X$ and distinct closed points $p, q \in X \backslash D$, there is a divisor $H(p,q) \subset X$ such that*

(i)     $H \cap D = H(p,q) \cap D$,

(ii)    $p \notin H(p,q)$, and

(iii)   $q \in H(p,q)$.

*Proof.* If $H$ is $\mathbb{Q}$-Cartier and ample then the restriction map

$$H^0\big(X, \mathcal{O}_X(mH)\big) \twoheadrightarrow H^0\big(D, \mathcal{O}_D(mH|_D)\big) + \mathcal{O}_X(mH) \otimes \big(k(p) + k(q)\big) \quad (9.4.15.1)$$

is surjective for some $m > 0$. We can thus find $s(p,q) \in H^0\big(X, \mathcal{O}_X(mH)\big)$ as needed.

Conversely, by (9.4.10) and (9.4.12) we can choose an ample divisor $D \subset X$ that detects linear similarity to $H$. Then assumption (9.4.15 (i)) guarantees that $H(p,q) \sim_{\mathrm{s}} H$. Assumption (9.4.15 (ii)) implies that $H$ is $\mathbb{Q}$-Cartier at $p$. Since $p, q$ are arbitrary points (if we also vary $D$), $H$ is $\mathbb{Q}$-Cartier and a multiple of it separates points.

Finally, for $m \in \mathbb{N}$ let $B_m \subset X \times X$ be the set of point pairs that are not separated by any member of $|mH|$. Then $B_m$ is closed, $B_{m_1} \subset B_{m_2}$ if $m_2 \mid m_1$, and we have just proved that $\cap_{m \in \mathbb{N}} B_m = \emptyset$. Thus $B_m = \emptyset$ for some $m$. $\square$

**Corollary 9.4.16.** *Let $X$ be a geometrically normal, projective variety over a field $k$ and $|H| \subset |X|$ a closed, irreducible subset of codimension 1. Assume that*

(1) *either $k$ is not locally finite and $\dim X \geq 2$,*

(2) *or $k$ is infinite, locally finite and $\dim X \geq 3$.*

*We can then decide, using only $|H| \subset |X|$, whether $|H|$ supports an ample* *divisor.* □

**Lemma 9.4.17.** *Let $X$ be a geometrically normal, projective variety over a field $k$ and $H_1, H_2$ irreducible, $\mathbb{Q}$-Cartier, ample divisors. Assume that*

(1) *either $k$ is not locally finite and $\dim X \geq 3$,*

(2) *or $k$ is infinite and $\dim X \geq 5$.*

*Then the following are equivalent.*

(3) $H_1 \sim_{\mathrm{sa}} H_2$.

(4) $|\mathbb{Q}H_1|^{\mathrm{irr}} = |\mathbb{Q}H_2|^{\mathrm{irr}}$.

(5) *Let $Z_1, Z_2 \subset X$ be any pair of disjoint, irreducible subvarieties of dimension $\geq 2$ if $k$ is locally finite and $\geq 1$ otherwise. Then there is an irreducible, $\mathbb{Q}$-Cartier, ample divisor $H'$ such that $\mathrm{Supp}(H' \cap Z_i) = \mathrm{Supp}(H_i \cap Z_i)$ for $i = 1, 2$.*

*Proof.* The implication (3) ⇔ (4) is clear. If (3) holds then choose $m_1, m_2 \gg 1$ such that $m_1 H_1 \sim m_2 H_2$ and

$$H^0\big(X, \mathcal{O}_X(m_1 H_1)\big) \to H^0\big(Z_1, \mathcal{O}_X(m_1 H_1)|_{Z_1}\big) + H^0\big(Z_2, \mathcal{O}_X(m_2 H_2)|_{Z_2}\big)$$

is surjective. As in (9.4.15 (ii)), the kernel separates points on $X \setminus (Z_1 \cup Z_2)$. Thus, by Bertini's theorem (11.1.15), we can then find an irreducible divisor $H' \in |m_1 H_1| = |m_2 H_2|$ whose restriction to $Z_i$ is $m_i H_i|_{Z_i}$.

Finally assume (5). By (9.4.10) and (9.4.12) we can choose both $Z_1, Z_2$ normal, disjoint and such that they detect linear similarity to the $H_i$. Then we have the chain of linear similarities

$$H_1 \overset{(by\ Z_1)}{\sim_{\mathrm{s}}} H' \overset{(by\ Z_2)}{\sim_{\mathrm{s}}} H_2. \qquad \square \qquad (9.4.17.1)$$

Putting together (9.4.15) and (9.4.17), we get the following.

**Corollary 9.4.18.** *Let $X$ be a normal, projective variety of dimension $\geq 3$ over a field $k$. Assume that $k$ is not locally finite. Then $|X|$ determines $\sim_{\mathrm{sa}}$.* □

**9.4.19** (Variants for reducible divisors). With $X$ as in (9.4.17), we can also get some results when the $H_i$ are reducible. For this, write $H_i = \sum_j a_{ij} H_{ij}$. We argue as above, except at the end, instead of (9.4.17.1) we get that

$$\sum_j a'_{1j} H_{1j} \overset{(by\ Z_1)}{\sim_{\mathrm{s}}} H' \overset{(by\ Z_2)}{\sim_{\mathrm{s}}} \sum_j a'_{2j} H_{2j},$$

where $a'_{ij} > 0$ if and only if $a_{ij} > 0$. Thus (9.4.17 (5)) is equivalent to the following.

(1) There are $\mathbb{Q}$-Cartier, ample, effective divisors $H'_1, H'_2$ such that $\mathrm{Supp}\, H'_i =$

Supp $H_i$ and $H_1' \sim_s H_2'$.

We also obtain the following.

(2) If the irreducible components of $H_2$ are $\mathbb{Q}$-Cartier and all but one of the irreducible components of $H_1$ are $\mathbb{Q}$-Cartier, then the remaining irreducible component of $H_1$ is also $\mathbb{Q}$-Cartier.

We can thus recognize irreducible $\mathbb{Q}$-Cartier divisors using the following criterion.

**Corollary 9.4.20.** *An irreducible divisor $D \subset X$ is $\mathbb{Q}$-Cartier if and only if there are irreducible, $\mathbb{Q}$-Cartier, ample divisors $A_1, A_2$ such that (9.4.19 (1)) holds for $B_1 := D + A_1$ and $B_2 := A_2$.*

**Remark 9.4.21.** Using (9.4.15) and (9.4.17) we get our first topological invariance claims. Namely, let $X_K, Y_L$ be normal, projective varieties such that $|X| \sim |Y|$. Assume that either $L$ is not locally finite and $\dim Y \geq 3$, or $\dim Y \geq 5$.

(1) If $X$ is $\mathbb{Q}$-factorial then so is $Y$.

(2) If $\operatorname{rank}_{\mathbb{Q}} \operatorname{Cl}(X) = 1$ then $\operatorname{rank}_{\mathbb{Q}} \operatorname{Cl}(Y) = 1$.

Note that by (10.3.2), $\mathbb{P}^2_{\overline{\mathbb{F}}_p}$ is homeomorphic to smooth surfaces with arbitrary large Picard number, so some restriction on the dimension is necessary in (9.4.17).

## 9.5   BERTINI-HILBERT DIMENSION

**9.5.1.** Let $X$ be a projective variety over a field $k$ and let $\mathscr{L}$ be an ample line bundle on $X$. We are looking for sections $s \in H^0(X, \mathscr{L})$ that satisfy three properties:

(1) The zero set $Z(s)$ is irreducible.

(2) The values of $s$ at some points $x_i \in X$ are specified. More generally, given a 0-dimensional subscheme $Z \subset X$, we would like to specify $s|_Z$.

(3) The zero set $Z(s)$ avoids a finite set of points $\Sigma \subset X$.

To formalize these, let $X$ be a proper scheme over a field $k$, $Z \subset X$ a subscheme, $\mathscr{L}$ an ample line bundle on $X$, and $s_Z \in H^0(Z, \mathscr{L}|_Z)$. Set

$$H^0(X, \mathscr{L}, s_Z) := \{ s \in H^0(X, \mathscr{L}) : s|_Z = c s_Z \text{ for some } c \in H^0(Z, \mathcal{O}_Z)^\times \}.$$
$$(9.5.1.1)$$

This is an open subset of a vector subspace of $H^0(X, \mathcal{L})$. If $X$ is integral then for the corresponding linear systems we use the notation $|\mathcal{L}, s_Z| \subset |\mathcal{L}|$. For a finite subset $\Sigma \subset X$, let

$$|\mathcal{L}, s_Z, \Sigma^c| := \{D \in |\mathcal{L}, s_Z| : D \cap \Sigma = \emptyset\} \tag{9.5.1.2}$$

denote the subset of those divisors that are disjoint from $\Sigma$. Finally we use

$$|\mathcal{L}, s_Z, \Sigma^c|^{\mathrm{irr}} := \{D \in |\mathcal{L}, s_Z, \Sigma^c| : D \text{ is irreducible}\}. \tag{9.5.1.3}$$

For our applications, we are free to replace $\mathcal{L}$ by $\mathcal{L}^m$. Thus we are most interested in the case when $H^0(X, \mathcal{L}) \to H^0(Z, \mathcal{L}|_Z)$ is surjective and the linear system $|\mathcal{L}, s_Z|$ is very ample on $X \setminus Z$. In this situation conditions (2) and (3) above are easy to satisfy and the key issue is the irreducibility of $Z(s)$.

Next we discuss three cases when we can guarantee irreducibility.

**Lemma 9.5.2.** *Let $X$ be an irreducible, projective variety of dimension $\geq 2$ over an infinite field $k$. Let $\Sigma \subset X$ be a finite subset and $Z \subset X$ a finite subscheme. Let $\mathcal{L}$ be an ample line bundle on $X$ and let $s_Z \in H^0(Z, \mathcal{L}|_Z)$ be a nowhere zero section over $Z$. Then,*

$$|\mathcal{L}^m, s_Z^m, \Sigma^c|^{\mathrm{irr}}$$

*contains an open and dense subset of $|\mathcal{L}^m, s_Z^m|$ for $m \gg 1$.*

*Proof.* The linear system $|\mathcal{L}^m, s_Z^m|$ is very ample on $X \setminus Z$, so this follows from the Bertini theorem (11.1.16). $\qquad\square$

Next we consider Hilbertian fields (8.6.6). Here $|\mathcal{L}^m, s_Z^m, \Sigma^c|^{\mathrm{irr}}$ need not be open, but it is still quite large. In light of (9.5.2), we need to primarily consider the case of curves.

**Lemma 9.5.3.** *Let $C$ be an irreducible, projective curve over a Hilbertian field $k$. Let $\Sigma \subset C$ be a finite subset and $Z \subset C$ a finite subscheme. Let $\mathcal{L}$ be an ample line bundle on $C$ and $s_Z \in H^0(Z, \mathcal{L}|_Z)^\times$. Then $|\mathcal{L}^m, s_Z^m, \Sigma^c|^{\mathrm{irr}}$ contains the complement of a thin subset (9.3.1) of $|\mathcal{L}^m, s_Z^m|$ for $m \gg 1$.*

*Proof.* As before, $|\mathcal{L}^m, s_Z^m, |$ is very ample on $C \setminus Z$, hence this follows from a basic property of Hilbertian fields (8.6.6). $\qquad\square$

For the following applications in Sections 9.6–9.7 we only need a weaker version of the conclusion in (9.5.3). Namely, we only need $|\mathcal{L}^m, s_Z^m, \Sigma^c|^{\mathrm{irr}}$ to be nonempty for some $m > 0$. This leads to the definition of *weakly Hilbertian* fields (9.9.1). The following is essentially the definition but we state it as a lemma to emphasize the similarity to (9.5.3).

**Lemma 9.5.4.** *Let $C$ be an irreducible, projective curve over a weakly Hilbertian field $k$. Let $\Sigma \subset C$ be a finite subset and $Z \subset C$ a finite subscheme. Let $\mathcal{L}$ be an ample line bundle on $C$. Then $|\mathcal{L}^m, s_Z^m, \Sigma^c|^{\mathrm{irr}}$ is nonempty for some $m > 0$. $\qquad\square$*

Note that although we ask for only one irreducible divisor, by enlarging $\Sigma$ we see that we get infinitely many. In fact, the sets

$$|\mathscr{L}^m, s_Z^m, \Sigma^c|^{\mathrm{irr}} \subset |\mathscr{L}^m, s_Z^m|$$

seem to be quite large, though we do not know how to formulate this precisely.

For most of the proofs we need to know the smallest dimension where linear systems are guaranteed to have many irreducible members. This leads to the following definition.

**Definition 9.5.5.** Let $k$ be a field that is not locally finite. We define the *Bertini-Hilbert dimension* of $k$—denoted by $\mathrm{BH}(k)$—by setting

(1) $\mathrm{BH}(k) = 1$ if $k$ is weakly Hilbertian (9.9.1), and

(2) $\mathrm{BH}(k) = 2$ otherwise.

**Remark 9.5.6.** In view of (9.5.2), the distinction is only about curves. If $k$ is Hilbertian then $\mathrm{BH}(k) = 1$ by (9.5.3).

**Remark 9.5.7.** We do not define $\mathrm{BH}(k)$ for locally finite fields. If $k$ is locally finite and $\mathscr{L}$ is an ample line bundle on an irreducible curve $C$, then every smooth point $p \in C$ is the zero set of a section of some $\mathscr{L}^m$. This would suggest that $\mathrm{BH}(k)$ should be 1, but in some applications setting $\mathrm{BH}(\mathbb{F}_q) = 2$ or even $\mathrm{BH}(\mathbb{F}_q) = \infty$ would seem the right choice.

## 9.6  LINKAGE OF DIVISORS AND RESIDUE FIELDS

**Definition 9.6.1.** Let $X$ be a normal, projective $k$-variety and $Z, W \subset X$ closed subsets. Two irreducible divisors $H_Z \sim_s H_W$ are (topologically, directly) *linked on* $Z \cup W$ if there is a third irreducible divisor $H \sim_s H_Z \sim_s H_W$ such that

$$H \cap Z = H_Z \cap Z \text{ and } H \cap W = H_W \cap W, \text{ as sets.}$$

This makes it clear that linkage depends only on $|X|$ and $\sim_s$, but it is technically simpler to work with the following equivalent line bundle version.

Let $\mathscr{L}$ be an ample line bundle on $X$. Then $H_Z, H_W \in |\mathbb{Q}\mathscr{L}|^{\mathrm{irr}}$ are (topologically, directly) $\mathscr{L}$-*linked on* $Z \cup W$ if the following equivalent condition holds.

(1) There is an $m > 0$ and a section $s \in H^0(X, \mathscr{L}^m)$ such that $(s = 0) \cap Z = H_Z \cap Z$ and $(s = 0) \cap W = H_W \cap W$ (as sets).

If $k$ is infinite, then, by Bertini's theorem this is equivalent to:

(2) There is an $H_X \in |\mathbb{Q}\mathscr{L}|^{\mathrm{irr}}$ such that $H_X \cap Z = H_Z \cap Z$ and $H_X \cap W = H_W \cap W$ (as sets).

Thus, $\mathcal{L}$-linking depends only on $(|X|, \sim_{\mathrm{sa}})$.

As we see below, this notion is not interesting if $Z \cap W = \emptyset$ and it has various problems if $\dim(Z \cap W) \geq 1$. Thus we focus on the case when $\dim(Z \cap W) = 0$.

A key observation is that linkage carries very significant information about

- residue fields of $Z \cap W$ in every characteristic, and
- the scheme structure of $Z \cap W$ in characteristic 0.

**Proposition 9.6.2.** *Let $X$ be a normal, projective $k$-variety and $Z, W \subset X$ closed subsets such that $\dim(Z \cap W) = 0$. Let $\mathcal{L}$ be an ample line bundle on $X$ and $H_Z, H_W \in |\mathbb{Q}\mathcal{L}|^{\mathrm{irr}}$. Then $H_Z, H_W$ are $\mathcal{L}$-linked on $Z \cup W$ if and only if (using the notation of (9.2.9))*

$$\mathcal{R}^Z_{Z \cap W}(H_Z, \mathcal{L}) \cap \mathcal{R}^W_{Z \cap W}(H_W, \mathcal{L}) \neq \emptyset.$$

*Proof.* Assume that $H = Z(s)$ gives the $\mathcal{L}$-linkage for some $s \in H^0(X, \mathcal{L}^m)$. Then $s|_Z \in \Gamma^{Z \cap H_Z}(Z, \mathcal{L}|_Z, m)$ and $s|_W \in \Gamma^{W \cap H_W}(W, \mathcal{L}|_W, m)$ have the same restriction to $Z \cap W$.

Conversely, if $s_Z \in H^0(Z, \mathcal{L}^m|_Z)$ and $s_W \in H^0(W, \mathcal{L}^m|_W)$ have the same image in $H^0(Z \cap W, \mathcal{L}^m|_{Z \cap W})$, then they glue to a section $s_{Z \cup W} \in H^0(Z \cup W, \mathcal{L}^m|_{Z \cup W})$, and then $s^{m'}_{Z \cup W}$ lifts to a section of $H^0(X, \mathcal{L}^{m'm})$ for some $m' > 0$. □

**9.6.3.** The conditions in (9.6.2) give the strongest restriction if

$$\mathcal{R}^Z_{Z \cap W}(H_Z, \mathcal{L})/k^\times \quad \text{and} \quad \mathcal{R}^W_{Z \cap W}(H_W, \mathcal{L})/k^\times \tag{9.6.3.1}$$

both have $\mathbb{Q}$-rank 1. However, in general these objects are essentially extensions of finite rank monoids by $k[Z]^\times$ (resp. $k[W]^\times$).

We get further interesting consequences if we relax these restrictions. The general situation seems rather complicated. In our applications it is advantageous to work with a non-symmetric situation:

(2) $H^0(Z, \mathcal{O}_Z) = k$, and

(3) $\mathcal{R}^W_{Z \cap W}(H_W, \mathcal{L})/k[W]^\times$ has $\mathbb{Q}$-rank 1.

Note that (2) holds if $Z$ is geometrically connected and reduced. In applications we achieve this by choosing $Z$ to be ample-ci (11.2.1).

We see in (9.2.10 (1)) that (3) holds if $\dim W \geq \mathrm{BH}(k)$ (with some additional mild genericity conditions).

Next we study the case when linking is always possible.

**Definition 9.6.4** (Free linking). Let $X$ be a normal, projective $k$-variety and $\mathcal{L}$ an ample line bundle on $X$. Let $Z, W \subset X$ be closed, integral subvarieties such that $\dim(Z \cap W) = 0$.

We say that $\mathscr{L}$-*linking is free* on $Z \cup W$ if there is a finite subset $\Sigma \subset X$ such that two divisors $H_Z, H_W \in |\mathbb{Q}\mathscr{L}|^{\mathrm{irr}}$ are $\mathscr{L}$-linked on $Z \cup W$ whenever they are disjoint from $\Sigma$. (In practice, any $\Sigma \supset \Sigma(Z \cup W)$ will work; cf. (9.2.2).)

As in (9.6.1), free $\mathscr{L}$-linking depends only on $(|X|, \sim_{\mathrm{sa}})$.

In the rest of the section we discuss various cases when the topological notion of free linking makes it possible to obtain information about the residue fields of closed points.

**9.6.5.** Let $X$ be a normal, projective $k$-variety and $\mathscr{L}$ an ample line bundle on $X$. Let $Z, W \subset X$ be closed, irreducible, positive dimensional subvarieties such that $\dim(Z \cap W) = 0$. We are interested in the following conditions:

(1) $\mathscr{L}$-linking is free on $Z \cup W$.

(2) $k[Z \cap W]^{\times}/k[Z]^{\times} \cdot k[W]^{\times}$ is a torsion group.

(3) One of the following holds.

   (a)  char $k = 0$, $Z \cap W$ is reduced, and either $k[Z \cap W] = k[Z]$ or $k[Z \cap W] = k[W]$.

   (b)  char $k > 0$, and either $k[\mathrm{red}(Z \cap W)]/k[Z]$ or $k[\mathrm{red}(Z \cap W)]/k[W]$ is purely inseparable.

   (c)  $k$ is locally finite.

**Remark 9.6.6.** Note that $(9.6.5\,(2))$ implies $(9.6.5\,(1))$ by $(9.6.2)$ and the equivalence of $(9.6.5\,(2))$ and $(9.6.5\,(3))$ is proved in $(11.3.32)$. We will prove, under additional assumptions, that these three conditions are equivalent – we do not know if they are equivalent in general.

Below we will show that $(9.6.5\,(1))$ implies $(9.6.5\,(3))$ if $W$ is geometrically connected and $\dim W \geq \mathrm{BH}(k)$. A careful study of the proof shows that the first assumption is not necessary, and the validity of $(10.6.7)$ would imply that $(9.6.5\,(1))$ always implies $(9.6.5\,(2))$.

**Proposition 9.6.7.** *Let $X$ be a normal, projective, geometrically irreducible $k$-variety and $\mathscr{L}$ an ample line bundle on $X$. Let $Z, W \subset X$ be closed, irreducible, positive dimensional subvarieties such that $\dim(Z \cap W) = 0$. Assume that $W$ is geometrically connected and $\dim W \geq \mathrm{BH}(k)$.*

   *Then conditions (1)-(3) in (9.6.5) are equivalent.*

*Proof.* By $(9.6.6)$ we need to show that $(9.6.5\,(1))$ implies $(9.6.5\,(3))$. Assuming $(9.6.5\,(1))$, choose $s_Z \in H^0(X, \mathscr{L}^m)$ with associated divisor $H_Z$ such that $\mathrm{Supp}(Z(s_Z)) = \mathrm{Supp}\, H_Z$ is disjoint from $\Sigma(Z \cup W)$.

By $(9.5.2)$ and $(9.5.4)$, for every $s_{Z \cap W} \in k[Z \cap W]^{\times}$, there is an $n > 0$ such that $s_{Z \cap W}^n$ extends to $s_W \in H^0(X, \mathscr{L}^n)$ and $W \cap H_W$ is irreducible, where $H_W := \mathrm{Supp}(Z(s_W))$.

If $H_Z, H_W$ are $\mathscr{L}$-linked, then there is an $s \in H^0(X, \mathscr{L}^m)$ as in (9.6.1). (We can use the same $m$, if we pass to suitable powers of $s, s_Z, s_W$.)

By (9.2.7), there are $u_Z \in k[Z]^\times$, $u_W \in k[W]^\times = k^\times$, a finitely generated subgroup $\Gamma_Z \subset k(Z)^\times$, $\gamma_Z \in \Gamma_Z$ and a natural number $r$ such that

$$s_W^r = s^r|_W \cdot u_W \text{ and } s_Z^r = s^r|_Z \cdot u_Z \cdot \gamma_Z. \tag{9.6.7.1}$$

Therefore

$$s_W^r|_{Z \cap W} \cdot s_Z^{-r}|_{Z \cap W} = u_W|_{Z \cap W} \cdot u_Z^{-1}|_{Z \cap W} \cdot \gamma_Z^{-1}|_{Z \cap W} \tag{9.6.7.2}$$

$$\in k^\times \cdot k[Z]^\times \cdot \Gamma_Z|_{Z \cap W}. = k[Z]^\times \cdot \Gamma_Z|_{Z \cap W}. \tag{9.6.7.3}$$

Next note that $s_W|_{Z \cap W} = s_{Z \cap W}^n$ where $s_{Z \cap W}$ is arbitrary. Thus $s_{Z \cap W}^n s_Z^{-1}|_{Z \cap W}$ is an arbitrary element of $k[Z \cap W]^\times$ (up to $n$-torsion). Therefore we get that

$$k[Z \cap W]^\times / \left( k[Z]^\times \cdot \Gamma_Z|_{Z \cap W} \right) \tag{9.6.7.4}$$

is torsion. Thus we obtain that

$$k[Z \cap W]^\times / k[Z]^\times \tag{9.6.7.5}$$

has finite $\mathbb{Q}$-rank. By (11.3.30) we are in one of four cases.

(1) $k$ is locally finite, giving (9.6.5 (3) (c)).

(2) char $k > 0$ and $k[Z] \hookrightarrow k[\text{red}(Z \cap W)]$ is a purely inseparable extension, giving (9.6.5 (3) (b)).

(3) char $k = 0$ and $k[Z] \cong k[Z \cap W]$; giving (9.6.5 (3) (c)).

(4) $\deg(k/\mathbb{Q}) < \infty$.

In the latter case $k$ is Hilbertian. Once $k$ is Hilbertian, at the beginning of the proof we can choose $Z \cap H_Z$ to be irreducible, in which case $\Gamma_Z = \{1\}$ by (9.2.7 (1)). Thus in this case we need to show that

$$k[Z \cap W]^\times / k[Z]^\times$$

is torsion and (11.3.30) implies that $Z \cap W$ is reduced.                                    $\square$

Using (9.6.7) we get a topological way of recognizing $k$-points.

**Corollary 9.6.8.** *Let $k$ be a perfect field that is not locally finite, and $X$ a normal, projective, geometrically irreducible $k$-variety of dimension $> 1 + \text{BH}(k)$. Let $\mathscr{L}$ be an ample line bundle and $p \in X$ a closed point. Assume that either char $k > 0$ or $p$ is a smooth point of $X$. The following are equivalent.*

(1) *$p$ is a $k$-point.*

(2) *There are ample-isci (11.2.1) subvarieties $Z, W$ such that*

    (a)  $\dim Z = 1$, $\dim W = \mathrm{BH}(k)$,

    (b)  $\mathrm{Supp}(Z \cap W) = \{p\}$ *and*

    (c)  $\mathcal{L}$*-linking is free on* $Z \cup W$.

*Proof.* Statement (2) implies (1) by (9.6.7). Conversely, we can take $Z, W$ to be general complete intersections of ample divisors containing $p$. ☐

**Remark 9.6.9.** If char $k = 0$ and (9.6.8 (2)) holds then $Z \cap W$ is a $k$-point, even if $X$ is singular there. However, for a singular $k$-point it may not be possible to find $Z, W$ such that $Z \cap W = \{p\}$ (as schemes). Thus the method does not yet provide a topological way of identifying singular $k$-points if char $k = 0$.

**Corollary 9.6.10.** *Let $k$ be a perfect field that is not locally finite, and $X$ a normal, projective, geometrically irreducible $k$-variety of dimension $> 1 + \mathrm{BH}(k)$. Let $\mathcal{L}$ be an ample line bundle and $p, q \in X$ closed points. Assume that either char $k > 0$ or $p$ is a smooth point. The following are equivalent.*

(1) *There is a $k$-embedding $k(p) \hookrightarrow k(q)$.*

(2) *There are irreducible subvarieties $Z, W$ such that*

    (a)  $\dim Z = 1, \dim W = \mathrm{BH}(k)$,

    (b)  $\mathrm{Supp}(Z \cap W) = \{p\}$,

    (c)  $q \in Z$,

    (d)  $W$ *is* $\mathcal{L}$*-isci, and*

    (e)  $\mathcal{L}$*-linking is free on* $Z \cup W$.

*Proof.* If (2) holds then $k(p) \cong k[Z]$ by (9.6.7) and (2) (c) gives an embedding $k[Z] \hookrightarrow k(q)$.

    Conversely, given $k(p) \hookrightarrow k(q)$, the required $Z$ is constructed in (11.2.8) and $W$ is then choosen to be a general complete intersection containing $p$. ☐

    Reversing the role of $p, q$ we the obtain a criterion to decide whether $k(p) \cong k(q)$. Note, however, that we get no information about $\deg(k(p)/k)$. Using (9.6.4) we see that the conditions (2.a–e) depend only on $(|X|, \sim_{\mathrm{sa}})$; thus, we obtain the following.

**Corollary 9.6.11** (Isomorphism of 0-cycles from $|X|$ and $\sim_{\mathrm{sa}}$). *Let $k$ be a perfect field that is not locally finite, and $X$ a normal, projective $k$-variety of dimension $> 1 + \mathrm{BH}(k)$. Let $Z_1, Z_2 \subset X$ be reduced 0-dimensional subschemes. Assume that either char $k > 0$ or $Z_1, Z_2 \subset X^{\mathrm{ns}}$. We can then decide, using only $(|X|, \sim_{\mathrm{sa}})$, whether $Z_1, Z_2$ are isomorphic as $k$-schemes.* ☐

**9.6.12** (Imperfect fields). If $k$ is an imperfect field, we can apply the above results to $k^{\mathrm{ins}}$. This results in the following changes in the statements.

In $(9.6.8\,(1))$ we characterize $k^{\mathrm{ins}}$-points.

In $(9.6.10\,(1))$ we characterize $k$-embeddings $k^{\mathrm{ins}}(p) \hookrightarrow k^{\mathrm{ins}}(q)$.

In $(9.6.11)$ we characterize isomorphisms $Z_1 \times_k k^{\mathrm{ins}} \cong Z_2 \times_k k^{\mathrm{ins}}$.

**Remark 9.6.13.** Let $X$ a normal, projective, geometrically irreducible $k$-variety of dimension $> 1 + \mathrm{BH}(k)$. Then char $k > 0$ if and only if the following holds.

There is an integral curve $C \subset X$ and a point $p \in C$, such that $\mathscr{L}$-linking is free on $C \cup W$ for every ample-sci subvariety $W$ of dimension $\mathrm{BH}(k)$ for which $\mathrm{Supp}(C \cap W) = \{p\}$.

Indeed, if char $k = 0$ then for any $p \in C$ we can choose $W$ such that $C \cap W$ is nonreduced, and then $\mathscr{L}$-linking is not free on $C \cup W$ by $(9.6.7)$.

Conversely, we use $(11.2.8)$ to get $p \in C$ such that $k(p)^{\mathrm{ins}} = k[C]^{\mathrm{ins}}$, and then $(9.6.7)$ and $(9.6.12)$ show that $\mathscr{L}$-linking is free on $C \cup W$ if char $k > 0$.


## 9.7   MINIMALLY RESTRICTIVE LINKING AND TRANSVERSALITY

**Definition 9.7.1.** Let $X$ be a normal, projective $k$-variety, let $\mathscr{L}$ be an ample invertible sheaf on $X$, and let $Z, W_1, W_2 \subset X$ be closed, irreducible, geometrically connected subvarieties such that $\dim(Z \cap W_i) = 0$. We say that $\mathscr{L}$-linking on $Z \cup W_2$ determines $\mathscr{L}$-linking on $Z \cup W_1$ if there is a finite subset $\Sigma \subset X$ such that the following holds.

Let $H_Z, H_W \in |\mathbb{Q}\mathscr{L}|^{\mathrm{irr}}$ be divisors disjoint from the $\Sigma$ such that $W_2 \cap H_W$ is irreducible. Then

$$\left( \begin{array}{c} H_Z, H_W \text{ are} \\ \text{linked on } Z \cup W_2 \end{array} \right) \Rightarrow \left( \begin{array}{c} H_Z, H_W \text{ are} \\ \text{linked on } Z \cup W_1 \end{array} \right).$$

In applying this notion we always assume that $\dim W_i \geq \mathrm{BH}(k)$, hence the above conditions are not empty.

We say that $\mathscr{L}$-linking is minimally restrictive on $Z \cup W_1$ if $\mathscr{L}$-linking on $Z \cup W_2$ determines $\mathscr{L}$-linking on $Z \cup W_1$, whenever $\mathrm{Supp}(Z \cap W_1) = \mathrm{Supp}(Z \cap W_2)$.

The key result—and rationale for the definition—is the following.

**Proposition 9.7.2.** *Let $k$ be a field of characteristic $0$, $X$ a normal, projective, geometrically irreducible $k$-variety, and $\mathscr{L}$ an ample line bundle on $X$. Let $Z, W_1, W_2 \subset X$ be closed, integral, geometrically connected subvarieties such that $\dim Z \geq 1$, $\dim W_i \geq \mathrm{BH}(k)$, and $\dim(Z \cap W_i) = 0$. Then the following are equivalent.*

(1) $Z \cap W_1 \subset Z \cap W_2$ *as schemes.*

(2) $\mathscr{L}$-linking on $Z \cup W_2$ determines $\mathscr{L}$-linking on $Z \cup W_1$.

*Proof.* Pick $H_Z = Z(s_Z)$ and $H_W = Z(s_W)$. By assumption $W_2 \cap H_W$ is irreducible and disjoint from $\Sigma(Z \cup W_2)$. Thus, by (9.2.10 (1)),

$$\mathscr{R}^{W_2}_{Z \cap W_2}(H_{W_2}, \mathscr{L}) = \langle s_W|_{Z \cap W_2}, k[W_2]^\times \rangle_{\mathbb{Q}} = \langle s_W|_{Z \cap W_2} \rangle_{\mathbb{Q}} \cdot k^\times,$$

where the last equality holds since $W_2$ is geometrically connected. So, by (9.6.?),
$H_Z, H_W$ are linked on $Z \cup W_2$ if and only if, for some $r > 0$,

$$s_W^r|_{Z \cap W_2} \in \mathscr{R}^Z_{Z \cap W_2}(H_Z, \mathscr{L}) \subset \coprod_m H^0(Z \cap W_2, \mathscr{L}^m|_{Z \cap W_2}). \qquad (9.7.2.1)$$

Now observe that if (1) holds then for $m$ sufficiently big the restriction map

$$H^0(Z \cap W_2, \mathscr{L}^m|_{Z \cap W_2}) \to H^0(Z \cap W_1, \mathscr{L}^m|_{Z \cap W_1})$$

is surjective. Thus if (1) holds then (9.7.2.1) implies that

$$s_W^r|_{Z \cap W_1} \in \mathscr{R}^Z_{Z \cap W_1}(H_Z, \mathscr{L}) \subset \coprod_m H^0(Z \cap W_1, \mathscr{L}^m|_{Z \cap W_1})), \qquad (9.7.2.2)$$

proving (2).

To see the converse, let $N$ be the kernel of

$$\rho \colon H^0(Z \cap W_1, \mathcal{O}_{Z \cap W_1})^\times \to H^0(Z \cap W_1 \cap W_2, \mathcal{O}_{Z \cap W_1 \cap W_2})^\times.$$

It is a direct sum of a commutative, unipotent group over $k$ and of the $k(p_i)^\times$
for every $p_i \in W_1 \setminus W_2$. $N$ is positive dimensional if and only if $Z \cap W_1 \not\subset Z \cap W_2$.
We distinguish between two cases, depending on whether $\mathrm{rank}_{\mathbb{Q}} N(k) = \infty$ or
not.

Next choose any $\sigma_2 \in H^0(Z \cap W_2, \mathscr{L}|_{Z \cap W_2})^\times$. If $\mathrm{rank}_{\mathbb{Q}} N(k) = \infty$ then the
restriction of $\sigma_2$ to $Z \cap W_1 \cap W_2$ can be lifted in two different ways to

$$\sigma_Z, \sigma_W \in H^0(Z \cap W_1, \mathscr{L}|_{Z \cap W_1})^\times$$

such that $\sigma_Z \sigma_W^{-1}$ is non-torsion in

$$\coprod_m H^0(Z \cap W_1, \mathscr{L}^m|_{Z \cap W_1})^\times / \mathscr{R}^Z_{Z \cap W_1}(H_Z, \mathscr{L}).$$

We can now glue $\sigma_2, \sigma_W$ to a section of $H^0(Z \cap (W_1 \cup W_2), \mathscr{L}|_{Z \cap (W_1 \cup W_2)})^\times$
and then lift (some power of) it to $s_W \in H^0(X, \mathscr{L})$ such that both $W_i \cap Z(s_W)$
are irreducible and disjoint from $\Sigma(Z \cup W_1 \cup W_2)$. Similarly, we can glue $\sigma_2, \sigma_Z$
to a section of $H^0(Z \cap (W_1 \cup W_2), \mathscr{L}|_{Z \cap (W_1 \cup W_2)})^\times$ and then lift (some power
of) it to $s_Z \in H^0(X, \mathscr{L})$ such that $Z \cap Z(s_Z)$ is irreducible and disjoint from
$\Sigma(Z \cup W_1 \cup W_2)$.

By construction, $s_Z|_{Z \cap W_2} = s_W|_{Z \cap W_2}$, hence $Z(s_Z)$ and $Z(s_W)$ are $\mathscr{L}$-linked on $Z \cup W_2$, but $s_Z|_{Z \cap W_1}$ and $s_W|_{Z \cap W_1}$ are multiplicatively independent, hence $Z(s_Z)$ and $Z(s_W)$ are not $\mathscr{L}$-linked on $Z \cup W_2$.

We are left with the case when $\mathrm{rank}_{\mathbb{Q}} N(k) < \infty$. In this case $\deg(k/\mathbb{Q}) < \infty$ by (11.3.23), hence $k$ is Hilbertian (8.6.6). We can thus choose $s_Z$ such that $Z \cap H_Z$ is irreducible and disjoint from $\Sigma(Z \cup W_2)$. So, by (9.2.10 (1)),

$$\mathscr{R}^Z_{Z \cap W_2}(H_Z, \mathscr{L}) = \langle s_Z|_{Z \cap W_2} \rangle_{\mathbb{Q}} \cdot k[Z]^\times = \langle s_Z|_{Z \cap W_2} \rangle_{\mathbb{Q}} \cdot k^\times.$$

This implies that $\mathscr{R}^Z_{Z \cap W_2}(H_Z, \mathscr{L})$ has trivial intersection with $N$. We can thus again choose $\sigma_Z, \sigma_W \in H^0(Z \cap W_1, \mathscr{L}|_{Z \cap W_1})^\times$ such that $\sigma_Z \sigma_W^{-1}$ is non-torsion, and complete the proof as before.                                                       □

**Corollary 9.7.3.** *Let $k$ be a field of characteristic 0, $X$ a normal, projective $k$-variety, and $\mathscr{L}$ an ample line bundle. Let $Z, W \subset X$ be closed, integral, geometrically connected subvarieties such that $\dim(Z \cap W) = 0$. Assume that $\dim X > \dim Z + \mathrm{BH}(k)$.*

*Then $\mathscr{L}$-linking is minimally restrictive on $Z \cup W$ if and only if $Z \cap W$ is reduced.*

*Proof.* Set $W_1 := W$ in (9.7.2) and let $W_2$ run through all ample-sci subvarieties of dimension $\mathrm{BH}(k)$ (11.2.1) that intersect $Z$ exactly along $Z \cap W_1$. Then apply the following (9.7.4).                                                                            □

**Lemma 9.7.4.** *Let $X$ be a projective $k$-variety. Fix $1 \leq r < \dim X$; let $Z \subset X$ be a subscheme of codimension $> r$ and $P \subset Z$ a reduced, finite subscheme. Let $\mathscr{W}(Z, P)$ be the set of all irreducible, $r$-dimensional, ample-sci (11.2.1) subvarieties $W \subset X$ for which $\mathrm{Supp}(Z \cap W) = P$. Then*

$$\bigcap_{W \in \mathscr{W}(Z, P)} W = P \quad (\text{scheme-theoretically}).$$

*Proof.* By definition we have $P \subseteq W$ for all $W \in \mathscr{W}(Z, P)$, so $P \subseteq \bigcap_{W \in \mathscr{W}(Z, P)}$. Thus it suffices to show the reverse inclusion.

If $\bigcap_{W \in \mathscr{W}(Z, P)} W$ is not contained in $P$, then it either contains a closed point not in $P$, or it contains a nonzero tangent vector at some point of $P$. So it suffices to show that for each point not in $P$ or nonzero tangent vector at a point in $P$, there exists some $W \in \mathscr{W}(Z, P)$ not containing that point or tangent vector.

We will show that, for $L$ sufficiently ample, and $s_1, \ldots, s_{\dim X - r}$ generic sections of $H^0(L)$ vanishing on $P$, the intersection of the vanishing loci of $s_1, \ldots, s_{\dim X - r}$ will suffice. Because we are looking for a generic solution, we can work freely over an algebraically closed field.

Assume $L$ is ample enough that the restriction map from $H^0(L)$ to $H^0(L \mid P')$ is surjective for each subscheme $P'$ of $X$ that consists of either the union of $P$ and another point $Q$ or $P$ with one of the points replaced by a double point.

Then for each $Q \in X \setminus P$, the condition that $s_i$ vanishes on $Q$ is a codimension 1 condition, so the condition that $s_1, \ldots, s_{\dim X - r}$ all vanish on $Q$ is a codimension $\dim X - r$ condition. Since $Z \setminus P$ has dimension $< \dim X - r$, the condition that all of $s_1, \ldots, s_r$ vanish on at least one point of $Z \setminus P$ is a codimension $> 0$ condition and thus does not hold generically, and hence the contrary condition that $\mathrm{Supp}(Z \cap W) = P$ holds generically.

Furthermore, for each $Q$ in $X \setminus P$, the condition that $Q \in W$ is a codimension $\dim X - r > 0$ condition and thus does not hold generically.

Finally, for each tangent vector at a point $p \in P$, $W$ contains that tangent vector if and only if it contains the scheme $P'$ obtained by replacing $p$ with the corresponding doubled point. This happens if and only if each of the $s_i$ vanish on $P'$, and these are all codimension 1 conditions, so this does not happen generically. $\qquad\square$

The following consequence of (9.7.3) allows us to understand intersection multiplicities topologically.

**Theorem 9.7.5** (Determining transversality from $|X|$ and $\sim_{\mathrm{sa}}$). *Let $k$ be a field of characteristic 0 and $X$ a normal, projective, geometrically irreducible $k$-variety of dimension $> 1 + \mathrm{BH}(k)$. Let $H \subset X$ be an irreducible, ample divisor and $C \subset X$ an irreducible, geometrically connected curve. Assume that $C \cap H \subset X^{\mathrm{ns}}$. The following are equivalent.*

(1) *All intersections of $C \cap H$ are transversal.*

(2) *$C \cap H$ is reduced.*

(3) *$\mathscr{L}$-linking is minimally restrictive on $C \cup H$ for some ample $\mathscr{L}$.*

*Proof.* The equivalence of (2) and (3) follows from (9.7.3) and the equivalence of (1) and (2) is a basic property of intersection mutiplicities; see, for example, [Ful98, 8.2]. $\qquad\square$

**Remark 9.7.6.** Note that there are several weaknesses of the current form of the above equivalences. First, we do not yet know how to decide which are the smooth points of $X$. We usually go around this by saying that some assertion holds outside some codimension $\geq 2$ subset. Second, we also do not yet know how to decide whether a curve $C$ is geometrically connected or not. However, if $C$ is ample-sci (11.2.1), then $C$ is geometrically connected (11.2.2).

The above arguments also show the following.

**Corollary 9.7.7.** *Let $k$ be a field of characteristic 0, $X$ a $k$-variety, $Z \subset X$ an irreducible, geometrically connected subvariety of codimension $r > \mathrm{BH}(k)$ and $p \in Z$ a closed point such that $X$ is smooth at $p$. Then $Z$ is smooth at $p$ if and only if there is an irreducible, ample-sci subvariety $W \subset X$ of dimension $r$ such that $p \in \mathrm{Supp}(Z \cap W)$ and $\mathscr{L}$-linking is minimally restrictive on $Z \cup W$.* $\qquad\square$

Interchanging the roles of $Z, W$ gives the following dual version.

**Corollary 9.7.8.** *Let $k$ be a field of characteristic 0, $X$ a $k$-variety, $W \subset X$ an irreducible, geometrically connected subvariety of dimension $r > \mathrm{BH}(k)$, and $p \in W$ a closed point such that $X$ is smooth at $p$. Then $W$ is smooth at $p$ if and only if there is an irreducible, ample, complete intersection subvariety $Z \subset X$ of codimension $r$ such that $p \in \mathrm{Supp}(Z \cap W)$ and $H$-linking is minimally restrictive on $Z \cup W$.* $\qquad\square$

**9.7.9.** The argument in (9.7.2) also applies in positive characteristic, except that then the kernel of

$$H^0\big(Z \cap W, \mathcal{O}_{Z \cap W}\big)^\times \to H^0\big(\mathrm{red}(Z \cap W), \mathcal{O}_{\mathrm{red}(Z \cap W)}\big)^\times$$

is $p$-power torsion. Thus multiplicative independence is not changed as we pass from $Z \cap W$ to $\mathrm{red}(Z \cap W)$. We get that, if $k$ is not locally finite, then the following are equivalent.

(1) $\mathrm{Supp}(Z \cap W_1) \subset \mathrm{Supp}(Z \cap W_2)$.

(2) $\mathscr{L}$-linking on $Z \cup W_1$ determines $\mathscr{L}$-linking on $Z \cup W_2$.

Thus, while $\mathscr{L}$-linking carries scheme-theoretic information in characteristic 0, it detects only the underlying reduced subscheme positive characteristic.

## 9.8   RECOVERING LINEAR EQUIVALENCE

**9.8.1.** In this section we discuss how to compute intersection numbers $(C \cdot D)$ of curves and divisors on a proper $k$-variety $X$. General theory tells us that we should write the scheme-theoretic intersection $C \cap D$ as the union of 0-dimensional subschemes $\{Z_i : i \in I\}$ and then

$$(C \cdot D) = \sum_{i \in I} \mathrm{length}_k \, \mathcal{O}_{Z_i}.$$

From the topology we see right away the points $p_i := \mathrm{red}\, Z_i$, but neither the nilpotent structure of $Z_i$ nor $\deg[k(p_i) : k]$ is visible to us.

We can use (9.7.5) to check that all intersections are transversal. To be precise, this works only if $X$ is smooth along $C \cap D$, and the latter is a problematic condition to check.

If the field $k$ is algebraically closed, then $\deg[k(p_i) : k] = 1$, and we are done. However, if the field is not algebraically closed, we would need to compute $\deg[k(p_i) : k]$, which we cannot do.

Nonetheless, by (9.6.11), we can determine when $\deg[k(p_1) : k] = \deg[k(p_2) : k]$ for two points. (This again with the caveat that $X$ should be smooth along the $p_i$.)

The end result says that, although we are not able to compute $(C \cdot D)$ itself, we can decide whether $(C_1 \cdot D) = (C_2 \cdot D)$ for two curves or $(C \cdot D_1) = (C \cdot D_2)$

for two divisors. This turns out to be sufficient for our applications.

The rest of the section is pure algebraic geometry, technically independent of previous results. However, the somewhat unusual assumptions and restrictions are dictated by the needs of (9.7.5) and (9.6.11).

**9.8.2.** Let $\mathscr{L}$ be a very ample line bundle on a reduced, projective curve over an algebraically closed field. The zero set of a general section of $\mathscr{L}$ consists of $\deg_C \mathscr{L}$ distinct points. However, if we work over a non-closed field $k$, then the zero set of a general section $s \in H^0(C, \mathscr{L})$ is a union of points of the form $\operatorname{Spec} k_i$ for some field extensions $k_i/k$, which depend on the choice of the section in a rather unpredictable way. We may thus aim to find sections $s \in H^0(C, \mathscr{L})$ whose zero set is arithmetically simple. If $k$ is Hilbertian, we can choose the zero set to be irreducible. Another direction would be to find zero sets that consist of low degree points. This is, however, impossible already for genus 1 curves over $\mathbb{Q}$.

Next we show that an intermediate result is possible. For any finite set of curves $C_i$ and line bundles $\mathscr{L}_i$, one can find sections whose zero sets consist entirely of points with a fixed (separable) residue field, at the expense of a lack of control over the particular field or its degree.

**Theorem 9.8.3.** *Let $C$ be a geometrically reduced, projective curve over a field $k$ with irreducible components $\{C_i : i \in I\}$. Let $\mathscr{L}$ be an ample line bundle on $C$ and $\Sigma \subset C$ a finite set. Then for infinitely many $m > 0$, there is a separable field extension $K/k$ and a section $s \in H^0(C, \mathscr{L}^m)$ such that*

(1) $(s = 0)$ *is disjoint from $\Sigma \cup \operatorname{Sing} C$, and*

(2) $C_i \cap (s = 0)$ *is isomorphic to the disjoint union of $(m/\deg(K/k)) \cdot \deg_{C_i} \mathscr{L}$ copies of $\operatorname{Spec} K$ for every $i$.*

*Proof.* For $m_1$ large enough there is a separable morphism $\pi \colon C \to \mathbb{P}^1$ such that $\mathscr{L}^{m_1} \cong \pi^* \mathcal{O}_{\mathbb{P}^1}(1)$. By (7.2.1) there is a separable point $p \in \mathbb{P}^1$ that is disjoint from $\pi(\Sigma \cup \operatorname{Sing} C)$ and such that $\pi^{-1}(p)$ is a reduced, disjoint union of copies of $p$. Let $s' \in H^0(\mathbb{P}^1, \mathcal{O}_{\mathbb{P}^1}(m_2))$ be a defining equation of $p$. Then $s := \pi^* s' \in H^0(C, \mathscr{L}^{m_1 m_2})$ has the required properties. $\square$

**Corollary 9.8.4.** *Let $X$ be a projective variety over a field $k$, $\mathscr{L}$ an ample line bundle, $\{C_i : i \in I\}$ a finite set of geometrically reduced curves and $\Sigma \subset X$ a finite subset. Then there is an $m > 0$, a section $s \in H^0(X, \mathscr{L}^m)$, and a separable field extension $K/k$ such that*

(1) $(s = 0)$ *is disjoint from $\Sigma \cup \operatorname{Sing}(\cup_i C_i)$, and*

(2) $C_i \cap (s = 0)$ *is isomorphic to the disjoint union of $(m/\deg(K/k)) \cdot \deg_{C_i} \mathscr{L}$ copies of $\operatorname{Spec}_k K$ for every $i$.* $\square$

**Theorem 9.8.5.** *Let $X$ be a projective variety over an infinite field, $H$ an ample, Cartier divisor, and $\{C_i : i \in I\}$ finitely many irreducible, geometrically* *reduced curves. The following are equivalent.*

(1) $(C_i \cdot H)$ *is independent of* $i$.

(2) *There is an irreducible divisor* $G \sim_{\mathrm{sa}} H$ *such that the (scheme-theoretic) intersections* $\{C_i \cap G : i \in I\}$ *are reduced and isomorphic to each other.*

*Moreover, we can choose* $G$ *such that*

(3) $G \in |mH|$ *for some* $m \gg 1$,

(4) $G$ *is disjoint from any given finite subset* $\Sigma \subset X$, *and*

(5) *the* $G \cap C_i$ *are disjoint from any given closed subset* $B \subsetneq X$ *that does not contain any of the* $C_i$.

*Proof.* Assume that (2) holds and $G \sim mH$. Then $m(C_i \cdot H) = \deg_k(C_i \cap G)$, proving (1). To see the converse, let $C$ denote the union of the $C_i$. Set $\mathscr{L} := \mathcal{O}_X(H)|_C$. Choose $m \gg 1$ such that $H^0(X, \mathcal{O}_X(mH)) \to H^0(C, \mathcal{O}_C(mH|_C))$ is surjective and there is a section $s \in H^0(C, \mathcal{O}_C(mH|_C))$ as in (9.8.3). Then $G := (s = 0)$ works.                                                                         □

**Corollary 9.8.6.** *Let* $X$ *be a normal, projective variety over an infinite field,* $H$ *an irreducible, ample, Cartier divisor,* $\{D_i : i \in I\}$ *finitely many irreducible, geometrically reduced divisors, and* $B \subset X$ *a closed subset of codimension* $\geq 2$ *containing* $\mathrm{Sing}\, X$. *The following are equivalent.*

(1) $(D_i \cdot H^{n-1})$ *is independent of* $i$.

(2) *There are irreducible,* $H$-*sci curves* $A$ *(11.2.1) that are disjoint from* $B$ *and such that the (scheme-theoretic) intersections* $A \cap D_i$ *are reduced and isomorphic to each other.*

*Proof.* As before, (2) $\Rightarrow$ (1) is clear. For the converse, we look for $A$ contained in a general complete intersection surface $S \subset X$. This reduces us to the special case when $\dim X = 2$. Then the $D_i$ are curves, so (9.8.6) follows from (9.8.5).  □

The main technical result is somewhat hard to state. To make it clearer, we list a series of six questions that one could ask about a variety $X$. The first four we already know how to answer using $|X|$ only. Then we show that we can also answer the remaining two.

**Queries 9.8.7.** Let $X$ be a normal, projective variety.

(1) Given an irreducible divisor $H \subset X$, is it $\mathbb{Q}$-Cartier and ample?

(2) Given two irreducible, ample divisors $H_1, H_2 \subset X$, are they linearly similar?

(3) Given an irreducible, geometrically connected curve $C \subset X$, and an irreducible, geometrically connected divisor $D \subset X$, is $C \cap D$ reduced or

$(C \cap D) \cap \operatorname{Sing} X \neq \emptyset$?

(4) Given two 0-dimensional, closed, reduced subschemes $Z_1, Z_2 \subset X$, is $Z_1 \cong Z_2$ or $(Z_1 \cup Z_2) \cap \operatorname{Sing} X \neq \emptyset$?

(5) Given two irreducible, ample divisors $H_1, H_2 \subset X$, are they numerically equivalent?

(6) Given two irreducible, ample divisors $H_1, H_2 \subset X$, are they $\mathbb{Q}$-linearly equivalent?

**Clarification 9.8.8.** In (3) and (4) we do not assume to know which part of the answer applies. That is, we get the correct answer if $X$ is nonsingular at the points $C \cap D$ (resp. $Z_1 \cup Z_2$) but we do not know whether the answer is correct or not otherwise. We also do not assume that we can decide which points are nonsingular.

In the applications we avoid this problem by working in the complement of an arbitrary codimension $\geq 2$ subset; see (9.8.12).

Note that if $X$ is a normal, projective variety over a field of characteristic 0 and $\dim X > 1 + \mathrm{BH}(k)$, then we know how to answer Queries (1)–(4) above by (9.4.15), (9.4.17), (9.7.5), and (9.6.11). The assumptions in these results dictated the Queries (9.8.7).

**Proposition 9.8.9.** *Let $X$ be a normal, projective variety of dimension $\geq 2$ over a field of characteristic 0. Assume that we know how to answer (9.8.7 (1))–(9.8.7 (4)). Then we can also answer (9.8.7 (5)) and (9.8.7 (6)).*

*Proof.* We need to string together the five claims below. In each of them the first part is the information we seek, the second shows how it can be obtained using (9.8.5) and (9.8.6), and the preceding Claims in the sequence.  $\square$

**Claim 9.8.10.** *Let $C_1, C_2$ be irreducible curves, not contained in $\operatorname{Sing} X$, and $H$ an ample divisor. Then*

(1) $(C_1 \cdot H) = (C_2 \cdot H)$ *if and only if*

(2) *for every finite subset $\Sigma \subset C_1 \cup C_2$ there is an irreducible divisor $G \sim_{\mathrm{s}} H$ disjoint from $\Sigma$, such that $C_i \cap G$ are reduced and $C_1 \cap G \cong C_2 \cap G$ (as $k$-schemes).*

*Proof.* This just restates (9.8.5). Here we use the answer to (9.8.7 (3)).  $\square$

**Claim 9.8.11.** *Let $C_1, C_2$ be irreducible curves, not contained in $\operatorname{Sing} X$. Then*

(1) $C_1 \equiv C_2$ *if and only if*

(2) $(C_1 \cdot H) = (C_2 \cdot H)$ *for every ample divisor $H$.*

*Proof.* This follows from (7.4.3) and (9.8.10).  $\square$

**Claim 9.8.12.** *Let $D_1, D_2$ be irreducible, geometrically connected divisors on $X$ and $H$ an ample divisor. Then*

(1) $(D_1 \cdot H^{n-1}) = (D_2 \cdot H^{n-1})$ *if and only if*

(2) *for every codimension $\geq 2$ subset $B \subset X$ there are irreducible, $H$-sci curves $A$ that are disjoint from $B$ and such that the $A \cap D_i$ are reduced, and $A \cap D_1 \cong A \cap D_2$ (as $k$-schemes).*

*Proof.* This uses (9.8.5) and the answers to (9.8.7 (3)) and (9.8.7 (4)). $\quad\square$

**Claim 9.8.13.** *Let $D_1, D_2$ be irreducible, geometrically connected divisors on $X$. Then*

(1) $D_1 \equiv D_2$ *if and only if*

(2) $(D_1 \cdot H^{n-1}) = (D_2 \cdot H^{n-1})$ *for every ample divisor $H$.*

*Proof.* This follows from (7.4.6) and the answer to (9.8.12). This gives the answer to (9.8.7 (5)). $\quad\square$

**Claim 9.8.14.** *Let $H_1, H_2$ be irreducible, ample divisors. Then*

(1) $H_1 \sim_{\mathbb{Q}} H_2$ *if and only if*

(2) $H_1 \sim_{\mathrm{sa}} H_2$ *and $H_1 \equiv H_2$.*

*Proof.* This needs the answer to (9.8.7 (2)) and (9.8.7 (4)). We get the answer to (9.8.7 (6)). $\quad\square$

The culmination of our work so far is the following.

**Theorem 9.8.15.** *Let $X$ be a normal, projective variety of dimension $> 1 + \mathrm{BH}(k)$ over a field $k$ of characteristic 0. Then we can decide using $|X|$ only when two irreducible, ample divisors are numerically equivalent.*

*Proof.* Under our assumptions, we know how to answer (9.8.7 (1))–(9.8.7 (4)) by (9.4.15), (9.4.17), (9.6.11), and (9.7.5). Thus, (9.8.9) implies the result. $\quad\square$

**Corollary 9.8.16.** *Let $X$ be a normal, projective variety of dimension $> 1 + \mathrm{BH}(k)$ over a field $k$ of characteristic 0. Let $H$ be an irreducible, ample divisor on $X$. Then we can decide using $|X|$ only when a t-pencil is algebraic and linearly similar to $H$.*

*Proof.* Let $\{D_\lambda : \lambda \in \Lambda\}$ be a t-pencil. By (7.3.4), it is algebraic and linearly similar to $H$ if and only if

(1) $D_\lambda$ is $\mathbb{Q}$-Cartier and ample for all but finitely many $\lambda \in \Lambda$.

(2) $D_\lambda \sim_{\mathrm{s}} H$ for all but finitely many $\lambda \in \Lambda$.

(3) There is an infinite subset $\Lambda^* \subset \Lambda$ such the $\{D_\lambda : \lambda \in \Lambda^*\}$ are numerically equivalent to each other.

The first of these we can decide by (9.4.15) and the second by (9.4.17). Once we know these, then all but finitely many of the $D_\lambda$ are ample. So their numerical equivalence is decided by (9.8.15). □

**Corollary 9.8.17.** *Let $X$ be a normal, projective variety of dimension $> 1 +$ $\mathrm{BH}(k)$ over a field $k$ of characteristic 0. Let $H$ be an irreducible, ample divisor on $X$. Then for each $1 \le r \le \dim X$ we can determine the similarity class of $\deg_H(\ )$ on $r$-cycles, using $|X|$.*

*Proof.* Using (9.8.16) we determine the set **HP** of all $\mathbb{Q}$-Cartier, algebraic pencils linearly similar to $H$.

By (7.4.11), **HP** is an ample, complete, and compatible set of algebraic pencils. Thus, by (7.4.15), **HP** determines the similarity class of $\deg_H(\ )$ on $r$-cycles for every $r$. □

We now come to the final result of our work in Chapters 7 to 9.

**Theorem 9.8.18.** *Let $X$ be a normal, projective variety of dimension $> 1 +$ $\mathrm{BH}(k)$ over a field $k$ of characteristic 0. Then linear equivalence of divisors is determined by $|X|$.*

*Proof.* Let $H$ be any irreducible, ample divisor on $X$. We get the similarity class of $\deg_H(\ )$ on divisors using (9.8.17). Once we have an ample degree function on divisors, linear equivalence is obtained by (7.5.8). □

## 9.9    APPENDIX: WEAKLY HILBERTIAN FIELDS

**Definition 9.9.1.** We call a field $k$ *weakly Hilbertian* if, for every irreducible, geometrically reduced, projective curve $C$ over $k$ and every ample line bundle $\mathscr{L}$ on $C$, there is an $n > 0$ and a nonzero section $s \in H^0(C, \mathscr{L}^n)$ such that $(s = 0) \subset C$ is irreducible. We discuss other versions of the definition in (9.9.3–9.9.4).

Having only one section with irreducible zero set is not very useful, but we show in (9.9.3) that there are infinitely many, and they can be chosen with much flexibility.

The Hilbertian field condition (8.6.6) requires that such an $s$ exists in every 2-dimensional, basepoint-free subspace of $H^0(C, \mathscr{L})$. Going from 2-dimensional subspaces to all of $H^0(C, \mathscr{L})$ is a minor change, but allowing powers of $\mathscr{L}$ gives much more flexibility.

A convenient aspect is that if $K/k$ is a finite extension then $K$ is weakly Hilbertian if and only if $k$ is; this does not hold for Hilbertian fields.

Most 'well known' fields are either Hilbertian or not even weakly Hilbertian, but there are many fields that are weakly Hilbertian but not Hilbertian, for example, $\mathbb{Q}^{\mathrm{solv}}$, the maximal, solvable, Galois extension of $\mathbb{Q}$; see (9.9.14).

**Notation 9.9.2.** For a field $k$ let $\mathbf{CL}(k)$ denote the set of all pairs $(C, \mathscr{L})$, where $C$ is a projective, irreducible, geometrically reduced curve over $k$ and $\mathscr{L}$ an ample line bundle on $C$. Such a pair is called nonsingular if and only if $C$ is nonsigular. We use $H^0(C, \mathscr{L})^{\circ} \subset H^0(C, \mathscr{L})$ to denote the nonzero sections.

If $Z \subset C$ is a finite subscheme and $s_Z \in H^0(Z, \mathscr{L}|_Z)$, then let $H^0(C, \mathscr{L}, s_Z) \subset H^0(C, \mathscr{L})$ be as in (9.5.1.1).

**Theorem 9.9.3.** *Let $k$ be a field that is not locally finite. The following are equivalent.*

(1) *For every $(C, \mathscr{L}) \in \mathbf{CL}(k)$ there is an $n > 0$ and $s \in H^0(C, \mathscr{L}^n)^{\circ}$ such that $(s = 0)$ is irreducible.*

(2) *For every $(C, \mathscr{L}) \in \mathbf{CL}(k)$ there are infinitely many $s_i \in H^0(C, \mathscr{L}^{n_i})^{\circ}$ such that the $(s_i = 0)$ are irreducible and distinct.*

(3) *For every nonsingular $(C, \mathscr{L}) \in \mathbf{CL}(k)$, closed subscheme $Z \subset C$ and nowhere zero $s_Z \in H^0(Z, \mathscr{L}|_Z)$, there is an $n > 0$ and $s \in H^0(C, \mathscr{L}^n, s_Z^n)^{\circ}$ such that $(s = 0)$ is irreducible.*

(4) *For every $(C, \mathscr{L}) \in \mathbf{CL}(k)$, closed subscheme $Z \subset C$ and nowhere zero $s_Z \in H^0(Z, \mathscr{L}|_Z)$, there are infinitely many $s_i \in H^0(C, \mathscr{L}^{n_i}, s_Z^n)^{\circ}$ such that the $(s_i = 0)$ are irreducible and distinct.*

*Proof.* (4) $\Rightarrow$ (2) $\Rightarrow$ (1) and (4) $\Rightarrow$ (3) are clear.

In order to show that (2) $\Rightarrow$ (4), we apply (9.9.5) to $Z \subset C$. We get a finite, birational morphism $\pi \colon C \to B$ and a line bundle $\mathscr{L}_B \subset \pi_* \mathscr{L}$ on $B$ with natural isomorphisms

$$H^0(B, \mathscr{L}_B^n) \cong H^0(C, \mathscr{L}^n, s_Z^n).$$

If $t_i \in H^0(B, \mathscr{L}_B^{n_i})^{\circ}$ has irreducible zero set that is different from $\pi(Z)$ then the corresponding $s_i \in H^0(C, \mathscr{L}^{n_i})^{\circ}$ also has irreducible zero set that is not contained in $Z$. Thus (4) holds for $(C, \mathscr{L}, s_Z)$ if and only if (2) holds for $(B, \mathscr{L}_B)$.

Next consider (1) $\Rightarrow$ (2). Assume that only finitely many points $c_i \in C$ occur as irreducible zero sets of sections of powers of $\mathscr{L}$. Since some of these points may be singular, let $p \colon \overline{C} \to C$ denote the normalization and $W \subset \overline{C}$ the union of all preimages of the $c_i$. Let

$$\Gamma \subset \amalg_{n>0} H^0(\overline{C}, p^* \mathscr{L}^n)^{\circ}/k^{\times}$$

be the subsemigroup of all sections whose zero sets are contained in $W$. Then the map from $\Gamma$ to $\mathbf{N}^{|W|}$ that sends a section to its order of vanishing at each point of $W$ is injective, so $\Gamma$ has finite rank.

Assume first that $k$ is not separably closed. Then there are closed points

$c \in C \backslash W$ such that $k(c)/k$ is a separable field extension of degree $> 1$. By (9.9.6) we get a finite homeomorphism $\pi \colon C \to B$ and an invertible sheaf $\mathscr{L}_B \subset \pi_* \mathscr{L}$ such that none of the sections in $\Gamma$ descend to $\mathrm{II}_{n>0} H^0(B, \mathscr{L}_B^n)^\circ$. Since $\pi$ is a homeomorphism, a closed subset of $B$ is irreducible if and only if its preimage in $C$ is irreducible. Thus $\mathscr{L}_B^n$ has no sections with irreducible zero set, proving $(1) \Rightarrow (2)$ in this case.

If $k$ is separably closed and not locally finite, we check in (9.9.8) and (9.9.19) that it does not satisfy (1).

Finally we show that $(3) \Rightarrow (1)$. Let $\pi \colon \overline{C} \to C$ denote the normalization and $\overline{Z} \subset \overline{C}$ the conductor subscheme of $\pi$.

Let $s_0$ be a nowhere zero section of $\mathscr{L}$ in a neighborhood of $\pi(\overline{Z})$. Set $s_{\overline{Z}} := \pi^* s_0|_{\overline{Z}}$. Then we have an inclusion $H^0(\overline{C}, \pi^* \mathscr{L}^n, s_{\overline{Z}}^n) \subset H^0(C, \mathscr{L}^n)$, so any section in $H^0(\overline{C}, \pi^* \mathscr{L}^n, s_{\overline{Z}}^n)$ with an irreducible zero set gives a required section in $H^0(C, \mathscr{L}^n)$. $\qquad\square$

**Remark 9.9.4.** Other versions of the properties (9.9.3 (1))–(9.9.3 (4)) are worth considering. The following variants of (9.9.3 (2)) are especially natural.

(1) For every nonsingular $(C, \mathscr{L}) \in \mathbf{CL}(k)$ there are infinitely many sections $s_i \in H^0(C, \mathscr{L}^{n_i})^\circ$ such that the $(s_i = 0)$ are irreducible and distinct.

(2) For every $(C, \mathscr{L}) \in \mathbf{CL}(k)$ there is an $n > 0$ and infinitely many sections $s_i \in H^0(C, \mathscr{L}^n)^\circ$ such that the $(s_i = 0)$ are irreducible, reduced, and distinct.

It is clear that $(2) \Rightarrow (9.9.3 (2)) \Rightarrow (1)$.

We see in (9.9.16) that the $p$-adic fields $\mathbb{Q}_p$ satisfy (1), but they are not weakly Hilbertian by (9.9.19). We do not know whether (2) is equivalent to (9.9.3 (2)). The examples in (9.9.13) all satisfy (2).

**9.9.5 (Pinching points).** Let $X$ be a $k$-scheme and $Z \subset X$ a closed subscheme that is finite over $k$. The universal pushout of $\operatorname{Spec} k \leftarrow Z \hookrightarrow X$ is a finite, birational morphism $\pi \colon X \to Y$ such that

$$\mathcal{O}_Y = k + \pi_* \mathcal{O}_X(-Z) \subset \pi_* \mathcal{O}_X.$$

The image of $Z$ is a point $z \in Y(k)$ and $X \setminus Z \cong Y \setminus \{z\}$.

Let $\mathscr{L}$ be an invertible sheaf on $X$ and $s_Z \in H^0(Z, \mathscr{L}|_Z)$ a nowhere zero section. Choosing any extension $\tilde{s}_Z$ to some open neighborhood of $Z \subset X$, we get an invertible subsheaf

$$\mathscr{L}(s_Z) := k \cdot \tilde{s}_Z + \pi_* \mathscr{L}(-Z) \subset \pi_* \mathscr{L}.$$

Note that $\mathscr{L}(s_Z)$ is independent of the choice of $\tilde{s}_Z$, and, for every $n > 0$, pushforward gives a natural isomorphism

$$H^0(Y, \mathscr{L}(s_Z)^n) \cong H^0(X, \mathscr{L}^n, s_Z^n).$$

**Lemma 9.9.6.** *Let $X$ be an irreducible $k$-scheme, $\mathscr{L}$ a line bundle on $X$ and $\Gamma \subset \amalg_{n>0} H^0(X, \mathscr{L}^n)^\circ/k^\times$ a (multiplicative) semigroup of finite $\mathbb{Q}$-rank. Let $x \in X$ be a closed point such that $k(x)/k$ is separable of degree $> 1$. Let $\pi\colon X \to Y$ be the pinching of $x$ as in (9.9.5). Then there is an invertible subsheaf $\mathscr{L}_Y \subset \pi_* \mathscr{L}$ such that*

$$\amalg_{n>0} H^0(Y, \mathscr{L}_Y^n)^\circ/k^\times \subset \amalg_{n>0} H^0(X, \mathscr{L}^n)^\circ/k^\times$$

*is disjoint from the saturation of $\Gamma$.*

*Proof.* Fixing some $s_0 \in H^0(\{x\}, \mathscr{L}_x)^\circ$ specifies an isomorphism

$$\amalg_{n\in\mathbb{Z}} H^0(\{x\}, \mathscr{L}_x^n)^\circ \cong \mathbb{Z} \times k(x)^\times.$$

(It is unfortunate notationwise that $\mathbb{Z}$ is additive but $k(x)^\times$ is multiplicative.) The restriction of $\Gamma$ to $\{x\}$ generates a finite rank subgroup; denote it by $\Gamma_x \subset \mathbb{Z} \times k(x)^\times/k^\times$. Its projection to the second factor is $\overline{\Gamma}_x \subset k(x)^\times/k^\times$.

If we choose another $s_x \in H^0(\{x\}, \mathscr{L}_x)^\circ$, then $t_x := s_x/s_0 \in k(x)^\times/k^\times$. By (9.9.7) we can choose $s_x$ such that $\overline{\Gamma}_x \cap (\langle t_x \rangle) = 1$. Thus $s_x^n \cdot k^\times$ is disjoint from $\Gamma \cap H^0(\{x\}, \mathscr{L}_x^n)/k^\times$ for $n > 0$. $\qquad\square$

**Lemma 9.9.7.** *Let $k$ be a field that is not locally finite. Let $T_1$ be a $k$-torus and $T_2 \subsetneq T_1$ a subtorus. Let $\Gamma \subset T_1(k)/T_2(k)$ be a subgroup of finite $\mathbb{Q}$-rank. Then there is a $t \in T_1(k)/T_2(k)$ such that*

$$\Gamma \cap \langle t \rangle = 1.$$

*Proof.* By (11.3.23 (4)), the rank of $T_1(k)/T_2(k)$ is infinite, so we can take $\bar{t} \in T_1/T_2$ such that $\langle \bar{t} \rangle$ intersects $\Gamma$ only at the origin. $\qquad\square$

Next we show that algebraically closed fields of characteristic 0 are not weakly Hilbertian; they do not even satisfy (9.9.4 (1)). Note that a much stronger variant of (9.9.8) could be true; see (10.6.9) and (9.9.10). The same methods work for real closed fields. We do not know any other subfield of $\overline{\mathbb{Q}}$ that does not satisfy (9.9.4 (1)) for smooth curves, though presumably there are many.

**Proposition 9.9.8.** *Let $K$ be an algebraically closed field of characteristic 0. There is a smooth projective curve $C$ and an ample line bundle $\mathscr{L}$, defined over $K$, such that every section of $\mathscr{L}^m$ has at least two distinct zeros for every $m > 0$.*

*Proof.* Let $\pi\colon C \to B$ be a nonconstant morphism between smooth, projective curves such that $g(C) \geq g(B) + 2$ and $g(B) \geq 1$. Assume that there is $b_0 \in B(K)$ such that $\pi^{-1}(b_0)$ is a single point $c_0$. Let $\Gamma \subset \mathrm{Pic}(B)$ be as in (9.9.9).

The $\mathbb{Q}$-rank of $\mathrm{Pic}(B)$ is infinite by (8.6.5 (4)), so there is an ample $\mathscr{L} \in \mathrm{Pic}(B)$ no power of which is in $\Gamma$. Then $\pi^* \mathscr{L}$ has the required property. $\qquad\square$

**Lemma 9.9.9.** *Let $\pi\colon C \to B$ be a nonconstant morphism between smooth, projective curves defined over $K$, such that $g(C) \geq g(B) + 2$. Assume that there* *is $b_0 \in B(K)$ such that $\pi^{-1}(b_0)$ is a single point $c_0$. Then*

$$\Gamma := \langle \mathscr{L} \in \mathrm{Pic}(B) : \pi^* \mathscr{L}^m \cong \mathcal{O}_C(n[c]) \text{ for some } c \in C(K), n, m > 0 \rangle \subset \mathrm{Pic}(B)$$

*has finite $\mathbb{Q}$-rank.*

*Proof.* Embed $C \hookrightarrow \mathbf{Jac}(C)$ sending $c_0$ to the origin. Let $A$ be the quotient $\mathbf{Jac}(C)/\pi^* \mathbf{Jac}(D)$, with the naturally induced map $\sigma : C \to A$. Let $\mathscr{L}$ be a line bundle of degree $d$ on $B$. If $\pi^* \mathscr{L}^m \sim \mathcal{O}_C(n[c])$ then $\sigma(c) \in A$ is an $n$-torsion point since $\pi^* \mathscr{L}(-d[b_0])$ maps to the origin in $A$. Thus there are only finitely many such $c \in C(K)$ by [Zha98], and the $\mathbb{Q}$-rank of $\Gamma$ is at most the number of such torsion points. $\square$

For nodal rational curves, there is an elementary proof; see (9.9.18) for a more advanced version of it.

**Example 9.9.10.** Nodal rational curves show that $\overline{\mathbb{Q}}$ is not weakly Hilbertian. In fact we show that for most line bundles, every section has at least as many zeros as the number of nodes.

A rational curve with $r$ nodes is obtained from $\mathbb{P}^1$ by identifying $r$ point pairs. Thus we start with $2r$ distinct points $a_1, \ldots, a_{2r} \in \mathbf{A}^1$ and identify $a_i$ with $a_{r+i}$ to get a nodal rational curve $C$.

A line bundle on $C$ is obtained by starting with some $\mathscr{L} = \mathcal{O}_{\mathbb{P}^1}(m)$ and specifying isomorphisms $\mathscr{L}|_{a_i} \cong \mathscr{L}|_{a_{r+i}}$. Thus sections of the resulting line bundle are given by polynomials $p(x)$ of degree $\leq m$ such that $p(a_i) = u_i p(a_{r+1})$ for every $i$, where $u_i \in \overline{\mathbb{Q}}^\times$ specify the line bundle.

A polynomial with zeros $\{z_j : j \in J\}$ is $s(x) = \gamma \prod_j (x - z_j)^{m_j}$. Thus for nonzero $u_1, \ldots, u_r$ we aim to solve the $r$ equations

$$\prod_{j \in J} \left( \frac{a_i - z_j}{a_{r+i} - z_j} \right)^{m_j} = u_i^n, \tag{9.9.10.1}$$

where $m_j, n \in \mathbb{Z}$ and $z_j \in \overline{\mathbb{Q}}$ are unknowns with $n \neq 0$.

For every $p$ choose an extension $v_p$ of the $p$-adic valuation to $\overline{\mathbb{Q}}$. The $v_p$-valuation of any $d \in \overline{\mathbb{Q}}^\times$ is 0 for all but finitely many $p$. We can thus choose $p$ such that

$$v_p(a_i) = v_p(a_i - a_j) = 0 \text{ for every } i \neq j. \tag{9.9.10.2}$$

Thus taking the valuation of (9.9.10.1) we get the equations

$$\sum_j m_j v_p \left( \frac{a_i - z_j}{a_{r+i} - z_j} \right) = n v_p(u_i). \tag{9.9.10.3}$$

Choose the $u_i$ such that $v_p(u_i) \neq 0$ for every $i$. By (9.9.11) below, for every $i$

we get a $\sigma(i) \in J$ such that

$$v_p(a_i - z_{\sigma(i)}) > 0 \quad \text{or} \quad v_p(a_{r+i} - z_{\sigma(i)}) > 0. \tag{9.9.10.4}$$

If $r > |J|$ then the same $z_j$ appears twice. Thus we have $v_p(a_{i_1} - z_j) > 0$ and $v_p(a_{i_2} - z_j) > 0$ for some $i_1 \neq i_2$ and $j$. Then

$$v_p(a_{i_1} - a_{i_2}) \geq \min\{v_p(a_{i_1} - z_j), v_p(a_{i_2} - z_j)\} > 0$$

gives a contradiction. $\qquad\qquad\qquad\qquad\qquad\qquad\qquad\qquad\qquad\qquad\qquad\square$

**Claim 9.9.11.** *Let $(R, v)$ be a valuation ring and $a, c \in R^\times$ such that $v(a) = v(c) = v(a - c) = 0$. Then*

$$v\left(\frac{a-z}{c-z}\right) > 0 \Leftrightarrow v(a-z) > 0 \quad \text{and} \quad v\left(\frac{a-z}{c-z}\right) < 0 \Leftrightarrow v(c-z) > 0. \quad \square$$

The weak Hilbertian property is well behaved in algebraic field extensions.

**Proposition 9.9.12.** *Let $K/k$ be a separable, algebraic field extension.*

(1) *If $K$ is weakly Hilbertian then so is $k$.*

(2) *If $k$ is weakly Hilbertian and $\deg(K/k) < \infty$, then so is $K$.*

*Proof.* For (1) we use (9.9.3 (3)). So we start with $C_k, \mathcal{L}_k, Z_k, s_{Z_k}$ and by base change we get $C_K, \mathcal{L}_K, Z_K, s_{Z_K}$. Let $s_K$ be a section in $H^0(C_K, \mathcal{L}_K^m, s_{Z_K}^m)$ with irreducible zero set.

Let $C'_K \subset C_K$ be one of the irreducible components. All these data are defined over a finite degree subextension $k \subset K_1 \subset K$. Since $C_k$ is non-singular, $C'_{K_1} \to C_k$ is flat, so $\mathrm{norm}_{K_1/k}$ sends sections of $(\mathcal{L}'_{K_1})^m$ to sections of $\mathcal{L}_k^{md}$, where $d = \deg(K_1/k)$. Thus $s_k := \mathrm{norm}_{K_1/k}(s'_{K_1})$ is a section in $H^0(C_k, \mathcal{L}_k^{md}, s_{Z_k}^{md})$ with irreducible zero set.

To prove (2) we use (9.9.3 (2)). Let $C_K$ be an irreducible curve over $K$. Since $K/k$ is finite, $C_K$ can be viewed as an irreducible curve $C_k$ over $k$. If $\mathcal{L}_K$ is a line bundle over $C_K$, it gives a line bundle $\mathcal{L}_k$ and a natural identification $H^0(C_K, \mathcal{L}_K) = H^0(C_k, \mathcal{L}_k)$. If $s_k \in H^0(C_k, \mathcal{L}_k^n)$ has an irreducible zero set, then so does the corresponding $s_K \in H^0(C_K, \mathcal{L}_K^n)$. $\qquad\square$

**Corollary 9.9.13.** *Let $k$ be a Hilbertian field and $K/k$ a Galois extension that is not separably closed. Then $K$ is weakly Hilbertian.*

*Proof.* By [Wei82], every nontrivial finite extension of such a field $K$ is Hilbertian. Thus, if $K$ is not separably closed, then $K$ is weakly Hilbertian by (9.9.12). $\qquad\qquad\qquad\qquad\qquad\qquad\qquad\qquad\qquad\qquad\qquad\qquad\qquad\qquad\square$

**Example 9.9.14.** $\mathbb{Q}^{\mathrm{solv}}$, the composite of all Galois extensions of $\mathbb{Q}$ with solvable Galois group, is weakly Hilbertian by (9.9.13), but not Hilbertian, as shown by the polynomial $y^2 - x$.

**Lemma 9.9.15.** *Let $K/k$ be a purely inseparable field extension. Then $K$ is weakly Hilbertian if and only if $k$ is.*

*Proof.* Going from $K$ to $k$ works as in the proof of $(9.9.12\,(1))$.

Conversely, assume that $k$ is weakly Hilbertian and let $(C_K, \mathscr{L}_K)$ be a curve. It is defined over a finite subextension $k \subset K_1 \subset K$. Since $K/K_1$ is purely inseparable, an irreducible subvariety of $C_{K_1}$ stays irreducible over $K$. Thus it is enough to show that $(C_{K_1}, \mathscr{L}_{K_1})$ satisfies the weak Hilbertian property.

Now note that $K_1^q \subset k$, where $q$ is a high enough power of the characteristic. Thus $K_1 \cong K_1^q$ is weakly Hilbertian by the first part. □

The behavior of $\mathbb{Q}_p$, $\mathbb{F}_p((t))$, and, more generally, quotient fields of Henselian valuation rings, is very interesting. Smooth curves have weak Hilbertian properties but singular curves do not.

**Proposition 9.9.16.** *Let $(R, m)$ be an excellent DVR with quotient field $K$ and locally finite residue field $k$. Let $C_K$ be a smooth, projective, irreducible curve over $K$ and $\mathscr{L}_K$ an ample line bundle on $C_K$. Then $|\mathscr{L}_K^n|$ has infinitely many irreducible members for $n$ sufficiently divisible.*

*Proof.* We extend $C_K$ to a flat morphism $C_R \to \operatorname{Spec} R$. We may assume that $C_R$ is regular by [Sha66]. Then $\mathscr{L}_K$ extends to a line bundle $\mathscr{L}_R$ on $C_R$.

Let $E_1, \ldots, E_r$ be the irreducible components of the central fiber $C_k = \sum m_i E_i$. The intersection matrix $(E_i \cdot E_j)$ is negative semidefinite with $[C_k]$ as the only null vector. Thus the intersection matrix of $E_2, \ldots, E_r$ is negative definite. We can thus find a divisor $F$ supported on $E_2, \ldots, E_r$ such that $\mathscr{L}_R^{n_1}(F)$ has degree 0 on $E_2, \ldots, E_r$.

By [Art62] the curves $E_2, \ldots, E_r$ can be contracted $C_R \to C_R^*$ and a suitable power of $\mathscr{L}_R^{n_1}(F)$ descends to a line bundle $\mathscr{L}^*$ on $C_R^*$.

Now we have a normal scheme with a flat morphism $\pi \colon C_R^* \to R$ whose generic fiber is $C_K$ and whose central fiber $C_k^*$ is an irreducible curve. Furthermore there is a line bundle $\mathscr{L}^*$ whose restriction to $C_K$ is a power of $\mathscr{L}_K$.

Set $E^* = \operatorname{red} C_k^*$ and pick any point $p \in E^*$ that is regular both on $E^*$ and on $C_R^*$. Since $\operatorname{rank}_{\mathbb{Q}} \operatorname{Pic}(E^*) = 1$, after passing to a power of $\mathscr{L}^*$, we may assume that

(1) $\mathscr{L}^*|_{E^*}$ has a section $\bar{s}_0$ that vanishes only at $p$,

(2) $\mathscr{L}^*(-E^*)/\mathscr{L}^*(-2E^*)$ has a section $\bar{s}_1$ that does not vanish at $p$, and

(3) $H^1\big(C_R^*, \mathscr{L}^*(-E^*)\big) = H^1\big(C_R^*, \mathscr{L}^*(-2E^*)\big) = 0$.

By (3) we can lift $\bar{s}_0$ and $\bar{s}_1$ to $s_0 \in H^0\big(C_R^*, \mathscr{L}^*\big)$ and $s_1 \in H^0\big(C_R^*, \mathscr{L}^*(-E^*)\big)$. For all but one residue value of $\lambda \in R$, $D_R(\lambda) := (s_0 + \lambda s_1 = 0)$ is regular at $p$. Since $p$ is its sole point over $k$, $D_R(\lambda)$ is irreducible and reduced. □

**Remark 9.9.17.** The proof uses excellence, but the result might hold without it. Note that a DVR of characteristic 0 is excellent [Sta22, 07QW]. In positive characteristic, local rings of smooth curves are excellent and so are power series rings $K[[t]]$. However, there are many non-excellent DVRs; see [DS18] for especially simple examples.

**Proposition 9.9.18.** *Let $(R, m)$ be a Henselian valuation ring with quotient field $K$ and residue field $k$. For $2g \leq |k|$ there are rational curves $C$ with $g$ nodes over $K$ such that, for 'most' ample line bundles $\mathcal{L}$ over $C$, every section of $\mathcal{L}^n$ has at least $g$ distinct zeros.*

*Proof.* We choose $a_i \in R \subset \mathbf{A}^1(K) \subset \mathbb{P}^1_K$ such that $\bar{a}_i \in k$ (their reduction mod $m$) are all distinct. As in (9.9.10), identifying $a_i$ with $a_{g+i}$ to get a nodal rational curve $C$ and a line bundle $\mathcal{L}$ on $C$ is achieved by starting with some $\mathcal{O}_{\mathbb{P}^1}(r)$ and specifying isomorphisms $\mathcal{O}_{\mathbb{P}^1}(r)|_{a_i} \cong \mathcal{O}_{\mathbb{P}^1}(r)|_{a_{g+i}}$. Thus sections of $\mathcal{L}^n$ are given by polynomials $f(x)$ of degree $\leq nr$ such that

$$f(a_i) = u_i^n f(a_{g+1}) \text{ for } i = 1, \ldots, g, \tag{9.9.18.1}$$

where the $u_i \in K^\times$ determine $\mathcal{L}$. We may assume that $f(x) \in R[x] \setminus m[x]$ and denote by $\bar{f}$ its image in $k[x]$. If the valuation of $u_i$ is not 0 for every $i$, this implies that

$$\bar{f}(\bar{a}_i) = 0 \text{ or } \bar{f}(\bar{a}_{g+i}) = 0 \text{ for } i = 1, \ldots, g. \tag{9.9.18.2}$$

Thus $\bar{f}$ has at least $g$ distinct zeros, so $f$ has at least $g$ distinct prime factors since $R$ is Henselian. □

**Corollary 9.9.19.** *The following fields are not weakly Hilbertian.*

(1) $\mathbb{Q}_p$ and $\overline{\mathbb{Q}} \cap \mathbb{Q}_p$.

(2) $\mathbb{F}_p((t))$ and $\overline{\mathbb{F}_p(t)} \cap \mathbb{F}_p((t))$.

(3) *Separably closed fields.*

*Proof.* The rings $\mathbb{Z}_p, \overline{\mathbb{Q}} \cap \mathbb{Z}_p, \mathbb{F}_p[[t]]$, and $\overline{\mathbb{F}_p(t)} \cap \mathbb{F}_p[[t]]$ are all Henselian, hence not weakly Hilbertian by (9.9.18).

Assume that $K$ is separably closed. If $K = \overline{\mathbb{F}}_p$ then $K$ is not weakly Hilbertian by definition. Otherwise $K$ has nontrivial valuations. The value ring is Henselian since $K$ is separably closed. □

**Remark 9.9.20** (Locally finite fields). Let $C$ be an integral, projective curve over a locally finite field $k$ and $\mathcal{L}$ an ample line bundle on $C$. Let $c \in C$ be any smooth point. Then $\mathcal{L}^{\deg c}(-\deg \mathcal{L} \cdot [c])$ has degree 0, hence is torsion since $k$ is locally finite. Thus there is an $m > 0$ and a section $s \in H^0(C, \mathcal{L}^m)$ such that $\mathrm{red}(s = 0) = \{c\}$.

Poonen explained to us that, using geometric class field theory and the function field Chebotarev density theorem, one can prove that $\mathscr{L}^m$ has a section with irreducible and reduced zero set for all $m \gg 1$. However, the probability that a random section has this property tends to 0 as $m \to \infty$.

# Chapter Ten

---

## Complements, counterexamples, and conjectures

In this chapter, we provide complements, counterexamples, and conjectures. We begin in Section 10.1 with a discussion of a topological form of the Gabriel-Rosenberg theorem: for varieties $X$ that are characterized by $|X|$, we show that the category of constructible étale sheaves also determines $X$. In Sections 10.2 to 10.5 we show that results analogous to our main results fail in other circumstances. In Section 10.6 we discuss a number of conjectures arising from our work.

## 10.1  A TOPOLOGICAL GABRIEL THEOREM

In this section, we prove the following analogues of Gabriel's reconstruction theorem.

**Notation 10.1.1.** Given a scheme $X$ and a ring $A$ that is either a finite field, a finite extension of $\mathbb{Z}_\ell$ for some prime $\ell$ (not necessarily invertible on $X$), or some field $\mathbb{Q}_\ell \subset A \subset \overline{\mathbb{Q}}_\ell$, we write $\mathscr{C}_{X,A}$ for the category of constructible étale sheaves of $A$-modules. For $\mathbb{Q}_\ell$-subfields of $\overline{\mathbb{Q}}_\ell$, we take the definition as in [Gro77, Exposé VI, 1.5.3].

Given a field $K$, call a $K$-scheme $X$ *recognizable* if $K$ and $X$ satisfy conditions (1), (2), or (3) of Main Theorem (1.2.1). The main result of this section is the following.

**Theorem 10.1.2.** *Fix a ring $A$ as in (10.1.1).*

(1) *For a scheme $X$, the Zariski topological space $|X|$ is determined by the category $\mathscr{C}_{X,A}$.*

(2) *Let $K$ and $L$ be fields of characteristic 0 and $X_K, Y_L$ recognizable varieties over $K$ and $L$, respectively. Any equivalence $\mathscr{C}_{X,A} \overset{\sim}{\to} \mathscr{C}_{Y,A}$ is induced by a unique isomorphism $X \to Y$ of schemes.*

The proof occupies the remainder of this section.

**10.1.3.** Note that by Main Theorem (1.2.1), to prove (10.1.2) it suffices to prove statement (1). We will thus work with a fixed category $\mathscr{C}_{X,A}$, which we will (temporarily) write as $\mathscr{C}$ in what follows, to ease notation.

Since we can pick out the $\mathbb{F}_\ell$-modules from the $\mathbb{Z}_\ell$-modules (using the criterion that the $\ell$-fold sum of the identity map vanishes), to prove (10.1.2) it suffices to prove it under the assumption that $A$ is either $\mathbb{F}_\ell$ or a subfield $\mathbb{Q}_\ell \subset A \subset \overline{\mathbb{Q}}_\ell$.

**Lemma 10.1.4.** *A constructible $A$-module $M$ is isomorphic to a sheaf of the form $\iota_* A$ for some $\iota \colon \operatorname{Spec} k \to X$ if and only if it is a (nonzero) simple object of $\mathscr{C}$.*

*Proof.* Since $M$ is constructible, its support is a constructible subset of $X$. Given a closed point $x$ of the support, we have from simplicity that the quotient map $M \to \iota_* M_x$ must be an isomorphism, where $\iota \colon x \to X$ is the inclusion map. Since $M$ is simple, $M_x$ must be a simple vector space, so it must have dimension 1. This gives the desired result. $\qquad\qquad\qquad\qquad\qquad\qquad\qquad\qquad\square$

**Definition 10.1.5.** We define some properties of objects of $\mathscr{C}$.

(1) An object $F \in \mathscr{C}$ is *irreducible* if

    (a)   for every simple object $s \in \mathscr{C}$ we have $\dim_A \operatorname{Hom}(F, s) \le 1$, and

    (b)   any pair of subobjects $F', F' \subset F$ have nonzero intersection.

(2) The *support* of an irreducible object $F$ is the set $\operatorname{Supp}(F)$ of simple quotients of $F$.

(3) Two irreducible objects $F$ and $F'$ are *equivalent* if $\operatorname{Supp}(F) = \operatorname{Supp}(F')$.

(4) An irreducible object $F$ is a *partial closure* of an irreducible object $G$ if there are nonzero subobjects $F' \subset F$ and $G' \subset G$ such that $F'$ is equivalent to $G'$.

(5) An irreducible object $F$ is *closed* if any partial closure of $F$ has the same support.

(6) An irreducible object $F$ is a *closure* of an irreducible object $F'$ if $F$ is a closed partial closure of $F'$.

We can associate two subsets of $X$ to an irreducible object $F$: the support of the sheaf $F$ and the set $\operatorname{Supp}(F)$ as defined above, which is identified with a subset of $X$ via Lemma 10.1.4. It is not hard to see that the support of the sheaf $F$ is the Zariski closure of the set of closed points, which is $\operatorname{Supp}(F)$. We will safely conflate these supports in what follows.

**Lemma 10.1.6.** *The following hold for irreducible objects.*

(1) *If $F$ is a closed irreducible object of $\mathscr{C}$ then the support of $F$ is a closed irreducible subset of $X$.*

(2) *The set of irreducible closed subsets of $X$ is in bijection with equivalence classes of closed irreducible objects of $\mathscr{C}$.*

(3) *An irreducible closed subset $Y \subset X$ lies in an irreducible closed subset $Z \subset X$ if and only if there is a closed irreducible sheaf $F$ with $\mathrm{Supp}(F) = Z(k)$, a closed irreducible sheaf $F'$ with $\mathrm{Supp}(F') = Y(k)$, and a surjection $F \to F'$.*

*Proof.* If the support of $F$ is not irreducible then there are two open subsets $U, V \subset \mathrm{Supp}(F)$ such that $U \cap V = \emptyset$. But then $(j_U)_! F_U$ and $(j_V)_! F_V$ are two nonzero subsheaves with trivial intersection. Suppose the support of $F$ is not closed. Consider the inclusion $\mathrm{Supp}(F) \hookrightarrow \overline{\mathrm{Supp}(F)}$. Since $F$ is constructible, there is an open subscheme $U \subset \mathrm{Supp}(F) \subset \overline{\mathrm{Supp}(F)}$. The constant sheaf on $\overline{\mathrm{Supp}(F)}$ is then a partial closure of $F$, since $j_! F|_U$ is equivalent to $i_! A$, where $j \colon U \to \mathrm{Supp}(F)$ and $i \colon U \to \overline{\mathrm{Supp}(F)}$ are the natural open immersions.

The second statement follows from the first statement and the fact that constant sheaves define all irreducible closed subsets.

The last statement follows from the fact that for any surjection $F \to F'$ we have $\mathrm{Supp}(F') \subset \mathrm{Supp}(F)$, combined with the fact that, if $i \colon Y \to Z$ is a closed immersion, the natural map $A_Z \to i_* A_Y$ is a surjection of irreducible sheaves. $\qquad\square$

**Proposition 10.1.7.** *The Zariski topological space $|X|$ is uniquely determined by the category $\mathscr{C}$.*

*Proof.* It suffices to reconstruct the Zariski topology on the set of closed points $X(k)$. First note that we can describe the set itself as the set of isomorphism classes of simple objects of $\mathscr{C}$, by (10.1.4). Given a sheaf $F$, we can thus describe the support $\mathrm{Supp}(F) \subset X(k)$. By (10.1.6), we can reconstruct the set of irreducible closed subsets $Z \subset X(k)$. This suffices to completely determine the topology, since closed subsets are precisely finite unions of irreducible closed subsets. $\qquad\square$

This completes the proof of (10.1.2). $\qquad\square$

**Remark 10.1.8.** It is natural to wonder if there are topological analogues of Balmer's monoidal reconstruction theorem [Bal05] or the theory of Fourier-Mukai transforms. These ideas will be pursued elsewhere.

**Remark 10.1.9.** In the above we work with the standard coefficient rings for étale sheaves. However, the same argument shows that (10.1.2) also holds with other coefficient rings such as $\mathbb{Z}$ (so $\mathscr{C}_{X,A}$ is the category of constructible sheaves of abelian groups) or $\mathbb{Q}$ (defined directly, in contrast with $\mathbb{Q}_\ell$).

## 10.2 EXAMPLES OVER FINITE FIELDS

Let $K$ be a locally finite field. The following example shows that, while $\left|\mathbb{P}^1_K \times \mathbb{P}^1_K\right|$ determines $K$, one cannot recover the field $K$ from the Zariski topology of $\mathbb{P}^2_K$.

**Example 10.2.1.** Let $K, L$ be locally finite fields. Then

(1) $\left|\mathbb{P}^1_K \times \mathbb{P}^1_K\right| \sim \left|\mathbb{P}^1_L \times \mathbb{P}^1_L\right| \quad \Leftrightarrow \quad K \cong L$.

(2) $\left|\mathbb{P}^2_K\right| \sim \left|\mathbb{P}^2_L\right|$.

Both assertions are special cases of more general results. Item (2) is essentially proved in [WK81]; we discuss a more general form of it in (10.3.1) below.

For (1), we show in (10.2.2) that, over any field $K$, one can recover $K^{\text{insep}}$ from $\left|\mathbb{P}^1_K \times \mathbb{P}^1_K\right|$. If $K$ is locally finite then $K = K^{\text{insep}}$, and we are done.

Note that the finite field case follows already from the simpler (10.2.5).

**Theorem 10.2.2.** Let $K$ be a perfect field. Then $\left|\mathbb{P}^1_K \times \mathbb{P}^1_K\right|$ determines $K$, up to isomorphism.

We present three methods to extract information from $\left|\mathbb{P}^1_K \times \mathbb{P}^1_K\right|$.

**10.2.3** (Lines in $\mathbb{P}^1_K \times \mathbb{P}^1_K$ and their intersections). Start with $K$ arbitrary. Consider all irreducible curves $C \subset \mathbb{P}^1_K \times \mathbb{P}^1_K$ that are disjoint from some other curve $C'$. These come in two families:

$$\mathbf{A} := \left\{\mathbb{P}^1_K \times \{p\} : p \in \mathbb{P}^1_K\right\} \text{ and}$$
$$\mathbf{B} := \left\{\{q\} \times \mathbb{P}^1_K : q \in \mathbb{P}^1_K\right\}.$$

Given $A \in \mathbf{A}$ corresponding to $p$ and $B \in \mathbf{B}$ corresponding to $q$, we see that

$$A \cap B \cong \operatorname{Spec}_K\left(K(q) \otimes_K K(p)\right).$$

**Lemma 10.2.4.** Let $K$ be a field and $L_1/K, L_2/K$ finite extensions. Write $L_1 \otimes_K L_2 \cong \oplus_{i \in I} A_i$, where the $A_i$ are local $K$-algebras. Then $|I|$ is at most the separable degree of $L_1/K$, and equality holds if $L_2$ is the normal closure of $L_1/K$.

*Proof.* If we replace $K$ by a purely inseparable field extension then $|I|$ does not change. We may thus assume that $L_i/K$ are separable. Each $A_i$ is an $L_2$-algebra, hence $\deg_K A_i$ is divisible by $\deg_K L_2$. Thus,

$$\deg_K L_1 \cdot \deg_K L_2 = \deg_K(L_1 \otimes_K L_2) \geq |I| \cdot \deg_K L_2.$$

If $L_2$ is the normal closure of $L_1/K$ then every $A_i \cong L_2$, hence we get that $|I| = \deg_K L_1$. $\qquad\square$

**Corollary 10.2.5.** Fix $A \in \mathbf{A}$ corresponding to $p \in \mathbb{P}^1_K$. Then the separable degree of $K(p)/K$ equals $\max\{|A \cap B| : B \in \mathbf{B}\}$.

In particular, we can determine—using $\left|\mathbb{P}^1_K \times \mathbb{P}^1_K\right|$ only—which $A \in \mathbf{A}$ corresponds to a purely inseparable point $p \in \mathbb{P}^1_K$.

*Proof.* The elements of $|A \cap B|$ correspond precisely to the factors of $K(p) \otimes K(q)$, where $B = \{q\} \times \mathbb{P}^1$. $\qquad\square$

The previous results determine $\mathbb{P}^1_K(K^{\text{insep}})$ as a point set. To go further, we use the following immediate consequence of (7.2.1).

**Lemma 10.2.6.** *Let $C \subset \mathbb{P}^1_K \times \mathbb{P}^1_K$ be an irreducible curve. Then the separable degree of the projection to the $\mathbf{B}$-factor equals $\max_{A \in \mathbf{A}, A \neq C} |C \cap A|$.* $\square$

**10.2.7** (Bidegree in characteristic 0). Let $C \subset \mathbb{P}^1_K \times \mathbb{P}^1_K$ be an irreducible curve. If char $K = 0$ then the coordinate projections $C \to \mathbb{P}^1_K$ are separable, hence (10.2.6) tells us that the bidegree of $C$ equals

$$\left( \max_{A \in \mathbf{A}, A \neq C} |C \cap A|, \ \max_{B \in \mathbf{B}, B \neq C} |C \cap B| \right).$$

In particular, the bidegree is determined by $\left| \mathbb{P}^1_K \times \mathbb{P}^1_K \right|$.

Since the bidegree determines the linear equivalence class, (1.3.1) shows that $\left| \mathbb{P}^1_K \times \mathbb{P}^1_K \right|$ determines $K$.

The above argument does not work in positive characteristic, since we do not know which projections are separable. However, we can tell which projections are purely inseparable, and this is what we exploit next.

**10.2.8** (Determining $\text{Aut}(\mathbb{P}^1_K)$). Here we show how to compute the abstract group structure of $\text{Aut}(\mathbb{P}^1_K)$ from $\left| \mathbb{P}^1_K \times \mathbb{P}^1_K \right|$ if $K$ is perfect. Once the abstract group structure is known, we use [BT73] to conclude that $\left| \mathbb{P}^1_K \times \mathbb{P}^1_K \right|$ determines $K$. This then concludes the proof of (10.2.2).

Let $\mathbf{C}$ denote the set of all curves $C \subset \mathbb{P}^1_K \times \mathbb{P}^1_K$ such that both coordinate projections are purely inseparable. By (10.2.6), the members of $\mathbf{C}$ are determined by $\left| \mathbb{P}^1_K \times \mathbb{P}^1_K \right|$.

Using the $\mathbf{A}, \mathbf{B}$ families from (10.2.3), any such $C$ determines a bijection

$$\sigma_C \colon \mathbf{A} \to \mathbf{B}.$$

Fix now one such curve $C_0$, giving $\sigma_0$. Then we get a subset

$$G_{\mathbf{A}} := \{ \sigma_0^{-1} \circ \sigma_C : C \in \mathbf{C} \} \subset S_{\mathbf{A}},$$

which depends only on $\left| \mathbb{P}^1_K \times \mathbb{P}^1_K \right|$ and $C_0$, where $S_{\mathbf{A}}$ denotes the group of permutations of $\mathbf{A}$.

Computing on $\mathbb{P}^1_K \times \mathbb{P}^1_K$ we see that $G_{\mathbf{A}}$ is actually a subgroup generated by $\text{Aut}(\mathbb{P}^1_K)$ and the Frobenius $F$. (Since $K$ is perfect, negative powers of the Frobenius also make sense.) Moreover, $G_{\mathbf{A}} \cong \text{Aut}(\mathbb{P}^1_K) \rtimes \mathbb{Z}$, so $\text{Aut}(\mathbb{P}^1_K)$ is the commutator subgroup of $G_{\mathbf{A}}$. Thus, $\text{Aut}(\mathbb{P}^1_K)$ is determined by $\left| \mathbb{P}^1_K \times \mathbb{P}^1_K \right|$, and so is $K$ by [BT73]. $\square$

## 10.3   SURFACES OVER LOCALLY FINITE FIELDS

The next result is a mild strengthening of [WK81].

**Theorem 10.3.1.** *Let $S_1, S_2$ be smooth, projective surfaces over locally finite fields $K_1$ and $K_2$. Assume that every effective divisor on the $S_i$ is ample. Then $|S_1| \sim |S_2|$.*

In view of (10.2.1 (1)), the assumption on every divisor being ample seems reasonable. By contrast, we do not know what happens with higher-dimensional varieties over finite or locally finite fields. For example, it is not known whether $|\mathbb{P}_K^3|$ determines $K$, or at least the characteristic.

The proof is given in (10.5.8) below; the key property that makes it work is the following.

**Proposition 10.3.2.** *Let $S$ be a normal, projective surface over a field $k$. The following are equivalent.*

(1) *$k$ is locally finite and any two curves in $S$ have a nonempty intersection.*

(2) *Let $D \subset S$ be any 1-dimensional, closed subset and $P \subset D$ a 0-dimensional subset. Then there is an irreducible curve $C \subset S$ such that $D \cap C = P$ if and only if $P \cap D_i \neq \emptyset$ for every 1-dimensional, irreducible component $D_i \subset D$.*

*Proof.* If $|S|$ satisfies (2) then $k$ is locally finite by (8.2.10). Applying (2) to an irreducible curve $D \subset S$ and $P = \emptyset$ shows that every irreducible curve $C \subset S$ has nonempty intersection with $D$.

For the converse we follow [WK81]. First we blow up $P$ and normalize to get $S_1 \to S$. Repeatedly blowing up points over $P$ we get $S_r \to S$ such that the intersection matrix of $D_r \subset S_r$ (the birational transform of $D$) is negative definite. By [Art62], $D_r \subset S_r$ can be contracted to get $\pi \colon S_r \to T$. By [CP16], there is an irreducible hypersurface section $C_T \subset T$ that is disjoint from $\pi(D_r)$. Let $C \subset S$ be its birational transform. $\qquad\qquad\qquad\qquad\qquad\square$

## 10.4   REAL ZARISKI TOPOLOGY

Let $X$ be an algebraic variety over a real closed field $\mathbb{R}$. It is then natural to consider its real Zariski topology—denoted by $|X_1|_\mathbb{R}$—which is the topology induced on the set of $\mathbb{R}$-points by the Zariski topology on $|X|$; see (10.4.2) for the precise definition.

It turns out that, over countable, real closed fields, the dimension is the only topological invariant. This applies to $\mathbb{R} = \overline{\mathbb{Q}} \cap \mathbb{R}$, but we do not know what happens over $\mathbb{R}$.

**Theorem 10.4.1.** *Let $X_1$ and $X_2$ be irreducible, quasi-projective varieties over countable, real closed fields $\mathbb{R}_1$ and $\mathbb{R}_2$. Assume that they both have smooth real points. Then $|X_1|_{\mathbb{R}_1} \sim |X_2|_{\mathbb{R}_2}$ if and only if $\dim X_1 = \dim X_2$.*

The proof is completed in (10.5.8).

**Definition 10.4.2.** Let $X$ be an algebraic variety defined over $\mathbb{R}$ and $X(\mathbb{R})$ its set of real points. We then get the *real Zariski topology* on $X(\mathbb{R})$, whose closed subsets are of the form $W(\mathbb{R})$, where $W \subset X$ is a closed subset. We denote this topological space by $|X|_{\mathbb{R}}$.

Note that, unlike in the complex case, $X(\mathbb{R})$ does not determine $X$. For example if $X = (x_1^2 + \cdots + x_n^2 = 0) \subset \mathbb{A}^n$, then $|X|_{\mathbb{R}}$ consists of a single point. More generally, if $X \subset \mathbb{A}^n$ has no smooth real points, then $X(\mathbb{R}) = (\mathrm{Sing}\, X)(\mathbb{R})$, so, $X(\mathbb{R})$ does not even detect the dimension of $X$. However, if $X \subset \mathbb{A}^n$ is irreducible and has smooth real points, then $X(\mathbb{R}) \subset \mathbb{R}^n$ determines $X$.

Let $X$ be an algebraic variety defined over $\mathbb{R}$. If $X$ has a smooth real point, then the underlying set of $|X|_{\mathbb{R}}$ is $X(\mathbb{R})$ and the irreducible subsets are given as $W(\mathbb{R})$, where $W \subset X$ is an irreducible closed subset that has a smooth real point.

More generally, one can work with any real closed field $\mathbb{R}$, and its algebraic closure $\mathbb{C} := \mathbb{R}(\sqrt{-1})$. We denote the real Zariski topology associated to an irreducible $\mathbb{R}$-variety $X$ by $|X|_{\mathbb{R}}$. Adopting a more scheme-theoretic view point, we can view $|X|_{\mathbb{R}}$ as a subset of $|X|$

$$|X|_{\mathbb{R}} := \{ Z \in |X| : Z \text{ has smooth } \mathbb{R}\text{-points.} \}.$$

The key property that makes (10.4.1) work is the following analog of (10.3.2).

**Lemma 10.4.3.** *Let $\mathbb{R}$ be a real closed field, $Y$ an irreducible $\mathbb{R}$-variety that has smooth real points, and $U, V \subsetneq Y$ closed subsets. Assume that $\dim U \leq \dim Y - 2$.*

*Then there is an irreducible $\mathbb{R}$-subvariety $Y_1 \subset Y$ of codimension 1 such that $U \subset Y_1$, $V \cap Y_1 = V \cap U$, and $Y_1$ has a smooth real point.*

Proof. We may assume that $Y$ is affine. Let $\{g_i : i \in I\}$ be defining equations for $U$ (over $\mathbb{C}$). Pick a smooth point $y \in Y \setminus (U \cup V)$ and let $\{h_j : j \in J\}$ be polynomials such that $h_j|_V \equiv 1$ and $y$ is an isolated smooth point of $(h_j = 0 : j \in J)$. Let $g$ be a general positive linear combination $\sum_{ij} c_{ij} g_i^2 h_j$. Then $Y_1 := (g = 0)$ works.  □

## 10.5  COUNTABLE NOETHERIAN TOPOLOGIES

Here we prove (10.3.1) and (10.4.1) using only basic properties of noetherian topologies and the key properties proved in (10.3.2) and (10.4.3).

The proofs rely on the observation that, if every closed subset is a complete intersection, then the Zariski topology carries relatively little information.

**10.5.1** (Noetherian spaces with a dimension function). We view a topological space as an underlying point set $|M|$, and a set of closed, irreducible subsets $\mathrm{Irr}(M) \subset 2^{|M|}$.

We assume that the noetherian property holds (so every closed subset is a finite union of irreducibles, and minimal unions are unique), and there is a dimension function $\dim : \mathrm{Irr}(M) \to \mathbb{Z}$ [Sta22, Tag 02I8]. That is, if $Z_1 \subsetneq Z_2$ are irreducible, and there is no irreducible subset satisfying $Z_1 \subsetneq Z_3 \subsetneq Z_2$, then $\dim Z_2 = \dim Z_1 + 1$. The dimension function is unique if we set $\dim \emptyset := -1$.

These conditions are satisfied by the Zariski topology of an algebraic variety.

Given a subset $I \subset \mathrm{Irr}(M)$, let $\mathscr{L}(I)$ be the lattice generated by the subvarieties $\{Z_i : i \in I\}$, that is, we repeatedly take intersections, irreducible decompositions, and finite unions. We write $\mathscr{L}(M)$ for $\mathscr{L}(\mathrm{Irr}(M))$. Note that $\mathscr{L}(M)$ determines $M$.

**Definition 10.5.2.** Let $M = (|M|, \mathrm{Irr}(M))$ be a topological space satisfying the conditions of (10.5.1). We say that $M$ satisfies the *complete intersection property* if the following holds.

(1) Let $U, V \subsetneq M$ be closed subsets and $\dim U < d < \dim M$. Then there is an irreducible $Z \subset M$ of dimension $d$ such that $U \subset Z$ and $V \cap Z = V \cap U$.

We saw in (10.4.3) that this is satisfied by the real Zariski topology.

We say that $M$ satisfies the *positive complete intersection property* if the following holds.

(2) Let $V \subsetneq M$ be closed subset and $U \subset V$ a finite subset. Then there is an irreducible curve $C \subset M$ such that $U \subset C$ and $V \cap C = U$ if and only if every positive dimensional irreducible component of $V$ has nonempty intersection with $U$.

Applying this to an irreducible $V \subsetneq M$ and $U = \emptyset$ shows that every irreducible curve $C \subset M$ has nonempty intersection with $V$. In particular, if $|X|$ satisfies (2) then $\dim X \leq 2$. We do not know how to formulate a meaningful variant of (2) for dimension $\geq 3$.

We saw in (10.3.2) that (2) is satisfied by the Zariski topology of a normal, projective surface over a locally finite field if every effective divisor is ample.

The abstract homeomorphism results that underlie (10.3.1) and (10.4.1) are the following.

**Proposition 10.5.3.** *Two countable, irreducible, topological spaces satisfying (10.5.1) and (10.5.2 (1)) are homeomorphic if and only if they have the same dimension.*

**Proposition 10.5.4.** *Any two countable, irreducible, 2-dimensional topological spaces satisfying (10.5.1) and (10.5.2 (2)) are homeomorphic to each other.*

By the seesaw argument (10.5.6), both of these are implied by the following.

**Lemma 10.5.5.** *Let $M_1, M_2$ be irreducible topological spaces satisfying (10.5.1) and (10.5.2 (1)) or (10.5.2 (2)). Let $I \subset J \subset \mathrm{Irr}(M_1)$ be finite subsets. Then every dimension preserving lattice embedding $\varphi_I : L(I) \hookrightarrow L(M_2)$ extends to a dimension preserving lattice embedding $\varphi_J : L(J) \hookrightarrow L(M_2)$.*

*Proof.* It is enough to prove this when $L(J)$ is obtained from $L(I)$ by adjoining a minimal element of $L(J) \setminus L(I)$. That is, an irreducible subset $Z_0 \subset X$ such that, for every $i \in I$, either $Z_0 \subset Z_i$ or $Z_i \cap Z_0 \in L(I)$.

If $Z_0 \subset Z_i$ for some $i$, then we can replace the $M_1$ by $|Z_i|$ and conclude by induction. Otherwise, we are looking for $W_0 \subset M_2$ such that $\dim W_0 = \dim Z_0$ and

$$W_0 \cap \cup_{i \in I} \varphi(Z_i) = \varphi\big(\cup_{i \in I}(Z_0 \cap Z_i)\big).$$

If the $M_i$ satisfy $(10.5.2\,(1))$ then the existence of such a $W_0$ follows from the definition.

If the $M_i$ satisfy $(10.5.2\,(2))$ then $Z_0$ has nonempty intersection with every positive dimensional irreducible component of $\cup_{i \in I} Z_i$, hence $\varphi\big(\cup_{i \in I}(Z_0 \cap Z_i)\big)$ has nonempty intersection with every positive dimensional irreducible component of $\cup_{i \in I} \varphi(Z_i)$. Thus, $W_0$ exists by definition. $\square$

**Lemma 10.5.6** (Seesaw isomorphism). *Two countable, abstract algebras $A, B$ are isomorphic if the following hold.*

(1) *For any two finitely generated subalgebras $A_1 \subset A_2 \subset A$, every embedding $\varphi_1 \colon A_1 \hookrightarrow B$ extends to an embedding $\varphi_2 \colon A_2 \hookrightarrow B$.*

(2) *For any two finitely generated subalgebras $B_1 \subset B_2 \subset B$, every embedding $\psi_1 \colon B_1 \hookrightarrow A$ extends to an embedding $\psi_2 \colon B_2 \hookrightarrow A$.*

*Proof.* Choose well-orderings $A = \{a_1, a_2, \dots\}$ and $B = \{b_1, b_2, \dots\}$. We start with the isomorphism $A_0 = \langle \emptyset \rangle = B_0$. Assume next that we already have subalgebras and an isomorphism $\varphi_i \colon A_i \cong B_i$. If $i$ is even, choose the smallest $a' \in A \setminus A_i$ and extend $\varphi_i$ to $\varphi_{i+1} \colon \langle A_i, a' \rangle \hookrightarrow B$. If $i$ is odd, choose the smallest $b' \in B \setminus B_i$ and extend $\varphi_i^{-1}$ to $\varphi_{i+1}^{-1} \colon \langle B_i, b' \rangle \hookrightarrow A$. $\square$

**Remark 10.5.7.** The assumption of countability seems essential here. For example, by the Baer-Specker theorem, $\mathbb{Z}^{\mathbb{N}}$ is not free, but every countable subgroup of it is free. (This does not follow directly from $(10.5.6)$. However, we can add a unary operation prim( ) that sends $\mathbf{u} \in \mathbb{Z}^{\mathbb{N}}$ to the smallest $\mathbf{u}/m \in \mathbb{Z}^{\mathbb{N}}$, where $m \in \mathbb{N}$.)

**10.5.8** (Proof of $(10.3.1)$ and $(10.4.1)$). In both cases, the topologies satisfy $(10.5.1)$. In view of $(10.5.3)$ and $(10.5.4)$ it remains to verify the following.

- The $|S_i|$ in $(10.3.1)$ satisfy the condition $(10.5.2\,(2))$.
- The $|X|_{\mathbb{R}}$ in $(10.4.1)$ satisfies the condition $(10.5.2\,(1))$.

As we already noted in $(10.5.2)$, $(10.5.2\,(2))$ is a restatement of $(10.3.2\,(2))$ and $(10.5.2\,(1))$ holds by $(10.4.3)$. $\square$

## 10.6   CONJECTURES

It is possible that with more work the methods used here can be extended to
weaken the normality assumption in Main Theorem (1.2.1):

**Question 10.6.1.** Let $K, L$ be fields and $X_K, Y_L$ seminormal, geometrically
irreducible varieties over $K$ (resp. $L$). Let $\Phi: |X_K| \simeq |Y_L|$ be a homeomorphism.
Assume that char $L = 0$ and dim $X_L \geq 2$. Then does the conclusion of Main
Theorem (1.2.1) hold? That is, $\Phi$ is the composite of a field isomorphism $\varphi:$
$K \cong L$ and an algebraic isomorphism of $L$-varieties $X_L^{\varphi} \cong Y_L$.

   As the following examples show, Main Theorem (1.2.1) is false as stated in
positive characteristic.

**Example 10.6.2.** Let $E$ be an elliptic curve over a field $k$ of characteristic
$p > 0$. Then the morphism

$$F_{E/k} \times \mathrm{id}\colon E \times_k E \to E^{(p)} \times_k E$$

is a homeomorphism that is not induced by an isomorphism of schemes

$$E \times_k E \to E^{(p)} \times_k E.$$

Indeed, such an isomorphism would have to respect the product structure, im-
plying that $E \simeq E^{(p)}$ over $k$. So we get examples of non-algebraizable homeomorphisms by choosing $E$ such that $E$ is not isomorphic to $E^{(p)}$ over $k$.

**Example 10.6.3.** For the second example, we assume that $k$ has characteristic
at least 5. Given a homogeneous polynomial $f(x, y, z)$ of degree $p$, let $D_f \subset \mathbb{P}^3$
denote the divisor given by the equation $w^p - f(x, y, z)$. The projection map

$$(x, y, z, w) \mapsto (x, y, z)\colon \mathbb{P}^3 \dashrightarrow \mathbb{P}^2$$

defines a morphism $\pi\colon D_f \to \mathbb{P}^2$ that realizes $D_f$ as obtained from the $p$th root
construction

$$D_f = \mathbf{Spec}_{\mathcal{O}_{\mathbb{P}^2}} \mathcal{O} \oplus \mathcal{O}(-1) \oplus \cdots \oplus \mathcal{O}(-p+1),$$

with the multiplication structure defined by the inclusion $\mathcal{O}(-p) \to \mathcal{O}$ associated
to the divisor $f(x, y, z) = 0$. In particular, $D_f$ is finite flat over $\mathbb{P}^2$. Since a general
such polynomial $f$ (for example, $f(x, y, z) = x^{p-1}y + y^{p-1}w + w^{p-1}x$) has a finite
set of critical points, we see that for such $f$ the scheme $D_f$ is a normal surface.
By adjunction, we have that $K_{D_f} \cong \mathcal{O}_{D_f}(p-3)$ is big. We will write $X_f \to D_f$
for a minimal resolution of $X_f$; the preceding considerations show that $X_f$ is a
smooth surface of general type.
   Since $\pi$ is purely inseparable, the map $\pi$ is a homeomorphism, but not an
isomorphism, as $D_f$ is not smooth. This gives counterexamples to Main Theorem (1.2.1) in positive characteristic.

Note that the example described here comes from a purely inseparable homeomorphism $X \to P$. In particular, $X$ and $P$ have isomorphic perfections. This leads naturally to the following question:

**Question 10.6.4.** Suppose $X$ and $Y$ are proper normal varieties of dimension at least 2 over uncountable algebraically closed fields with perfections $X^{\mathrm{perf}}$ and $Y^{\mathrm{perf}}$. Is the map

$$\mathrm{Isom}(X^{\mathrm{perf}}, Y^{\mathrm{perf}}) \to \mathrm{Isom}(|X|, |Y|) \quad \text{a bijection?}$$

In the spirit of Grothendieck and Voevodsky, it is also natural to ask the following question.

**Question 10.6.5.** Is the perfection of a normal scheme of positive dimension over an uncountable algebraically closed field uniquely determined by its proétale topos?

The following is probably old; [Ter85] attributes a version of it to T. Shioda.

**Conjecture 10.6.6.** *Let $k$ be a field that is not locally finite, $X$ a normal projective $k$-variety of dimension $\geq 3$ and $H$ an ample Cartier divisor. Then, for $m \gg 1$ and for 'most' $k$-divisors $D \in |mH|(k)$, the restriction maps*

$$\mathrm{Pic}(X) \to \mathrm{Pic}(D) \quad \text{and} \quad \mathrm{Cl}(X) \to \mathrm{Cl}(D) \quad \text{are isomorphisms.}$$

The traditional statement of the Noether-Lefschetz theorem says that the conclusion holds outside a countable union of proper, closed subvarieties of $|mH|$; see [Gro05] for the Picard group and [RS06, RS09, Ji21] for the class group. This gives a positive answer to the conjecture whenever $k$ is uncountable. Using the ideas of [And96, MP12, Amb18, Chr18], this implies the claim for all algebraically closed fields, save the locally finite ones; see [Ji21, Sec.4.2]. See also [Ter85] for similar results over $\mathbb{Q}$ for complete intersections in $\mathbb{P}^n$.

For independence of intersection points, we start with an example. Let $E \subset \mathbb{P}^2$ be an elliptic curve, and $L \subset \mathbb{P}^2$ a very general line intersecting $E$ at three points $p_1, p_2, p_3$. Let $[H] \in \mathrm{Pic}(E)$ denote the hyperplane class. One can see that $m_1 p_1 + m_2 p_2 + m_3 p_3 \sim nH$ holds only for $m_1 = m_2 = m_3 = n$. This is easy over $\mathbb{C}$, but gets quite a bit harder over $\mathbb{Q}$ if we want the line to be also defined over $\mathbb{Q}$. The following conjecture says that a similar claim holds for all smooth curves.

**Conjecture 10.6.7.** *Let $k$ be a field that is not locally finite. Let $C$ be a smooth, projective curve of genus $\geq 1$ over $k$ and $L$ a very ample line bundle on $C$. For a section $s \in H^0(C, L)$ write $\{p_i(s) : i \in I\}$ (resp. $\{\bar{p}_i(s) : i \in \bar{I}\}$) for the closed points (resp. $\bar{k}$-points) of $(s = 0)$. Then, for 'most' sections, we have injections*

(1) $\oplus_{i \in I} \mathbb{Z}[p_i(s)] \hookrightarrow \mathrm{Pic}(C)$ *(weak form),*

(2) $\oplus_{i \in \bar{I}} \mathbb{Z}[\bar{p}_i(s)] \hookrightarrow \mathrm{Pic}(C_{\bar{k}})$ *(strong form).*

It is not clear what 'most' should mean. It is possible that this holds outside a field-locally thin set (9.3.1), but some heuristics suggest otherwise. For $k = \mathbb{C}$, a proof due to Voisin is explained in [Ji21, 4.5].

For the proof of (9.6.5) we would need the following stronger variant. If true, it would allow us to prove Main Theorem (1.2.1) for 3-folds as well.

**Conjecture 10.6.8.** *Using the notation of (10.6.7), let $A \subsetneq \mathbf{Pic}^{\circ}(C)$ be an abelian subvariety and $\Gamma \subset \mathrm{Pic}(C)$ a finitely generated subgroup that contains $[L]$. Then, for 'most' sections, we have injections*

$$\oplus_{i \in I} \mathbb{Z}[p_i(s)] / \textstyle\sum_{i \in I}[p_i(s)] \;\hookrightarrow\; \mathrm{Pic}(C)/\langle A(k), \Gamma \rangle \;\text{(weak form)},$$

$$\oplus_{i \in \overline{I}} \mathbb{Z}[\overline{p}_i(s)] / \textstyle\sum_{i \in \overline{I}}[\overline{p}_i(s)] \;\hookrightarrow\; \mathrm{Pic}(C_{\overline{k}})/\langle A(\overline{k}), \Gamma \rangle \;\text{(strong form)}.$$

The next two conjectures posit that, for 'most' ample line bundles, every section has many zeros.

**Conjecture 10.6.9.** *Let $K$ be an algebraically closed field other than $\overline{\mathbb{F}}_p$. Let $C$ be a smooth, projective curve over $K$. Then, for 'most' ample line bundles $L$, every section of $L^m$ has at least $g(C)$ zeros for every $m \geq 1$.*

Line bundles of degree $d$ that have a section with fewer than $g$ zeros form a closed subset of dimension $g - 1$ of $\mathbf{Pic}^d(C)$ obtained as the image of the maps

$$\varphi_{\mathbf{m}} \colon C^{g-1} \to \mathbf{Pic}^d(C) \;\text{ given by }\; (c_1, \ldots, c_{g-1}) \mapsto \mathcal{O}_C\left(\textstyle\sum_i m_i[c_i]\right),$$

where $\mathbf{m} := (m_1, \ldots, m_{g-1})$ such that $\sum m_i = d$. Thus, (10.6.9) is true if $K$ is uncountable. The most interesting open case is probably $\overline{\mathbb{Q}}$. By (9.9.9), there is a curve $C$ and a line bundle $L$ over $\overline{\mathbb{Q}}$, such that every section of $L^m$ has at least two zeros for every $m \geq 1$.

We prove the nodal rational curve cases of (10.6.9) in (9.9.8).

Thinking of the curve $C$ as a subvariety of its Jacobian leads to the following stronger form.

**Conjecture 10.6.10.** *Let $K$ be an algebraically closed field other than $\overline{\mathbb{F}}_p$. Let $A$ be an abelian variety over $K$ and $Z_i \subset A$ subvarieties such that $\sum_i \dim Z_i < \dim A$. Then, for 'most' $p \in A(K)$, the equation*

$$n[p] = \textstyle\sum_i m_i[z_i] \quad n, m_i \in \mathbb{Z}, z_i \in Z_i(K),$$

*has only the trivial solution $n = m_i = 0$.*

Next we give an example with only one $Z_i$ where this holds.

**Example 10.6.11.** Let $k$ be any field. Assume that $A = B \times E$, where $B$ is a simple abelian variety and $E$ an elliptic curve, and we have only one $Z = Z_1 \subset A$ of dimension $\leq \dim A - 2$. Assume also that $Z$ does not contain any translate of $E$.

Let $\pi \colon A \to B$ be the coordinate projection. If $p \in E(k)$, $z \in Z(k)$, and $n[p] = m[z]$, then $m[\pi(z)] = 0$, that is, $\pi(z)$ is a torsion point in $\pi(Z)$. By

[Zha98] there are only finitely many such $z$, so there are only finitely many $\{z_j \in Z : j \in J\}$ for which there is an $m_j > 0$ such that $m_j[z_j] \in E$.

Thus, if $p \in E(k)$ then $n[p] = m[z]$ has a nontrivial solution if and only if $p$ is in the saturation of $m_j[z_j]$ for some $j \in J$.

If $\text{rank}_{\mathbb{Q}} E(k) \geq 2$, then finitely many subgroups of $\mathbb{Q}$-rank 1 do not cover $E(k)$.

To get such Jacobian examples, fix an elliptic curve $E$ over $\overline{\mathbb{Q}}$ and let $C$ be a sufficiently general member of a very ample linear system on $E \times \mathbb{P}^1$. Then, by [Koc18, 1.6], $\mathbf{Jac}(C)$ is isogeneous to the product of $E$ and of a simple abelian variety $B$.

# Chapter Eleven

## Appendix

In this appendix we collect various results that do not fit neatly within the main part of the exposition.

### 11.1  BERTINI-TYPE THEOREMS

The basic Bertini theorem says that if $X \subset \mathbb{P}^n$ is an irreducible variety of dimension $\geq 2$ over an algebraically closed field and $H \subset \mathbb{P}^n$ is a general hyperplane, then $X \cap H$ is irreducible and smooth outside $\operatorname{Sing} X$. There are many similar Bertini-type theorems saying that if $X$ has a property $\mathscr{P}$, then a general member of a linear system also has property $\mathscr{P}$.

We start with the geometric versions and then discuss some variants that explore the difference between irreducibility and geometric irreducibility.

**Definition 11.1.1.** Fix an algebraically closed field $K$. Let $X$ be a quasi-projective $K$-variety and $|H_i|$ finite-dimensional (possibly incomplete) linear systems on $X$. Assume that $X$ satisfies a property $\mathscr{P}$. We say that $\mathscr{P}$ is *inherited by general complete $|H_i|$-intersections* if there is an open, dense subset $U \subset \prod_i |H_i|$ such that if $(D_1, \ldots, D_r) \in U$ then $Z := D_1 \cap \cdots \cap D_r$ also satisfies $\mathscr{P}$.

We, of course, need to assume that the $|H_i|$ are reasonably large. The following assumptions are close to optimal in most cases.

**Theorem 11.1.2** (Bertini smoothness theorem). *Let $X$ be a quasi-projective variety over an algebraically closed field $K$ and $|H_i|$ finite-dimensional (possibly incomplete) linear systems on $X$. Assume that*

(1) *either* $\operatorname{char} K = 0$ *and the* $|H_i|$ *are basepoint-free,*

(2) *or the* $|H_i|$ *are very ample.*

*The following properties are inherited by general complete $|H_i|$-intersections:*

(3) *smoothness,*

(4) *Serre's condition $S_2$,*

(5) *normality.*

*Proof.* This follows from [FOV99, 3.4.9 and 3.4.13a]. Alternatively, these are proved in [Har77, II.8.18, III.10.9] and [Jou83], but for statements (4) and (5) one needs to use that a general hypersurface section of an $S_2$ scheme is also $S_2$; see [Gro60, IV.12.1.6]. $\qquad\square$

Applying this to subvarieties, we get the following.

**Corollary 11.1.3.** *Assume that $K, X$, and the $|H_i|$ are as in (11.1.2). Let $W_j \subset X$ be a finite set of locally closed subvarieties. Then for a general, complete $|H_i|$-intersection $Z := D_1 \cap \cdots \cap D_r$ we have*

(1) $\operatorname{codim}(Z \cap W_j, Z) = \operatorname{codim}(W_j, X)$ *or the intersection is empty;*

(2) *if the $W_j$ are smooth, then so are $Z \cap W_j$;*

(3) *if the $W_j$ are $S_2$, then so are $Z \cap W_j$;*

(4) *if the $W_j$ are normal, then so are $Z \cap W_j$.* $\qquad\square$

In many applications, we need complete intersections that are special in some respects but general in some others. Here we deal with local conditions.

**11.1.4.** A set LC of *local conditions* consists of finitely many (not necessarily closed) points $p_j \in X$ with maximal ideals $m_j \subset \mathcal{O}_{p_j, X}$, natural numbers $n_{ij}$ and $g_{j1}, \ldots, g_{jr} \in \mathcal{O}_{p_j, X}$.

We always assume that none of the $p_j$ are generic points of $X$, and $p_i \notin \overline{p_j}$ for $i \neq j$. The *base locus* of LC is $B(\mathrm{LC}) := \cup_j \overline{p_j} \subset X$.

We say that a divisor $D_i$ *satisfies* LC if $g_{ji}$ is a local equation for $D_i$ at $p_j$ modulo $m_j^{n_{ji}}$ for every $j$. (Thus if $n_{ji} = 0$ then no condition is imposed on $D_i$ at $p_j$.)

For example, if $g_{ji}(p_j) \neq 0$ and $n_{ji} \geq 1$ then this just says that $p_j \notin D_i$. If $g_{ji} \in m_j$ and $n_{ji} \geq 1$ then we get that $\overline{p_j} \subset D_i$. If $p_j$ is a smooth point, $g_{ji} \in m_j \setminus m_j^2$, and $n_{ji} \geq 2$ then $D_i$ is generically smooth along $\overline{p_j}$.

Given a linear system $|H_i|$, the set of $D_i \in |H_i|$ that satisfy LC form an open subset of a linear subspace of $|H_i|$; we denote the latter by $|H_i, \mathrm{LC}|$.

We say that $D_1 \cap \cdots \cap D_r$ satisfies LC if every $D_i$ satisfies LC. Such intersections are parametrized by an open, dense subset of $\prod_i |H_i, \mathrm{LC}|$, which is a product of projective spaces (but may be empty).

The following combines the local conditions with (11.1.2) (2)–(5). We assume for simplicity that $X$ is projective, so the complete linear systems $|m_i H_i|$ are finite-dimensional.

**Corollary 11.1.5.** *Let $X$ be a projective variety over an algebraically closed field $K$ and $H_i$ ample divisors on $X$. Let LC be a set of local conditions with* *base locus $B = B(\mathrm{LC})$. Then, for $d_i \gg 1$, the restrictions*

$$|d_i H_i, \mathrm{LC}||_{X \setminus B(\mathrm{LC})}$$

*are very ample. In particular, if $X \setminus B$ is smooth (resp. normal) then $(D_1 \cap \cdots \cap D_r) \setminus B$ is also smooth (resp. normal) for general $(D_1, \ldots, D_r) \in \prod_i |d_i H_i, \mathrm{LC}|$.*
□

**Remark 11.1.6.** Note that per our conventions in (2.5.2), the linear systems $|d_i H_i|$ are complete, but the restrictions may be incomplete. Also, if the codimension of $B$ is $\geq r + 1$, then $D_1 \cap \cdots \cap D_r$ is the closure of $(D_1 \cap \cdots \cap D_r) \setminus B$. In these cases $D_1 \cap \cdots \cap D_r$ is a true complete intersection.

**11.1.7** (General fields). Assume now that $X$ and the other data are defined over a field $k$. Applying (11.1.5) over its algebraic closure, we get $U \subset \prod_i |d_i H_i, \mathrm{LC}|$, and the $k$-points of $U$ correspond to complete intersections defined over $k$ with the desired properties. Since $U$ is an open subset of a rational variety, this implies that the $k$-points are dense if $k$ is infinite, and we get the above results also over such fields. If $k$ is finite, however, then a dense open subset of $\mathbb{P}^n$ may be disjoint from $\mathbb{P}^n(k)$. Nonetheless, the results of [Poo08, CP16] say that the open sets $U$ above should have many $k$-points as $d_i$ grows, though not all cases have been worked out.

Our proofs have other problems with finite fields, so we will be able to make only very limited use of these cases.

Next we discuss a Bertini-type irreducibilty theorem over infinite fields. The key point is to understand the relationship between irreducibility and geometric irreducibility. We use mainly (11.1.16), which does not seem to be treated in the standard reference books, so we give details.

**Lemma 11.1.8.** *Let $K/k$ be a normal, algebraic field extension with Galois group $G = \mathrm{Aut}(K/k)$. Let $X$ be a $k$-scheme of finite type.*

(1) *$Z \mapsto \mathrm{red}(Z_K)$ provides a one-to-one correspondence between*

    (a) *closed subsets of $X$, and*

    (b) *closed, $G$-invariant subsets of $X_K$.*

(2) *$Z$ is irreducible if and only if $G$ acts transitively on the set of irreducible components of $Z_K$.*

(3) *Assume that $X$ is irreducible and let $X_K^1 \subset X_K$ be an irreducible component. If $Z_K \cap X_K^1$ is irreducible then $Z$ is irreducible.*

*Proof.* The first two claims are clear. Let $\{X_K^i \subset X_K : i \in I\}$ be the irreducible components. Since $X$ is irreducible, $G$ acts transitively on the $X_K^i$. Thus, it also acts transitively on the $Z_K \cap X_K^i$. We conclude that (2) implies (3) since $Z_K = \cup_i (Z_K \cap X_K^i)$. □

**Lemma 11.1.9.** *Let $X$ be a normal scheme over a field $k$ and $K/k$ a finite field extension. Then every connected component of $X_K$ is irreducible.*

*Proof.* If $k$ is perfect then $X_K$ is also normal and the claim is clear. The general case needs more work; see [Sta22, Tag 0BQ1]. □

**Lemma 11.1.10.** *Let $X$ be an irreducible scheme over a field $k$. Assume that $X$ has a normal $k$-point $x_0$. Then $X$ is geometrically irreducible.*

*Proof.* Let $K/k$ a finite, normal field extension and $X_K^i \subset X_K$ the irreducible components. Since $X$ is irreducible, the Galois group $\mathrm{Gal}(\overline{k}/k)$ acts transitively on the $X_K^i$ by (11.1.8 (1) (a)).

By (11.1.9), $x_{0,K}$ is contained in a unique irreducible component of $X_K$; call it $X_K^1$. Since $x_0$ is a $k$-point, every Galois conjugate of $X_K^1$ also contains $x_0$. So $X_K^1$ is the unique irreducible component of $X_K$. □

**Lemma 11.1.11.** *Let $p\colon Y \to S$ be a morphism of irreducible $k$-schemes of finite type. Assume that there is a section $\sigma\colon S \to Y$ such that $Y$ is normal at the generic point of $\sigma(S)$. Then there is a dense, open $S^\circ \subset S$ such that all fibers over $S^\circ$ are geometrically irreducible.*

*Proof.* The generic fiber is irreducible and the generic point of $\sigma(S)$ is a normal $k(S)$-point on it. Thus, the generic fiber is geometrically irreducible by (11.1.10). The rest follows from [Sta22, Tag 0559]. □

**Lemma 11.1.12.** *Let $X$ be an irreducible $k$-variety of dimension $\geq 2$ and $|H| = |H_1, H_2|$ a mobile, linear pencil with base locus $B = H_1 \cap H_2$. Assume that there is a point $x \in B(k)$ that is smooth on both $B$ and $X$. Then all but finitely many members of $|H|_{\overline{k}}$ are irreducible.*

*Proof.* Let $\pi\colon X' \to X$ denote the blowup of $B$. The birational transform of $|H|$ defines a morphism $p\colon X' \to \mathbb{P}^1$. Note that $\pi^{-1}(p) \cong \mathbb{P}^1$ gives a section of $p$. The rest follows from (11.1.11). □

**Lemma 11.1.13.** *Let $K$ be an algebraically closed field, $S$ an irreducible $K$-surface and $p\colon S \to \mathbb{P}^2$ a finite morphism. Then there is a dense, open subset $\check{U} \subset \check{\mathbb{P}}^2$ in the dual projective space such that $p^{-1}(L) \subset S$ is irreducible for every $[L] \in \check{U}$.*

*Proof.* See [Jou83, 6.10.3]. □

**Theorem 11.1.14** (Bertini irreducibility theorem I). *Let $K$ be an algebraically closed field, $X$ an irreducible $K$-variety and $|H|$ a finite-dimensional, mobile, (possibly incomplete) linear system such that the image $X \dashrightarrow |H|^\vee$ has dimension $\geq 2$. Then there is a dense, open subset $U \subset |H|$ such that $H_\lambda \subset X$ is irreducible for $[H_\lambda] \in U$.*

*Proof.* It is enough to show that $H_\lambda \subset X$ is irreducible for a dense set. For a dense set of linear projections $|H|^\vee \dashrightarrow \mathbb{P}^2$, the composite map $X \dashrightarrow \mathbb{P}^2$ is dominant. Let $X'$ be the normalization of the graph of $X \dashrightarrow \mathbb{P}^2$ and $p' \colon X' \to S$ and $p \colon S \to \mathbb{P}^2$ the Stein factorization. By (11.1.13) $p^{-1}(L) \subset S$ is irreducible for a general line $L \subset \mathbb{P}^2$, and so is $p'^{-1}(p^{-1}(L))$. $\qquad\square$

**Theorem 11.1.15** (Bertini irreducibility theorem II). *Let $k$ be an infinite field, $X$ an irreducible $k$-variety, and $|H|$ a finite-dimensional, mobile, (possibly incomplete) linear system such that the image $X \dashrightarrow |H|^\vee$ has dimension $\geq 2$. Then there is a dense, open subset $U \subset |H|$ such that $H_\lambda \subset X$ is irreducible for $[H_\lambda] \in U(k)$.*

*Proof.* As in (11.1.14), we can reduce to the case when $S$ is an irreducible $k$-surface, $p \colon S \to \mathbb{P}^2$ a finite morphism, and $|H|$ the pullback of $|\mathcal{O}_{\mathbb{P}^2}(1)|$. Let $K \supset k$ be an algebraic closure.

Let $\{S_K^i \subset S_K : i \in I\}$ be the irreducible components and $p^i \colon S_K^i \to \mathbb{P}^2$ the restriction of $p_K$. For each $i$ we have a dense, open subset $U^i \subset \check{\mathbb{P}}^2$ as in (11.1.13); set $U := \cap_{i \in I} U^i$. If $[L] \in U(k)$ then $p^{-1}(L)$ is irreducible by (11.1.8 (1) (b)). $\qquad\square$

We mostly use the following special case, obtained by combining (11.1.15) and (11.1.5).

**Corollary 11.1.16.** *Let $k$ be an infinite field, $X$ a projective $k$-variety, and $H$ an ample, Cartier divisor on $X$. Let $\{Z_i \subset X : i \in I\}$ be a finite collection of irreducible subvarieties such that $\dim Z_i \geq 2$ for every $i$. Let $W \subset X$ be a closed subscheme.*

*Then, for $m \gg 1$, there is a dense, open subset $U_m \subset |mH|(-W)$, where $|mH|(-W)$ denotes the subspace of $|mH|$ of divisors containing $W$ scheme-theoretically, such that for every $[H_\lambda] \in U_m(k)$ and $i \in I$, the following hold.*

(1) $H_\lambda \cap (Z_i \setminus W)$ *is irreducible.*

(2) *If $Z_i$ is geometrically reduced, then so is $H_\lambda \cap (Z_i \setminus W)$.*

(3) $H_\lambda \cap (Z_i^{\mathrm{sm}} \setminus W)$ *is smooth, where $Z_i^{\mathrm{sm}} \subset Z_i$ denotes the smooth locus.*

*Furthermore, if $X$ and $W$ are regular at the generic points of $W$, then*

(4) $H_\lambda$ *is regular at the generic points of $W$.* $\qquad\square$

## 11.2   COMPLETE INTERSECTIONS

In this section we collect various results on complete intersections that we need.

**11.2.1.** Let $X$ be an irreducible variety. A subscheme $Z \subset X$ of codimension $r$ is a *complete intersection* (resp. *set-theoretic complete intersection*) if there are effective Cartier divisors $D_1, \ldots, D_r$ such that $Z = D_1 \cap \cdots \cap D_r$ scheme-theoretically (resp. $\mathrm{Supp}\, Z = \mathrm{Supp}(D_1 \cap \cdots \cap D_r)$).

If the $D_i$ are ample, we call $Z$ an *ample (set-theoretic) complete intersection*, usually abbreviated as *ample-ci* (resp. *ample-sci.*)

If $H$ is a Cartier divisor and $D_i \in |m_i H|$ for every $i$, then we say that $Z$ is a *complete $H$-intersection*. We usually abbreviate this as $H$-*ci*, and the set-theoretic version as $H$-*sci*.

If $Z$ and the $D_i$ are furthermore irreducible then we say that $Z$ is *irreducible-sci* or *isci*. For us $H$-*isci* subvarieties are especially useful.

Ample complete intersections inherit many properties of a variety, but the strongest results are for general complete intersections, that is, when the $D_i \in |m_i H|$ are sufficiently general.

Let $Z$ be a scheme. Connectedness and irreducibility of $Z$ depends only on the topological space $|Z|$, but geometric connectedness and geometric irreducibility cannot be determined using $|Z|$ only.

We frequently need to guarantee that certain schemes are geometrically connected. The next criterion can be proved by repeatedly using [Har77, II.7.8]; see also [Har62].

**Claim 11.2.2.** *Let $X$ be a normal, projective, geometrically irreducible variety and $Z \subset X$ a positive dimensional ample-sci. Then $Z$ is geometrically connected.*                    $\square$

Note that a proper $k$-scheme $Y$ is geometrically connected if and only if $H^0(Y, \mathcal{O}_Y)$ is a local, Artin $k$-algebra such that $H^0(Y, \mathcal{O}_Y)/\sqrt{0}$ is a purely inseparable field extension of $k$. We can thus restate (11.2.2) as follows.

**Claim 11.2.3.** *Let $X$ be a normal, projective, geometrically irreducible variety and $Z \subset X$ a positive dimensional, reduced ample-sci. Then $k[Z]/k[X]$ is purely inseparable.*                    $\square$

We also use the following variant of the Lefschetz hyperplane theorem, essentially due to [Nér52b]. For the Picard variety, this is proved in [Gro05]. For normal varieties in characteristic 0, the class group version is proved in [RS06, RS09]. For positive characteristic see [Ji21, Prop.3.1].

**Theorem 11.2.4.** *Let $X$ be a geometrically normal, projective variety and $|H|$ an ample linear system on $X$. Let $Z \subset X$ be a normal, $\geq 2$-dimensional, general complete $H$-intersection. Then the restriction map*

$$\mathrm{Cl}(X) \to \mathrm{Cl}(Z) \ \text{ is injective.}$$

**11.2.5** (Disjointness of conjugates). Let $K/k$ be a finite separable field extension, let $V_k$ be a $k$-vector space of dimension $n$, and let $W_k$ be a $k$-vector space of dimension $r$. Denote by $V_K$ and $W_K$ their scalar extensions to $K$. Fix also an algebraic closure $k \hookrightarrow \bar{k}$ of $k$. The space

$$H := \mathrm{Hom}_K(V_K, W_K)$$

has a natural structure of a $k$-scheme $\mathbb{H}$. Its functor of points sends a $k$-algebra $R$ to the set of $K \otimes_k R$-linear maps

$$V_K \otimes_k R \to W_K \otimes_k R.$$

In particular, since $K/k$ is separable we have

$$\mathbb{H}(\overline{k}) = \prod_{\sigma: K \hookrightarrow \overline{k}} \mathrm{Hom}_{\overline{k}}(V_{\overline{k}}, W_{\overline{k}}), \qquad (11.2.5.1)$$

where the product is taken over embeddings $\sigma: K \hookrightarrow \overline{k}$ over $k$.

Let

$$A \in \mathbb{H}(k) = \mathrm{Hom}_K(V_K, W_K)$$

be a map over $K$, and for an embedding $\sigma: K \hookrightarrow \overline{k}$ let

$$A^\sigma \in \mathrm{Hom}_{\overline{k}}(V_{\overline{k}}, W_{\overline{k}})$$

be the map induced by scalar extension along $\sigma$. So under the identification (11.2.5.1) the image of $A$ under the map

$$\mathbb{H}(k) \to \mathbb{H}(\overline{k})$$

is the vector of maps $(A^\sigma)_{\sigma: K \hookrightarrow \overline{k}}$. For $A \in \mathbb{H}(k)$ and $\sigma: K \hookrightarrow \overline{k}$ let $L_{A,\sigma} \subset V_{\overline{k}}$ denote the kernel of $A^\sigma$. If $A^\sigma$ is surjective then $L_{A,\overline{\sigma}}$ has codimension $r$ in $V_{\overline{k}}$.

**Claim 11.2.6.** *There exists a nonempty Zariski open subset $U \subset \mathbb{H}$ such that for any $k$-point $A \in U(k)$ and distinct embeddings*

$$\sigma_1, \sigma_2: K \hookrightarrow \overline{k}$$

*the codimension of $L_{A,\sigma_1} \cap L_{A,\sigma_2}$ in $V_{\overline{k}}$ is equal to $\min\{2r, n\}$.*

*Proof.* We first reduce to the critical case when $n = 2r$. Consider an inclusion of $k$-vector spaces

$$V_k \hookrightarrow V_{k'} \qquad (11.2.6.1)$$

and let $\mathbb{H}'$ be the $k$-scheme of maps $V_K' \to W_K$. Restriction defines a smooth surjective morphism of $k$-schemes

$$\pi: \mathbb{H}' \to \mathbb{H}.$$

In fact, a splitting of (11.2.6.1) identifies $\mathbb{H}'$ with a product of $\mathbb{H}$ with a smooth affine scheme.

If $n < 2r$, choose an injection of $V_k$ into $V_{k'}$ of dimension $2r$. Assuming the case $n = 2r$ we get a dense open subset $U' \subset \mathbb{H}'$ such that for $A' \in U'(k)$ and two embeddings $\sigma_1, \sigma_2$ we have

$$0 = L_{A',\sigma_1} \cap L_{A',\sigma_2} \subset V_{k'}'.$$

It follows that if $U := \pi(U') \subset \mathbb{H}$ then for $A \in U(k)$ we also have $L_{A,\sigma_1} \cap L_{A,\sigma_2} = 0$ as asserted in the claim. This reduces the proof to the case $n \geq 2r$.

Next consider the case when $n > 2r$ and choose a subspace $V'_k \subset V_k$ of dimension $2r$. Then as in the preceding paragraph we get the scheme of maps $\mathbb{H}'$ for $V'_k$ and a smooth surjection

$$\pi : \mathbb{H} \to \mathbb{H}'.$$

Again assuming the case $n = 2r$ we get an open subset $U' \subset \mathbb{H}'$ with the desired properties. Let $U \subset \mathbb{H}$ be the preimage of $U'$. Then for $A \in U(k)$ we have

$$L_{A,\sigma_1} \cap L_{A,\sigma_2} \cap V'_{\overline{k}} = 0,$$

which implies that the codimension of $L_{A,\sigma_1} \cap L_{A,\sigma_2}$ in $V_k$ is at least $2r$. Since this is also the maximum possible codimension, this gives the claim for $n > 2r$.

So to complete the proof of the claim it suffices to consider the case $n = 2r$. Let $K' \subset \overline{k}$ be the Galois closure of $\sigma_1(K) \subset \overline{k}$ so that $\sigma_2$ and $\sigma_1$ differ by an automorphism $\sigma$ of $K'$. Let $\mathbb{H}'$ be the scheme of maps defined using $K'$ instead of $K$. For a $k$-algebra $R$ and $R$-point $A \in \mathbb{H}'(R)$ corresponding a $K' \otimes_k R$-linear map

$$V_{K'} \otimes_k R \to W_{K'} \otimes_k R$$

we can consider the associated twist $A^\sigma$ obtained by post-composing with the automorphism

$$W_{K'} \otimes_k R \to W_{K'} \otimes_k R$$

obtained by applying $\sigma$ to $K'$. Taking the determinant of the induced map

$$(A, A^\sigma) : V_{K'} \otimes_k R \to (W_{K'} \oplus W_{K'}) \otimes_k R$$

we obtain an element of $K' \otimes_k R$, well defined up to units. If $R$ is furthermore a $K'$-algebra then composing with the induced map $K' \otimes_k R \to R$ we obtain an element of $R$. The locus where this element is zero is an open subset $\widetilde{U}'_{K'} \subset \mathbb{H}'_{K'}$, where $\mathbb{H}_{K'}$ denotes the base change of $\mathbb{H}$ from $k$ to $K'$. Let $U'_{K'} \subset \mathbb{H}'_{K'}$ denote the intersection of the Galois conjugates of $\widetilde{U}'_{K'}$. Then $U'_{K'}$ is $G_{K'/k}$-invariant and therefore descends to an open subset $U' \subset \mathbb{H}'$. A point $A \in \mathbb{H}'(k)$ lies in $U'$ if and only if for every embedding $\lambda : K' \to \overline{k}$ we have (here we write $L'$ instead of $L$ to emphasize that we are working with $K'$)

$$L'_{A,\lambda} \cap L'_{A,\lambda\sigma} = 0.$$

Scalar extension from $K$ to $K'$ defines an inclusion

$$i : \mathbb{H} \hookrightarrow \mathbb{H}'.$$

To complete the proof of the claim it suffices to show that $U := i^{-1}(U')$ is nonempty.

Choose bases $V_k \simeq k^{2r}$ and $W_k \simeq k^r$, and let $A, C \in \mathrm{Mat}_{r\times r}(k)$ be matrices over $k$ and set $B = \alpha C$ for $\alpha \in K$ a primitive element for $K/k$ (which exists since $K/k$ is separable). Then the map

$$(A, B)\colon K^{2r} \to K^r$$

defines an element of $\mathbb{H}(k)$ that lies in $U$. Indeed, if $\lambda\colon K \hookrightarrow \overline{k}$ is an embedding then the induced map

$$\overline{k}^{2r} \to \overline{k}^{2r}$$

given by the two embeddings $\lambda$ and $\sigma\lambda$ is given by the matrix

$$\begin{pmatrix} A & \lambda(\alpha)C \\ A & \sigma\lambda(\alpha)C \end{pmatrix}.$$

By row reduction the determinant of this matrix is the same as the determinant of

$$\begin{pmatrix} A & \lambda(\alpha)C \\ 0 & \sigma\lambda(\alpha) - \lambda(\alpha))C \end{pmatrix},$$

which is $(\sigma\lambda(\alpha)^\sigma - \lambda(\alpha))^r \det A \det C$. This is nonzero since $\alpha$ is a primitive element. $\qquad\square$

The scheme $\mathbb{H}$ can be viewed as the Weil restriction, denoted by $\mathfrak{R}_k^K(\ )$, from $K$ to $k$ of the scheme of maps $V_K \to W_K$ over $K$; see [BLR90, Sec. 7.6]. We can thus globalize (11.2.6) first to projective spaces and then to their subvarieties as follows.

**Claim 11.2.7.** *Let $X$ be a $k$-variety of pure dimension $n$ and $|M_1|, \ldots, |M_r|$ basepoint-free linear systems. Let $K/k$ be a finite, separable field extension. Then there is a dense, Zariski open subset*

$$U \subset \mathfrak{R}_k^K |M_1|^{\mathrm{var}} \times \cdots \times \mathfrak{R}_k^K |M_r|^{\mathrm{var}},$$

*such that, if $(D_1, \ldots, D_r) \in U$, then*

$$\mathrm{codim}_X\left(D_1^{\sigma_1} \cap \cdots \cap D_r^{\sigma_1} \cap D_1^{\sigma_2} \cap \cdots \cap D_r^{\sigma_2}\right) = \min\{2r, n+1\},$$

*for all pairs of distinct d distinct $k$-embeddings $\sigma_1, \sigma_2\colon K \hookrightarrow \overline{k}$.* $\qquad\square$

*Proof.* For any variety $V$ over $K$, $R$-points of $(\mathfrak{R}_k^K V)_{\overline{k}}$ for a ring $R$ containing $\overline{k}$ are the same thing as $R \otimes_k K$-points of $V$. Because $R \otimes_k K$ is isomorphic to $R^{[K:k]}$, with the factors indexed by embeddings $\sigma\colon K \to \overline{k}$, $(\mathfrak{R}_k^K V)_{\overline{k}}$ is $V^{[K:k]}$, with the factors indexed by embeddings $\sigma$. The projection onto the $\sigma$th factor arises from the map of rings $R \otimes_k K \to R$ that is the identity on $R$ and $\sigma$ composed with the map $\overline{K} \to R$ on $K$.

Given a $k$-point of $\mathfrak{R}_k^K V$, base-changed to a $\overline{k}$-point of $(\mathfrak{R}_k^K V)_{\overline{k}}$, the projection onto the $\sigma$th factor is given by the embedding $\sigma$.

Applying this to $V = |M_i|^{\mathrm{var}}$, we obtain an isomorphism

$$\left(\mathscr{R}_k^K \, |M_1|^{\mathrm{var}} \times \cdots \times \mathscr{R}_k^K \, |M_r|^{\mathrm{var}}\right)_{\overline{k}} \to \left(|M_1|^{\mathrm{var}} \times \cdots |M_r|^{\mathrm{var}}\right)_{\overline{k}}^{[K:k]}.$$

Let $U'$ be the open subset of $\left(|M_1|^{\mathrm{var}} \times \cdots |M_r|^{\mathrm{var}}\right)_{\overline{k}}^{[K:k]}$ consisting of tuples of divisors $D_{i,\sigma}$ indexed by $i$ from 1 to $r$ and $\sigma \colon K \to \overline{k}$ such that for all distinct pairs $\sigma_1, \sigma_2 \colon K \to \overline{K}$ we have

$$\mathrm{codim}_X \left(D_{1,\sigma_1} \cap \cdots \cap D_{r,\sigma_1} \cap D_{1,\sigma_2} \cap \cdots \cap D_{r,\sigma_2}\right) = \min\{2r, n+1\}.$$

Then $U'$ is a nonempty Zariski open set.

The action of $\mathrm{Gal}(\overline{k}/k)$ is by permuting the embeddings $\sigma \colon K \to \overline{k}$, and so $U'$ is stable under $\mathrm{Gal}(\overline{k}/k)$, hence descends to an open set $U$ over $k$.

A $k$-point lies in $U$ if and only if the associated $\overline{k}$-point lies in $U'$, which happens if and only if the codimension condition holds.                    □

**Lemma 11.2.8.** *Let $k$ be an infinite field and $X$ a normal, projective $k$-variety of dimension $n > 2r$. Let $p, q \in X$ be closed points such that there are embeddings $k \subset k(p) \subset k(q)^{\mathrm{ins}} \subset \overline{k}$.*

*Then there is an irreducible, $r$-dimensional $k$-variety $W \subset X$ such that*

(1) *$p, q \in W$ and*

(2) *$k(p)/k[W]$ is purely inseparable.*

*Furthermore, if $p$ is a smooth, separable point of $X$ then we can also assume that*

(3) *$p$ is a smooth point of $W$.*

*Proof.* Let $k \subset K_p \subset k(p)$ and $k \subset K_q \subset k(q)$ be maximal separable subextensions. After base change to $K_p$, we have a degree 1 point $\overline{p}$ lying over $p$ and a degree $= \deg(K_q/K_p)$ point $\overline{q}$ lying over $q$. Let $\{\sigma\}$ be the set of all $k$-embeddings $\sigma \colon K_p \hookrightarrow \overline{k}$. Thus, $\overline{p}$ and $\overline{q}$ each have $\deg(K_p : k)$ conjugates over $k$ and these are disjoint from each other.

Next take a general ample-ci variety $W_1 \subset X_{K_p}$ that contains $\overline{p}$ and $\overline{q}$. By (11.2.7) the $W_1^\sigma$ are disjoint from each other. Thus, their union $W_{K_p} = \cup_\sigma W_1^\sigma$ descends to a $k$-subvariety $W \subset X$ with the required properties.                    □

## 11.3   PICARD GROUP, CLASS GROUP, AND ALBANESE VARIETY

Here we review various results on the Picard group, class group and Albanese variety that we need elsewhere. For the Picard group and Picard scheme, [Gro62,

Lects.V–VI], [Mum66, Sec.19], [Mum70], and [BLR90] contain proofs; for these we just fix our notation. Modern references for the class group and Albanese variety are harder to find; about these we give longer explanations.

We start with a summary of the basic facts about abelian varieties. A very good introduction is [Mil08]; more detailed treatments and further results can be found in [Mum70] and [BL04]. The abelian varieties that we encounter are either Picard varieties or Albanese varieties of normal, projective varieties.

**11.3.1** (Abelian varieties). Let $k$ be a field. An *abelian variety* over $k$ is a smooth, proper algebraic group over $k$. We denote the identity element by $e_A$ or $0_A$. The group operation is usually written additively since $A$ is commutative (11.3.1.4).

Historically the first examples were Jacobians of smooth, projective curves $\mathbf{Jac}(C)$. It turns out that every abelian variety is isomorphic to a closed subgroup of $\mathbf{Jac}(C)$ for some $C$; this almost follows from (11.3.14). This is quite useful conceptually at the beginning, but $C$ is not unique, and there does not seem to be any optimal way of choosing $C$, so it is not always helpful in proving theorems.

For the convenience of the reader we list a few basic facts about abelian varieties that we use.

In the following, we say that a sequence of abelian varieties

$$0 \to A_1 \xrightarrow{p} A_2 \xrightarrow{q} A_3 \to 0 \tag{11.3.1.1}$$

is *exact modulo torsion* if $p$ is a closed embedding, $q$ is surjective and $p(A_1) = (\ker q)^\circ$.

Recall also the following terminology. A morphism $p\colon A_1 \to A_2$ of abelian varieties is called an *isogeny* if $p$ is surjective and $\dim A_1 = \dim A_2$. Isogeny is an equivalence relation on abelian varieties and the dual of an isogeny is an isogeny. An abelian variety is called *simple* if it has no positive dimensional abelian subvarieties (other than itself).

(1) Every map $\pi\colon \mathbb{P}^1 \to A$ from the projective line to an abelian variety is constant.

(2) Let $g\colon X \dashrightarrow A$ be a rational map from a smooth variety $X$ to an abelian variety $A$. Then $g$ is a morphism. More generally, this holds for any variety $A$ that contains no rational curves; see [Kol96, VI.1.9]. A different proof is in [Mil08, 3.2].

(3) Let $g\colon A_1 \dashrightarrow A_2$ be a rational map of abelian varieties that sends the unit $e_1$ to $e_2$. Then $g$ is a morphism and a group homomorphism. In particular, $(A, e_A)$ determines the group structure. This is the main reason why the multiplication $\mu\colon A \times A \to A$ is usually suppressed in the notation.

(4) Every abelian variety is commutative.

(5) For abelian varieties, $A \mapsto \mathbf{Pic}^\circ(A)$ is a duality that preserves the dimension. The dual is frequently denoted by $\hat{A}$ or $A^t$. Over $\mathbb{C}$, this is a special case of the Appell-Humbert theorem [Mum70, pp. 21–22]. In general, see [Mil08, Sec.8] or [Mum70, Sec.13].

(6) Let $A$ be an abelian variety and $B \subset A$ a closed subgroup scheme. Then there is a unique abelian variety $A_3$ and an exact sequence

$$B \xrightarrow{p} A \xrightarrow{q} A_3 \to 0.$$

We call $A_3 := A/B$ the *quotient* of $A$ by $B$.

(7) $p \colon A_1 \to A_2$ is a closed embedding if and only if $\hat{p} \colon \hat{A}_2 \to \hat{A}_1$ is its own Stein factorization. (That is, $\hat{p}_* \mathcal{O}_{\hat{A}_2} = \mathcal{O}_{\hat{A}_1}$.) More generally, if (11.3.1.1) is exact (resp. exact modulo torsion) then so is its dual sequence

$$0 \to \hat{A}_3 \xrightarrow{\hat{q}} \hat{A}_2 \xrightarrow{\hat{p}} \hat{A}_1 \to 0.$$

See [Mil08, Sec.9] or [Mum70, Sec.15].

(8) (Poincaré reducibility theorem) Let $A$ be an abelian variety and $A_1 \hookrightarrow A$ an abelian subvariety. Then there is an abelian subvariety $A_3 \hookrightarrow A$ such that $A_1 + A_3 \to A$ is an isogeny.

(9) (Poincaré reducibility theorem, dual version) Let $A$ be an abelian variety and $A \twoheadrightarrow A_3'$ an abelian quotient. Then there is an abelian quotient $A \twoheadrightarrow A_1'$ such that $A \to A_1' + A_3'$ is an isogeny. It is worth noting that, unlike in the previous results, Poincaré reducibility fails for compact, complex analytic groups.

(10) Every abelian variety is isogenous to a product of simple abelian varieties. The simple factors are unique, up to isogeny.

**11.3.2** (Picard group of a normal variety). The group of line bundles on a scheme $X$ is the *Picard group* of $X$, denoted by $\mathrm{Pic}(X)$. If $X$ is proper then $\mathrm{Pic}^\circ(X) \subset \mathrm{Pic}(X)$ denotes the subgroup of divisors that are algebraically equivalent to 0. The quotient $\mathrm{NS}(X) := \mathrm{Pic}(X)/\mathrm{Pic}^\circ(X)$ is the *Néron-Severi* group of $X$. It is a finitely generated abelian group. Its $\mathbb{Q}$-rank is the *Picard number* of $X$, denoted by $\rho(X)$.

If $X$ is proper over a field $k$ with algebraic closure $\bar{k}$ then $\mathrm{Pic}(X_{\bar{k}})$ has a natural $k$-scheme structure, denoted by $\mathbf{Pic}(X)$. The identity component is denoted by $\mathbf{Pic}^\circ(X)$; it is a commutative algebraic group. If $X$ is geometrically normal and char $k = 0$ then $\mathbf{Pic}^\circ(X)$ is an abelian variety. If $X$ is geometrically normal and $k$ is perfect then red $\mathbf{Pic}^\circ(X)$ is an abelian variety. The nonreduced structure of $\mathbf{Pic}^\circ(X)$ will pay no role in our questions.

There is a natural inclusion $\mathrm{Pic}(X) \hookrightarrow \mathbf{Pic}(X)(k)$, which is an isomorphism if $X$ has a $k$-point. In general, the quotient $\mathbf{Pic}(X)(k)/\mathrm{Pic}(X)$ is a torsion group.

**11.3.3** (Class group of a normal variety). For a normal $k$-variety $X$, let $\mathrm{Cl}(X)$ denote the group of Weil divisors modulo linear equivalence. It is also isomorphic to the group of rank 1 reflexive sheaves, where the product is the double dual of the tensor product.

Let $\mathrm{Cl}^\circ(X) \subset \mathrm{Cl}(X)$ be the subgroup of divisors that are algebraically equivalent to 0. If $X$ is proper, we call the quotient $\mathrm{NS}^{\mathrm{cl}}(X) := \mathrm{Cl}(X)/\mathrm{Cl}^\circ(X)$ the *Néron-Severi class group*[*] of $X$, and its $\mathbb{Q}$-rank the *class rank* of $X$, denoted by $\rho^{\mathrm{cl}}(X)$.

Note that we have natural inclusions

$$\mathrm{Pic}^\circ(X) \subset \mathrm{Cl}^\circ(X) \ \ \text{and} \ \ \mathrm{NS}(X) \subset \mathrm{NS}^{\mathrm{cl}}(X), \tag{11.3.3.1}$$

which are isomorphisms if and only if every Weil divisor is Cartier, for example, when $X$ is smooth.

Basic results about these groups are the following.

**Lemma 11.3.4.** *Let* $p\colon Y \to X$ *be a birational morphism of normal, proper varieties over a perfect field* $k$. *Then*

(1) $p_*\colon \mathrm{Cl}^\circ(Y) \to \mathrm{Cl}^\circ(X)$ *is an isomorphism and*

(2) $p_*\colon \mathrm{NS}^{\mathrm{cl}}(Y) \twoheadrightarrow \mathrm{NS}^{\mathrm{cl}}(X)$ *is onto.*

**Lemma 11.3.5.** *Let* $X$ *be a normal, proper variety over a perfect field* $k$. *Then there is a normal, proper variety* $Y$ *and a birational morphism* $p\colon Y \to X$ *such that* $\mathrm{Cl}^\circ(Y_K) = \mathrm{Pic}^\circ(Y_K)$ *for every* $K \supset k$.

It is quickest to prove these by using the Albanese variety; see (11.3.9.5–6). As a consequence, we can define the scheme structure of $\mathbf{Cl}^\circ$ by

$$\mathbf{Cl}^\circ(X) \cong \mathbf{Cl}^\circ(Y) \cong \mathrm{red}\,\mathbf{Pic}^\circ(Y). \tag{11.3.5.1}$$

In the complex case these results are due to Picard [Pic1895] and Severi [Sev1906], but the most complete references may be the papers of Matsusaka [Mat52] and Néron [Nér52a]; see also [Kol18, Sec.3] for some discussions.

More recent results on various aspects of the class group of singular varieties are discussed in [BVS93, BVRS09, RS09].

**Definition 11.3.6.** Let $X$ be a normal, proper $k$-variety and $\Sigma \subset X$ a subset. Let $\mathrm{WDiv}(X,\Sigma) \subset \mathrm{WDiv}(X)$ and $\mathrm{Cl}(X,\Sigma) \subset \mathrm{Cl}(X)$ denote the subgroup of those Weil divisors that are Cartier at every point $x \in \Sigma$.

Note that $\mathrm{Cl}(X,\Sigma)$ is isomorphic to the group of those rank 1 reflexive sheaves that are locally free at every point $x \in \Sigma$.

We see in (11.3.7) that $\mathrm{Cl}(X_{\bar{k}}, \Sigma_{\bar{k}})$ is naturally identified with a closed $k$-subgroup $\mathbf{Cl}(X,\Sigma) \subset \mathbf{Cl}(X)$. We denote its identity component by $\mathbf{Cl}^\circ(X,\Sigma)$.

---

[*]The literature seems inconsistent. Frequently this is called the Néron-Severi group.

Note that in general $\mathbf{Cl}(X,\Sigma)\cap \mathbf{Cl}^\circ(X)$ may be disconnected.

The quotient $\mathrm{NS}^{\mathrm{cl}}(X,\Sigma) := \mathbf{Cl}(X,\Sigma)/\mathbf{Cl}^\circ(X,\Sigma)$ is finitely generated.

**Lemma 11.3.7.** *Let $X$ be a normal, proper variety over a perfect field $k$ and $\Sigma \subset X$ an arbitrary subset. Then there is a closed, algebraic $k$-subgroup*

$$\mathbf{Cl}(X,\Sigma) \subset \mathbf{Cl}(X)$$

*such that*

$$\mathbf{Cl}(X_{\bar{k}},\Sigma_{\bar{k}}) = \mathbf{Cl}(X,\Sigma)(\bar{k}).$$

*Proof.* Assume first that $\Sigma = \{x\}$ is a closed point and there is a universal family $\mathbf{L}$ on $X \times \mathbf{Cl}^\circ(X)$ that is flat over $\mathbf{Cl}^\circ(X)$. The set of points $V \subset X \times \mathbf{Cl}^\circ(X)$ where $\mathbf{L}$ is not locally free is closed. Since $\mathbf{Cl}(X,\{x\})$ is the complement of the image of $V \cap (\{x\} \times \mathbf{Cl}^\circ(X))$, it is constructible. It is also a subgroup and a constructible subgroup is closed.

In general such an $\mathbf{L}$ does not exist, but we check in (11.3.8) that a flat universal family exists after a finite field extension and a constructible subdivision $\tau\colon \amalg_j W_j \to \mathbf{Cl}^\circ(X)$. The argument above then shows that $\mathbf{Cl}(X,\{x\})$ is constructible, hence closed as before.

If $\Sigma$ is any set of closed points then $\mathbf{Cl}(X,\Sigma) = \cap_{x\in\Sigma}\mathbf{Cl}(X,\{x\})$.

If $\eta$ is a non-closed point, then $\mathbf{Cl}(X,\{\eta\})$ is the union of all $\mathbf{Cl}(X,\Sigma_U)$, where $U$ runs through all open subsets of $\bar{\eta}$ and $\Sigma_U$ denotes the set of closed points of $U$. By the noetherian property, $\mathbf{Cl}(X,\{\eta\}) = \mathbf{Cl}(X,\Sigma_U)$ for some $U$.

Finally, $\mathbf{Cl}(X,\Sigma) = \cap_{x\in\Sigma}\mathbf{Cl}(X,\{x\})$ holds for any set of points $\Sigma$.  □

**Lemma 11.3.8.** *Let $X$ be a normal, proper variety over an algebraically closed field $K$. There is a locally closed decomposition $\tau\colon \amalg_j W_j \to \mathbf{Cl}^\circ(X)$ such for every $j$ there is a universal family $\mathbf{L}_j$ on $X \times W_j$ that is flat over $W_j$ and whose fiber over $w \in W_j$ is the reflexive sheaf corresponding to $\tau(w) \in \mathbf{Cl}^\circ(X)$.*

*Proof.* By (11.3.5), there is a proper, birational morphism from a normal variety $p\colon Y \to X$ such that $\mathbf{Cl}^\circ(X) = \mathbf{Pic}^\circ(Y)$. Let $\mathbf{L}$ be the universal line bundle over $Y \times \mathbf{Pic}^\circ(Y)$. Pushing it forward we get a rank 1 sheaf

$$\mathbf{L}_X := (\pi_* \mathbf{L})^{[**]} \quad \text{over} \ X \times \mathbf{Cl}^\circ(X).$$

In general, $\mathbf{L}_X$ is not flat over $\mathbf{Cl}^\circ(X)$. However, by generic flatness, $\mathbf{L}_X$ is flat with reflexive fibers over a dense, open subset $W_1 \subset \mathbf{Cl}^\circ(X)$. Repeating this with $\mathbf{Cl}^\circ(X) \setminus W_1$ we get the required locally closed decomposition.  □

**11.3.9** (Albanese variety). Let $X$ be a proper, normal variety over a perfect field $k$. There are two different notions of the *Albanese variety* of $X$ in the literature. In [Gro62, VI.3.3] it is the target of the universal *morphism* from $X$ to an abelian torsor, that is, a principal homogeneous space under an abelian variety. We denote this by

$$\mathrm{alb}^{\mathrm{gr}}_X : X \to \mathbf{Alb}^{\mathrm{gr}}(X). \tag{11.3.9.1}$$

Pullback by $\mathrm{alb}^{\mathrm{gr}}_X$ gives an isomorphism

$$\mathbf{Pic}^{\circ}\big(\mathbf{Alb}^{\mathrm{gr}}(X)\big) \cong \mathrm{red}\,\mathbf{Pic}^{\circ}(X). \tag{11.3.9.2}$$

If $X$ has a $k$-point then $\mathbf{Alb}^{\mathrm{gr}}(X)$ an abelian variety. $\mathbf{Alb}^{\mathrm{gr}}(X)$ is a birational invariant for smooth, proper varieties, but not a birational invariant for normal varieties.

In the pre-EGA literature, for example, [Mat52, Ser59], the Albanese map is the universal *rational map* from $X$ to an abelian torsor, called the *classical Albanese variety*,

$$\mathrm{alb}_X : X \dashrightarrow \mathbf{Alb}(X). \tag{11.3.9.3}$$

More precisely, $\mathbf{Alb}(X)$ is the unique abelian torsor $A$ together with a rational map $\mathrm{alb}_X : X \dashrightarrow A$ such that for any abelian torsor $B$ and rational map $a \colon X \dashrightarrow B$, there exists a unique map $j \colon A \to B$ with $j \circ \mathrm{alb}_X = a$.

If $X$ has a smooth $k$-point then so does $\mathbf{Alb}(X)$, and then it is an abelian variety.

$\mathbf{Alb}(X)$ is a birational invariant of $X$ (for normal, proper varieties) and the two versions coincide if $X$ is smooth. Therefore, if $X' \to X$ is a resolution then $\mathbf{Alb}(X) = \mathbf{Alb}(X') = \mathbf{Alb}^{\mathrm{gr}}(X')$. In any case, by (11.3.1 (2)) we get a morphism over the smooth locus

$$\mathrm{alb}_X : X^{\mathrm{sm}} \to \mathbf{Alb}(X). \tag{11.3.9.4}$$

Let $X'$ be the normalization of the closure of the graph of $\mathrm{alb}_X$. Then we have a commutative diagram

$$\tag{11.3.9.5}$$

where $\mathrm{alb}_{X'}$ is a morphism. In particular, (11.3.9.2) gives that

$$\mathbf{Cl}^{\circ}(X') = \mathrm{red}\,\mathbf{Pic}^{\circ}(X') \cong \mathbf{Pic}^{\circ}\big(\mathbf{Alb}(X')\big). \tag{11.3.9.6}$$

Therefore

$$\mathbf{Cl}^{\circ}(X) \cong \mathbf{Pic}^{\circ}\big(\mathbf{Alb}(X)\big). \tag{11.3.9.7}$$

Let $p \colon Y \dashrightarrow X$ be a map of normal varieties. As long as $p(Y)$ is not contained in the singular locus of $X$, the composite $\mathbf{Alb}_X \circ p \colon Y \dashrightarrow \mathbf{Alb}(X)$ is defined, hence we get a morphism

$$\text{alb}_p : \textbf{Alb}(Y) \to \textbf{Alb}(X). \qquad (11.3.9.8)$$

**11.3.10.** Let $p\colon Y \to X$ be a morphism of normal varieties. Let

$$\text{alb}^{\text{gr}}_{Y\setminus X} : X \to \textbf{Alb}^{\text{gr}}(Y\setminus X) \qquad (11.3.10.1)$$

denote the universal morphism from $X$ to an abelian torsor that maps every irreducible component of $Y$ to a point. Thus, we get an exact sequence

$$\textbf{Alb}^{\text{gr}}(Y) \to \textbf{Alb}^{\text{gr}}(X) \to \textbf{Alb}^{\text{gr}}(Y\setminus X) \to 0. \qquad (11.3.10.2)$$

**Claim 11.3.11.** *The induced sequence*

$$0 \to \textbf{Pic}^{\circ}\big(\textbf{Alb}^{\text{gr}}(Y\setminus X)\big) \to \textbf{Pic}^{\circ}(X) \to \textbf{Pic}^{\circ}(Y)$$

*is exact.*

*Proof.* To see this, let $K$ denote the kernel of $\textbf{Pic}^{\circ}(X) \to \textbf{Pic}^{\circ}(Y)$. It is clear that $\textbf{Pic}^{\circ}\big(\textbf{Alb}^{\text{gr}}(Y\setminus X)\big) \subset K$. To see the converse, we may assume that $X$ has a $k$-point. By $(11.3.1\,(5))$ and $(11.3.1\,(8))$ we get an exact sequence

$$\textbf{Alb}^{\text{gr}}(Y) \to \textbf{Alb}^{\text{gr}}(X) \to \textbf{Pic}^{\circ}(K) \to 0.$$

The resulting $X \to \textbf{Alb}^{\text{gr}}(X) \to \textbf{Pic}^{\circ}(K)$ maps every irreducible component of $Y$ to a point, so it factors through $\textbf{Alb}^{\text{gr}}(Y\setminus X)$. By duality $(11.3.1\,(5))$ we get $K \to \textbf{Pic}^{\circ}\big(\textbf{Alb}^{\text{gr}}(Y\setminus X)\big)$. $\square$

We would like to know when $\text{alb}_p$ is dominant. Lefschetz theory suggests that this should hold if $p(Y)$ is ample-ci (11.2.1). This is, however, not always true. For example, let $X \subset \mathbb{P}^3$ be the cone over a smooth, cubic, plane curve $C$ and $Y \subset X$ the line over an inflection point of $C$. Then $\textbf{Alb}(Y) = 0$ but $\textbf{Alb}(X) \cong \text{Jac}(C)$.

However, the next results show that $\text{alb}_p$ is usually dominant.

**Lemma 11.3.12.** *Let $X, Y$ be normal, projective varieties and $p\colon Y \to X$ a morphism. Assume that $p(Y)$ has nonempty intersection with every nonzero divisor in $X$ and $\text{alb}_X$ is a morphism along $p(Y)$. Then $\text{alb}_p : \textbf{Alb}(Y) \to \textbf{Alb}(X)$ is surjective.*

*Proof.* If $\text{alb}_p$ is not surjective then the quotient $\textbf{Alb}(X)/\,\text{alb}_p(\textbf{Alb}(Y))$ is positive dimensional. Hence, there is a nonzero, effective divisor

$$D \subset \textbf{Alb}(X)/\,\text{alb}_p(\textbf{Alb}(Y))$$

whose pullback to $\textbf{Alb}(X)$ is disjoint from $\text{alb}_p(\textbf{Alb}(Y))$. Then its pullback to $X$ is a divisor which is disjoint from $p(Y)$. $\square$

**Corollary 11.3.13.** *Let $X$ be a normal, projective variety and $C \subset X^{\mathrm{ns}}$ an irreducible ample-sci curve (11.2.1). Then $\mathbf{Cl}^\circ(X) \to \mathbf{Jac}(\overline{C})$ has finite kernel.*

*Proof.* Note that $\mathrm{alb}_X$ is a morphism along $C$ by (11.3.9.4) and $C$ has nonempty intersection with every nonzero divisor. Thus, (11.3.12) applies. ☐

**Corollary 11.3.14.** *Let $A$ be an abelian variety and $C \subset \hat{A}$ an irreducible ample-sci curve (11.2.1). Then $A \to \mathbf{Jac}(\overline{C})$ has finite kernel.* ☐

**Lemma 11.3.15.** *Let $X$ be a normal, proper variety. Then there is a finite subset $\Sigma \subset X$ such that the following holds.*

*Let $Y \subset X$ be an irreducible divisor that is disjoint from $\Sigma$. Assume that $Y$ has nonempty intersection with every nonzero divisor in $X$. Let $\overline{Y} \to Y$ be the normalization. Then $\mathbf{Alb}(\overline{Y}) \to \mathbf{Alb}(X)$ is surjective.*

*Proof.* Consider the normalization of the closure of the graph of $\mathrm{alb}_X$

$$X \xleftarrow{\pi} X' \xrightarrow{\mathrm{alb}_{X'}} \mathbf{Alb}(X).$$

Let $E_i' \subset X'$ be the $\pi$-exceptional divisors. Choose $\Sigma$ to contain the generic point of each $\pi(E_i')$ and every non-Cartier center (11.3.19).

Then $Y$ is a Cartier divisor and $\pi^{-1}(Y) = \pi_*^{-1}(Y)$. Therefore, if $D' \subset X'$ is a divisor then

$$\pi\big(\pi_*^{-1}(Y) \cap D'\big) = \pi\big(\pi^{-1}(Y) \cap D'\big) = Y \cap \pi(D') \neq \emptyset.$$

Let $\overline{Y}'$ denote the normalization of $\pi_*^{-1}(Y)$. Then $\mathbf{Alb}(\overline{Y}') \to \mathbf{Alb}(X)$ is surjective by (11.3.12) and $\mathbf{Alb}(\overline{Y}') \cong \mathbf{Alb}(\overline{Y})$. ☐

**Definition 11.3.16** (Partial Albanese variety). Let $X$ be a proper, normal variety over a perfect field $k$ and $\Sigma \subset X$ a subset.

Define the *Albanese map* of $(X, \Sigma)$ as the universal rational map from $X$ to an abelian torsor that is regular along $\Sigma$

$$\mathrm{alb}_{X,\Sigma} : X \dashrightarrow \mathbf{Alb}(X, \Sigma). \tag{11.3.16.1}$$

If $\Sigma \subset X^{\mathrm{sm}}$ then $\mathbf{Alb}(X, \Sigma) = \mathbf{Alb}(X)$. In general $\mathbf{Alb}(X, \Sigma)$ is a quotient of $\mathbf{Alb}(X)$.

**Theorem 11.3.17.** *Let $X$ be a normal, proper variety over a perfect field $K$ and $\Sigma \subset X$ a subset. Then pullback by $\mathrm{alb}_{X,\Sigma}$ gives an isomorphism*

$$\mathrm{alb}_{X,\Sigma}^* : \mathbf{Pic}^\circ\big(\mathbf{Alb}(X, \Sigma)\big) \cong \mathbf{Cl}^\circ(X, \Sigma). \tag{11.3.17.1}$$

*Proof.* Consider $\mathrm{alb}_{X,\Sigma} : X \dashrightarrow \mathbf{Alb}(X, \Sigma)$. By assumption, it is a morphism along $\Sigma$, thus the pullback of a line bundle on $\mathbf{Alb}(X, \Sigma)$ is locally free along

$\Sigma$. That is,

$$\mathrm{alb}^*_{X,\Sigma}\,\mathbf{Pic}^\circ\big(\mathbf{Alb}(X,\Sigma)\big)\subset \mathbf{Cl}^\circ(X,\Sigma). \tag{11.3.17.2}$$

For the converse, assume first that $\Sigma=\{x\}$ is a closed point. As in (11.3.9.5) we have

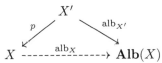

where $\mathrm{alb}_{X'}$ is a morphism. Let $Y'$ be the normalization of $p^{-1}(x)$. Since $X'\to\mathbf{Alb}^{\mathrm{gr}}(Y'\backslash X')$ contracts every irreducible component of $Y'$ to a point, the composite $X\dashrightarrow X'\to\mathbf{Alb}^{\mathrm{gr}}(Y'\backslash X')$ is a morphism at $x$ by Zariski's main theorem. This gives $\mathbf{Alb}(X,\{x\})\to\mathbf{Alb}^{\mathrm{gr}}(Y'\backslash X')$. Thus, we get a commutative diagram

$$
\begin{array}{ccc}
X' & \longrightarrow & \mathbf{Alb}^{\mathrm{gr}}(Y'\backslash X') \\
\downarrow & & \uparrow \\
X & \longrightarrow & \mathbf{Alb}(X,\{x\})
\end{array}
\tag{11.3.17.3}
$$

If $L\in\mathbf{Cl}^\circ(X,\{x\})(\bar{k})$ then its pullback to $X'$ is trivial along $Y'$, hence it is obtained as the pullback of a line bundle on $\mathbf{Alb}^{\mathrm{gr}}(Y'\backslash X')$ by (11.3.11). Factoring through $\mathbf{Alb}(X,\{x\})$ shows that $\mathbf{Cl}^\circ(X,\{x\})\subset\mathrm{alb}^*_{X,\{x\}}\,\mathbf{Pic}^\circ\big(\mathbf{Alb}(X,\{x\})\big)$.

The same argument works for any finite number of closed points. If $\Sigma$ is an infinite set of closed point then, by the noetherian property, $\mathbf{Cl}^\circ(X,\Sigma)=\mathbf{Cl}^\circ(X,\Sigma')$ for every large enough finite subset $\Sigma'\subset\Sigma$.

Finally, assume that $y$ is a non-closed point. Then $\mathbf{Cl}^\circ(X,\{y\})$ is the union of all $\mathbf{Cl}^\circ(X,\Sigma_U)$ where $\Sigma_U$ is the set of all closed points in some open subset $U\subset\overline{\{y\}}$. By the noetherian property, we have equality $\mathbf{Cl}^\circ(X,\{y\})=\mathbf{Cl}^\circ(X,\Sigma_U)$ for some fixed $U$. $\qquad\square$

**Corollary 11.3.18.** *Let $X$ be a normal, projective variety over a perfect field and $Z\subset X$ a closed, reduced subscheme with generic points $g_Z$. Then there is a normal, projective variety $X'$, a birational morphism $p\colon X'\to X$, and a closed, reduced subscheme $Z'\subset X'$ with generic points $g_{Z'}$ such that*

(1) *$p$ is a local isomorphism at all generic points of $Z'$,*

(2) *$Z=p(Z')$,*

(3) *$\mathrm{alb}_{X',g_{Z'}}$ is a morphism along $Z'$ and*

(4) *$\mathbf{Cl}^\circ(X,g_Z)=\mathbf{Cl}^\circ(X',g_{Z'})=\mathbf{Cl}^\circ(X',Z')$.*

*If either $\dim Z=1$ or the characteristic is $0$, we can also achieve that*

(5) $Z'$ *is smooth.*

*Proof.* We can take $X'$ to be the normalization of the closure of the graph of $\mathrm{alb}_{X,g_Z}$. Then we can resolve the singularities of $Z'$ if desired. $\square$

**Definition 11.3.19.** Let $X$ be a reduced scheme and $D$ an effective Weil divisor. There is a unique largest open subscheme $X_D^{\mathrm{car}} \subset X$, called the *Cartier locus* of $D$, such that the restriction of $D$ to $X_D^{\mathrm{car}}$ is Cartier. The complement $X \setminus X_D^{\mathrm{car}}$ is the *non-Cartier locus* of $D$. A point $x \in X$ is a *non-Cartier center* of $X$ if there is a Weil divisor $D$ such that $x$ is the generic point of an irreducible component of the non-Cartier locus of $D$.

For example, let $X = (xy = 0) \subset \mathbb{A}^3_{xyz}$ and set $D_c := (x = z - c = 0)$. Its non-Cartier locus is the point $(x = y = z - c = 0)$. Thus, every closed point of the $z$-axis is a non-Cartier center of $X$. The generic point of the $z$-axis is also a non-Cartier center of $X$ for the divisor $(x = y = 0)$.

In direct analogy one can define the notions of $\mathbb{Q}$-*Cartier locus* and *non-$\mathbb{Q}$-Cartier center*.

The next result of [BGS11, 6.7] shows that the situation is quite different for normal varieties. (Note that [BGS11] works over an algebraically closed field, but this is not necessary.)

**Theorem 11.3.20.** *A geometrically normal variety has only finitely many non-Cartier or non-$\mathbb{Q}$-Cartier centers.*

*Proof.* We may assume that $X$ is proper and irreducible. Let $U \subset X$ be an open subset such that $X$ has only finitely many non-$(\mathbb{Q}\text{-})$Cartier centers in $U$. We show that there is a strictly larger open subset $U \subsetneq U' \subset X$ such that $X$ has only finitely many non-$(\mathbb{Q}\text{-})$Cartier centers in $U'$. We can start with the smooth locus $U = X^{\mathrm{sm}}$, since it is disjoint from every non-$(\mathbb{Q}\text{-})$Cartier center. Noetherian induction then gives that $X$ has only finitely many non-$(\mathbb{Q}\text{-})$Cartier centers.

Let $Z \subset X \setminus U$ be an irreducible component. By (11.3.21) there is a dense, open subset $Z^0 \subset Z$ such that if a Weil divisor $D$ is $(\mathbb{Q}\text{-})$Cartier at the generic point $g_Z \in Z$ then it is $(\mathbb{Q}\text{-})$Cartier along $Z^0$. We may assume that $Z^0$ is disjoint from every other irreducible component of $X \setminus U$. Then $U' := U \cup Z^0$ is open in $X$ and $g_Z$ is the only possible new non-$(\mathbb{Q}\text{-})$Cartier center in $U'$. $\square$

**Lemma 11.3.21.** *Let $X$ be a normal, proper variety over an algebraically closed field and $Z \subset X$ an irreducible subvariety. Then there is a dense, open subset $Z^0 \subset Z$ such that the following holds.*

*Let $D$ be a Weil divisor that is $(\mathbb{Q}\text{-})$Cartier at the generic point $g_Z \in Z$. Then it is $(\mathbb{Q}\text{-})$Cartier everywhere along $Z^0$.*

*Proof.* As in (11.3.6), let $\mathbf{Cl}(X, g_Z) \subset \mathbf{Cl}(X)$ be the subgroup of those divisors that are Cartier at the generic point of $Z$ and $\mathbf{Cl}^\circ(X, g_Z) \subset \mathbf{Cl}^\circ(X)$ the identity component.

As we noted in (11.3.6), the quotient $\mathbf{Cl}(X, g_Z)/\mathbf{Cl}^\circ(X, g_Z)$ is finitely generated; say by the divisors $D_i$. There is a dense, open subset $Z_1^0 \subset Z$ such that every $D_i$ is Cartier along $Z_1^0$, hence the same holds for every linear combination of the $D_i$.

Next we show that there is a dense, open subset $Z_2^0 \subset Z$ such that every divisor in $\mathbf{Cl}^\circ(X, g_Z)$ is Cartier along $Z_2^0$. Consider the Albanese map $\mathrm{alb}_{X,Z} : X \dashrightarrow \mathbf{Alb}(X, Z)$. By (11.3.16) it is defined at $g_Z$, hence on a dense, open subset $Z_2^0 \subset Z$. By (11.3.17), $\mathbf{Cl}^\circ(X, g_Z)$ is the pullback of $\mathbf{Pic}^\circ\big(\mathbf{Alb}(X, g_Z)\big)$, hence every member of $\mathbf{Cl}^\circ(X, g_Z)$ is Cartier along $Z_2^0$. Finally, $Z^0 = Z_1^0 \cap Z_2^0$ is the dense, open subset that we need.

If $D$ is $\mathbb{Q}$-Cartier at $g_Z$ then $mD$ is Cartier at $g_Z$ for some $m > 0$, hence $D$ is $\mathbb{Q}$-Cartier along $Z^0$ by the previous results.                                                    □

By an *algebraic group* we mean a finite type group scheme over a field. Such a group scheme is called a *linear algebraic group* if it is, in addition, affine.

**11.3.22** (Structure of commutative algebraic groups). Let $A$ denote a commutative algebraic group over a perfect field $k$, $A^\circ \subset A$ the identity component, and $A^{\mathrm{lin}} \subset A$ the largest connected linear algebraic subgroup. Then $A/A^\circ$ is a finite étale group scheme and $A^\circ/A^{\mathrm{lin}}$ is an abelian variety. This is a consequence of the Barsotti-Chevalley theorem [Mil17, Theorem 8.27].

Let $A^{\mathrm{unip}} \subset A^{\mathrm{lin}}$ be the largest unipotent subgroup and $A^{\mathrm{tor}} \subset A^{\mathrm{lin}}$ the largest subgroup of multiplicative type. By [Mil17, Theorem 16.13 (b)] we have

$$A^{\mathrm{lin}} = A^{\mathrm{unip}} \times A^{\mathrm{tor}}.$$

If $A$ is furthermore assumed reduced, and hence smooth since $k$ is perfect, then $A^{\mathrm{tor}}$ is a *torus* (that is, isomorphic to $\mathbb{G}_m^r$ over $k^{\mathrm{sep}}$ for some $r$).

A reduced algebraic group $A$ is called *semi-abelian* if $A^{\mathrm{unip}} = 0$.

Let $A^{\mathrm{prop}} \subset A$ denote the largest proper, connected subgroup. Then $A^{\mathrm{prop}} \cap A^{\mathrm{lin}}$ is finite but usually $A^{\mathrm{prop}} + A^{\mathrm{lin}}$ does not equal $A$.

See [Bor91], [Mil17, Chap. 8] or [Bri17b] for details and proofs.

**11.3.23** ($\mathbb{Q}$-rank). For abelian varieties, the $\mathbb{Q}$-rank of $A(k)$ is a subtle invariant of $A$ and $k$; see, for example, (8.6.3) and (8.6.5). By contrast the $\mathbb{Q}$-rank of a linear algebraic group is easy to compute.

(1) $G(k)$ is torsion for every algebraic group $G$ over a locally finite field $k$.

(2) $U(k)$ is $p^\infty$-torsion for every unipotent algebraic group $U$ over a field $k$ of characteristic $p > 0$.

(3) $\mathrm{rank}_{\mathbb{Q}} U(k) = \dim U \cdot \deg(k/\mathbb{Q})$ for every unipotent algebraic group $U$ over a field $k$ of characteristic 0.

(4) $\mathrm{rank}_{\mathbb{Q}} T(k) = \infty$ for every positive dimensional torus over a field that is not locally finite.

Of these, only the last claim is nontrivial. It was proved in (7.2.8).

**11.3.24** (Jacobians of curves). Let $C$ be a proper scheme of dimension 1 over a field $k$. Then $\mathbf{Pic}^{\circ}(C)$ is called the *Jacobian* or *generalized Jacobian* of $C$ and denoted by $\mathbf{Jac}(C)$.

Let $C^{\mathrm{wn}} \to C$ denote the *weak normalization* and $\overline{C} \to C^{\mathrm{wn}} \to C$ the *normalization*. Pullback induces maps

$$\mathbf{Jac}(C) \to \mathbf{Jac}(C^{\mathrm{wn}}) \to \mathbf{Jac}(\overline{C}).$$

The kernel of $\mathbf{Jac}(C) \to \mathbf{Jac}(C^{\mathrm{wn}})$ is $\mathbf{Jac}(C)^{\mathrm{unip}}$ and the kernel of $\mathbf{Jac}(C) \to \mathbf{Jac}(\overline{C})$ is $\mathbf{Jac}(C)^{\mathrm{lin}}$. Thus, (11.3.23) gives the following.

(1) If $k$ is locally finite then $\mathbf{Jac}(C)(k)$ is torsion.

(2) If $\operatorname{char} k > 0$ but $k$ is not locally finite and $C$ is geometrically integral, then $\mathbf{Jac}(C)(\overline{k})$ is torsion if and only if $C$ is rational and $C^{\mathrm{wn}} = \overline{C}$.

(3) If $\operatorname{char} k = 0$ then $\mathbf{Jac}(C)(\overline{k})$ is torsion if and only if $h^1(C, \mathcal{O}_C) = 0$. This implies that every irreducible component of $C_{\overline{k}}$ is smooth and rational.

**11.3.25.** Let $P$ be a 0-cycle on a smooth algebraic group $A$ and write $P_{\overline{k}} = \cup_i m_i[p_i]$. Set

$$\operatorname{tr}_A P := \sum_i m_i[p_i] \quad \text{(summation in } A\text{)}. \tag{11.3.25.1}$$

Note that $\operatorname{tr}_A P \in A(k)$. (When the residue field is inseparable over $k$, this uses the fact that the multiplication by $p$ map on $A$ always factors through Frobenius.) If $A$ is the additive group $\mathbb{G}_a$ then this is the usual trace, but for the multiplicative group $\mathbb{G}_m$ this is the norm. Since we usually use additive notation, trace seems a better choice. For $Z \subset A$ set

$$\operatorname{tr}_A Z := \{\operatorname{tr}_A P : P \text{ is a 0-cycle on } Z\}. \tag{11.3.25.2}$$

Let $C$ be a smooth, projective curve. There is a natural embedding $j \colon C \hookrightarrow \mathbf{Jac}_1(C) \subset \mathbf{Pic}(C)$. If $P$ is a 0-cycle on $C$ then

$$\operatorname{tr}_{\mathbf{Pic}(C)}\big(j(P)\big) = [\mathcal{O}_C(P)] \in \mathbf{Jac}_{\deg P}(C). \tag{11.3.25.3}$$

The following is a restatement of [Bri17a, 4.9].

**Lemma 11.3.26.** *The association $A \mapsto A(k) \otimes \mathbb{Q}$ defines an exact functor on the category of commutative algebraic groups.*

*Proof.* The only nontrivial claim is that if $g \colon A \to B$ is a dominant morphism then $g(k) \colon A(k) \to B(k)$ is surjective modulo torsion. To see this pick $b \in B(k)$ and let $P$ be a 0-cycle on the fiber $A_b$. Then $\operatorname{tr}_A P \in A(k)$ and $g(\operatorname{tr}_A P) = \deg P \cdot b$. $\square$

We need some facts about the multiplicative group of Artin algebras

**11.3.27.** Let $A$ be an Artinian $k$-algebra. The group of units $A^\times$ is the $k$-points of a commutative algebraic group $\mathcal{R}_k^A \mathbb{G}_m$ called the *Weil restriction* of $\mathbb{G}_m$ from $A$ to $k$. The algebraic group $\mathcal{R}_k^A \mathbb{G}_m$ represents the pushforward (in the sense of big étale sheaves) $f_* \mathbb{G}_m$ of the multiplicative group along the morphism $f \colon \mathrm{Spec}(A) \to \mathrm{Spec}(k)$. More concretely we have

$$(\mathcal{R}_k^A \mathbb{G}_m)(B) = (A \otimes_k B)^\times \text{ for a } k\text{-algebra } B.$$

Note that $\dim \mathcal{R}_k^A \mathbb{G}_m = \dim_k A$.

For example, if $K/k$ is a field extension of degree $n$, choose a basis $e_i \in K$. As a variety, $\mathcal{R}_k^K \mathbb{G}_m$ is $\mathbb{A}^n \setminus \big(\mathrm{norm}_{K/k}(\sum x_i e_i) = 0\big)$.

**11.3.28.** Let $(A, m)$ be a local, Artinian $k$ algebra with residue field $K = A/m$. There is an exact sequence

$$1 \to U \to A^\times \to K^\times \to 1, \qquad (11.3.28.1)$$

where the map $a \mapsto 1 + a$ identifies $m$ with $U$. Note that $a \mapsto 1 + a$ is a group isomorphism if $m^2 = 0$ but not otherwise. In characteristic $0$ one can correct this by taking $a \mapsto \exp(a)$. We will think of $(11.3.28.1)$ as the $k$-points of an exact sequence of algebraic $k$-groups

$$1 \to U \to \mathcal{R}_k^A \mathbb{G}_m \to \mathcal{R}_k^K \mathbb{G}_m \to 1, \qquad (11.3.28.2)$$

where $U$ is a unipotent group. In positive characteristic the algebraic groups $(m, +)$ and $(U, \cdot)$ need not be isomorphic.

We also use the following variant of Hensel's lemma.

**Claim 11.3.29.** *Let $k \subset k' \subset K$ be a subfield that is separable over $k$. Then there is a unique lifting $k' \to A$.* □

Combining the above with the previous discussion on algebraic groups yields the following lemmas.

**Lemma 11.3.30.** *Let $k$ be a field and $A \to B$ a homomorphism of Artin $k$-algebras. Then $\mathrm{coker}[A^\times \to B^\times]$ is torsion if and only if one of the following holds.*

(1) *$A \to B$ is surjective.*

(2) *char $k > 0$ and $B/\sqrt{0}$ is purely inseparable over $A/\sqrt{0}$.*

(3) *$k$ is locally finite.*

*Moreover, $\mathrm{coker}[A^\times \to B^\times]$ has finite $\mathbb{Q}$-rank in one additional case:*

(4) *$\deg(k/\mathbb{Q}) < \infty$ and $A/\sqrt{0} \to B/\sqrt{0}$ is surjective.* □

**Lemma 11.3.31.** *Let $k$ be a field and $A \to B$ a homomorphism of Artin $k$-algebras. Then $\ker[A^\times \to B^\times]$ is torsion if and only if one of the following holds.*

(1) $A \to B$ *is injective.*

(2) char $k > 0$ *and $A/\sqrt{0} \to B/\sqrt{0}$ is injective.*

(3) $k$ *is locally finite.*

*Moreover, $\ker[A^\times \to B^\times]$ has finite $\mathbb{Q}$-rank in one additional case:*

(4) $\deg(k/\mathbb{Q}) < \infty$ *and $A/\sqrt{0} \to B/\sqrt{0}$ is injective.* □

Much of the following is proved in [CTGW96].

**Lemma 11.3.32.** *Let $k$ be a field that is not locally finite. Let $A \supset k$ be a finite, reduced $k$-algebra. Let $k \subset L_1, L_2 \subset A$ be subfields. The following are equivalent.*

(1) $A^\times/(L_1^\times \cdot L_2^\times)$ *is torsion.*

(2) $A^\times/(L_1^\times \cdot L_2^\times)$ *has finite $\mathbb{Q}$-rank.*

(3) $A$ *is a field and $A/L_i$ is purely inseparable for some $i = 1, 2$.*

*Proof.* If $A/L_i$ is purely inseparable then $A^q \subset L_i$ for some power $q$ of char $k$, hence $A^\times/L_i^\times$ is torsion. This proves $(3) \Rightarrow (1)$ and $(1) \Rightarrow (2)$ is clear.

Assume (2). We may replace $A$ by its maximal separable subalgebra. Thus assume that $A/k$ is separable. If $A = L_i$ for some $i$ then we are done. Otherwise, $\dim_k A = \dim_{L_i} A \cdot \dim_k L_i \geq 2 \dim_k L_i$.

If $B$ is a reduced, separable $k$-algebra, then $B^\times$ is identified with the $k$-points of the $k$-torus $\mathcal{R}_k^B \, \mathbb{G}_m$. Thus, $L_1^\times \cdot L_2^\times \to A^\times$ can be viewed as the $k$-points of a morphism of $k$-tori

$$\mu : \mathcal{R}_k^{L_1} \, \mathbb{G}_m \times \mathcal{R}_k^{L_2} \, \mathbb{G}_m \longrightarrow \mathcal{R}_k^A \, \mathbb{G}_m.$$

Both of the $L_i$ contain $k$, thus

$$\dim_k \mathrm{Im}(\mu) \leq \dim_k L_1 + \dim_k L_2 - 1 < \dim A.$$

Thus, $\mathrm{coker}(\mu)$ is a positive dimensional $k$-torus, hence $\mathrm{rank}_\mathbb{Q}\big(\mathrm{coker}(\mu)(k)\big) = \infty$ by (11.3.23 (4)). Finally, (11.3.26) shows that

$$\mathrm{rank}_\mathbb{Q}\big(A^\times/(L_1^\times \cdot L_2^\times)\big) = \mathrm{rank}_\mathbb{Q}\big(\mathrm{coker}(\mu)(k)\big) = \infty.$$ □

# Bibliography

[AGV73]     Michael Artin, Alexandre Grothendieck, and Jean-Louis Verdier
            (eds.), *Théorie des topos et cohomologie étale des schémas. Tome
            3*, Lecture Notes in Mathematics, Vol. 305, Springer-Verlag, Berlin-
            New York, 1973, Séminaire de Géométrie Algébrique du Bois-Marie
            1963–1964 (SGA 4), Dirigé par M. Artin, A. Grothendieck et J. L.
            Verdier. Avec la collaboration de P. Deligne et B. Saint-Donat.

[Amb18]     Emiliano Ambrosi, *Specialization of Néron-Severi groups in posi-
            tive characteristic*, arXiv e-prints (2018), https://arxiv.org/abs/
            1810.06481v2.

[And96]     Yves André, *Pour une thèorie inconditionnelle des motifs*, Inst.
            Hautes Etudes Sci. Publ. Math. **83** (1996), 5–49.

[Art57]     Emil Artin, *Geometric algebra*, Interscience Publishers, Inc., New
            York-London, 1957. MR 0082463

[Art62]     Michael Artin, *Some numerical criteria for contractability of curves
            on algebraic surfaces*, Amer. J. Math. **84** (1962), 485–496.

[Bae52]     Reinhold Baer, *Linear algebra and projective geometry*, Academic
            Press, Inc., New York, N.Y., 1952. MR 0052795

[Bal05]     Paul Balmer, *The spectrum of prime ideals in tensor triangulated
            categories*, J. Reine Angew. Math. **588** (2005), 149–168.

[BGH18]     Clark Barwick, Saul Glasman, and Peter Haine, *Exodromy*, https:
            //arxiv.org/abs/1807.03281v7, 2018.

[BGS11]     Samuel Boissière, Ofer Gabber, and Olivier Serman, *Sur le pro-
            duit de variétés localement factorielles ou Q-factorielles*, https:
            //arxiv.org/abs/1104.1861v4, April 2011.

[BK12]      Alina Bucur and Kiran S. Kedlaya, *The probability that a com-
            plete intersection is smooth*, J. Théor. Nombres Bordeaux **24** (2012),
            no. 3, 541–556.

[BKT10]     Fedor Bogomolov, Mikhail Korotiaev, and Yuri Tschinkel, *A Torelli
            theorem for curves over finite fields*, Pure Appl. Math. Q. **6** (2010),

no. 1, Special Issue: In honor of John Tate. Part 2, 245–294.

[BL04]       Christina Birkenhake and Herbert Lange, *Complex abelian varieties*, second ed., Grundlehren der Mathematischen Wissenschaften [Fundamental Principles of Mathematical Sciences], vol. 302, Springer-Verlag, Berlin, 2004.

[BLR90]      Siegfried Bosch, Werner Lütkebohmert, and Michel Raynaud, *Néron models*, Ergebnisse der Mathematik und ihrer Grenzgebiete (3), vol. 21, Springer-Verlag, Berlin, 1990.

[BLR93]      Manuel Blum, Michael Luby, and Ronitt Rubinfeld, *Self-testing/correcting with applications to numerical problems*, JCSS **47** (1993), 549–595.

[BM20]       Daniel Barlet and Jón Magnússon, *Cycles Analytiques Complexes II: L'espace des Cycles*, Cours Specialises, vol. 27, Société Mathématique de France, 2020.

[Bor91]      Armand Borel, *Linear algebraic groups*, second ed., Graduate Texts in Mathematics, vol. 126, Springer-Verlag, New York, 1991.

[BPS16]      Fedor A. Bogomolov, Alena Pirutka, and Aaron Michael Silberstein, *Families of disjoint divisors on varieties*, Eur. J. Math. **2** (2016), no. 4, 917–928. MR 3572551

[Bri17a]     Michel Brion, *Commutative algebraic groups up to isogeny*, Doc. Math. **22** (2017), 679–725.

[Bri17b]     _____, *Some structure theorems for algebraic groups*, Algebraic groups: structure and actions, Proc. Sympos. Pure Math., vol. 94, Amer. Math. Soc., Providence, RI, 2017, pp. 53–126.

[BRT19]      F. Bogomolov, M. Rovinsky, and Y. Tschinkel, *Homomorphisms of multiplicative groups of fields preserving algebraic dependence*, Eur. J. Math. **5** (2019), no. 3, 656–685. MR 3993257

[BT73]       Armand Borel and Jacques Tits, *Homomorphismes "abstraits" de groupes algébriques simples*, Ann. of Math. (2) **97** (1973), 499–571. MR 316587

[BT09]       Fedor Bogomolov and Yuri Tschinkel, *Milnor $K_2$ and field homomorphisms*, Surveys in differential geometry. Vol. XIII. Geometry, analysis, and algebraic geometry: Forty years of the Journal of Differential Geometry, Surv. Differ. Geom., vol. 13, Int. Press, Somerville, MA, 2009, pp. 223–244. MR 2537087

[BT11]       _____, *Reconstruction of higher-dimensional function fields*, Mosc. Math. J. **11** (2011), no. 2, 185–204, 406. MR 2859233

[BT12]        ———, *Introduction to birational anabelian geometry*, Current developments in algebraic geometry, Math. Sci. Res. Inst. Publ., vol. 59, Cambridge Univ. Press, Cambridge, 2012, pp. 17–63. MR 2931864

[BT13]        ———, *Galois theory and projective geometry*, Comm. Pure Appl. Math. **66** (2013), no. 9, 1335–1359. MR 3078692

[BVRS09]   Luca Barbieri-Viale, Andreas Rosenschon, and Vasudevan Srinivas, *The Néron-Severi group of a proper seminormal complex variety*, Math. Z. **261** (2009), no. 2, 261–276.

[BVS93]     Luca Barbieri Viale and Vasudevan Srinivas, *On the Néron-Severi group of a singular variety*, J. Reine Angew. Math. **435** (1993), 65–82.

[Čes21]      Kęstutis Česnavičius, *Reconstructing a variety from its topology (after Kollár, building on earlier work of Lieblich and Olsson)*, 2021, Séminaire Bourbaki, 73ème année, no 1175.

[Cho49]     Wei-Liang Chow, *On the geometry of algebraic homogeneous spaces*, Annals of Mathematics **50** (1949), no. 1, 32–67.

[Chr18]     Atticus Christensen, *Specialization of Néron-Severi groups in characteristic p*, arXiv e-prints (2018), https://arxiv.org/abs/1810.06550v2.

[CP16]      François Charles and Bjorn Poonen, *Bertini irreducibility theorems over finite fields*, J. Amer. Math. Soc. **29** (2016), no. 1, 81–94.

[CP18]      Anna Cadoret and Alena Pirutka, *Reconstructing function fields from Milnor K-theory*, 2018.

[CTGW96] Jean-Louis Colliot-Thélène, Robert M. Guralnick, and Roger Wiegand, *Multiplicative groups of fields modulo products of subfields*, Journal of Pure and Applied Algebra **106** (1996), no. 3, 233–262.

[DS18]      Rankeya Datta and Karen E. Smith, *Excellence in prime characteristic*, Local and global methods in algebraic geometry, Contemp. Math., vol. 712, Amer. Math. Soc., Providence, RI, 2018, pp. 105–116.

[dS22]       Juan B. Sancho de Salas, *The fundamental theorem of affine and projective geometries*, 2022.

[Eis95]      David Eisenbud, *Commutative algebra*, Graduate Texts in Mathematics, vol. 150, Springer-Verlag, New York, 1995, With a view toward algebraic geometry. MR 1322960 (97a:13001)

[Fal94]    Gerd Faltings, *The general case of S. Lang's conjecture*, Barsotti Symposium in Algebraic Geometry (Abano Terme, 1991), Perspect. Math., vol. 15, Academic Press, San Diego, CA, 1994, pp. 175–182.

[Fan1892]  Gino Fano, *Sui postulati fondamentali della geometria proiettiva*, Giornale di Matematiche **30** (1892), 106–132.

[FJ08]     Michael D. Fried and Moshe Jarden, *Field arithmetic*, third ed., Ergebnisse der Mathematik und ihrer Grenzgebiete. 3. Folge. A Series of Modern Surveys in Mathematics, vol. 11, Springer-Verlag, Berlin, 2008, Revised by Jarden.

[FKL16]    Mihai Fulger, János Kollár, and Brian Lehmann, *Volume and Hilbert function of $\mathbb{R}$-divisors*, Michigan Math. J. **65** (2016), no. 2, 371–387. MR 3510912

[FOV99]    Hubert Flenner, Liam O'Carroll, and Wolfgang Vogel, *Joins and intersections*, Springer Monographs in Mathematics, Springer-Verlag, Berlin, 1999.

[FP10]     Arno Fehm and Sebastian Petersen, *On the rank of abelian varieties over ample fields*, Int. J. Number Theory **6** (2010), no. 3, 579–586.

[FP19]     _____, *Ranks of abelian varieties and the full Mordell-Lang conjecture in dimension one*, Abelian varieties and number theory, Contemp. Math., vol. 767, Amer. Math. Soc., Providence, RI, 2021, pp. 13–24

[Ful98]    William Fulton, *Intersection theory*, second ed., Ergebnisse der Mathematik und ihrer Grenzgebiete. 3. Folge. A Series of Modern Surveys in Mathematics, vol. 2, Springer-Verlag, Berlin, 1998.

[Gab62]    Pierre Gabriel, *Des catégories abéliennes*, Bull. Soc. Math. France **90** (1962), 323–448. MR 232821

[Gro60]    Alexander Grothendieck, *Éléments de géométrie algébrique. I–IV.*, Inst. Hautes Études Sci. Publ. Math., no. 4,8,11,17,20,24,28,32, IHES, 1960.

[Gro62]    _____, *Fondements de la géométrie algébrique. [Extraits du Séminaire Bourbaki, 1957–1962.]*, Secrétariat mathématique, Paris, 1962.

[Gro77]    _____, *Cohomologie l-adic et Fonctions L.*, Lecture Notes in Mathematics, vol. 589, Springer Verlag, Heidelberg, 1977.

[Gro05]    _____, *Cohomologie locale des faisceaux cohérents et théorèmes de Lefschetz locaux et globaux (SGA 2)*, Documents Mathématiques (Paris), vol. 4, Société Mathématique de France, Paris, 2005, Séminaire de Géométrie Algébrique du Bois Marie, 1962, Augmenté d'un exposé de Michèle Raynaud. With a preface and edited by Yves Laszlo. Revised reprint of the 1968 French original.

[Har62]   Robin Hartshorne, *Complete intersections and connectedness*, Amer. J. Math. **84** (1962), 497–508.

[Har77]   ———, *Algebraic geometry*, Springer-Verlag, New York, 1977, Graduate Texts in Mathematics, No. 52.

[Hoc69]   Mel Hochster, *Prime ideal structure in commutative rings*, Trans. Amer. Math. Soc. **142** (1969), 43–60. MR 251026

[Hua49]   Loo-Keng Hua, *Geometry of symmetric matrices over any field with characteristic other than two*, Ann. of Math. (2) **50** (1949), 8–31. MR 28296

[IL19]   Bo-Hae Im and Michael Larsen, *Abelian varieties and finitely generated Galois groups*, Abelian varieties and number theory, Contemp. Math., vol. 767, Amer. Math. Soc., Providence, RI, 2021, pp. 1–12

[Jac85]   Nathan Jacobson, *Basic algebra. I*, second ed., W. H. Freeman and Company, New York, 1985.

[Ji21]   Lena Ji, *The Noether–Lefschetz theorem*, https://arxiv.org/pdf/2107.12962.pdf.

[Jou83]   Jean-Pierre Jouanolou, *Théorèmes de Bertini et applications*, Progress in Mathematics, vol. 42, Birkhäuser Boston, Inc., Boston, MA, 1983.

[Kle1874]   Felix Klein, *Nachtrag zu dem zweiten Aufsatze über Nicht-Euklidische Geometrie*, Math. Ann. **7** (1874), 531–537.

[Kle66]   Steven L. Kleiman, *Toward a numerical theory of ampleness*, Ann. of Math. (2) **84** (1966), 293–344. MR 206009

[KLOS20]   János Kollár, Max Lieblich, Martin Olsson, and Will Sawin, *Topological reconstruction theorems for varieties*, https://arxiv.org/abs/2003.04847v1, 2020.

[KM09]   János Kollár and Frédéric Mangolte, *Cremona transformations and diffeomorphisms of surfaces*, Adv. Math. **222** (2009), no. 1, 44–61.

[Kob06]   Emi Kobayashi, *A remark on the Mordell-Weil rank of elliptic curves over the maximal abelian extension of the rational number field*, Tokyo J. Math. **29** (2006), no. 2, 295–300.

[Koc18]   Fabian Koch, *Jacobians of curves on surfaces*, http://www.math.

uni-bonn.de/people/huybrech/KochMaster.pdf, 2018.

[Kol96]    János Kollár, *Rational curves on algebraic varieties*, Ergebnisse der Mathematik und ihrer Grenzgebiete. 3. Folge., vol. 32, Springer-Verlag, Berlin, 1996.

[Kol13]    ———, *Singularities of the minimal model program*, Cambridge Tracts in Mathematics, vol. 200, Cambridge University Press, Cambridge, 2013, With the collaboration of Sándor Kovács.

[Kol18]    ———, *Mumford divisors*, https://arxiv.org/abs/1803.07596v2, March 2018.

[Kol20]    ———, *What determines a variety?*, https://arxiv.org/abs2002.12424v2, 2020.

[Kra96]    Hanspeter Kraft, *Challenging problems on affine n-space*, Séminaire Bourbaki: volume 1994/95, exposés 790-804, Astérisque, no. 237, Société mathématique de France, 1996. MR 1423629

[KSC04]    János Kollár, Karen E. Smith, and Alessio Corti, *Rational and nearly rational varieties*, Cambridge Studies in Advanced Mathematics, vol. 92, Cambridge University Press, Cambridge, 2004.

[Lan62]    Serge Lang, *Diophantine geometry*, Interscience Tracts in Pure and Applied Mathematics, No. 11, Interscience Publishers (a division of John Wiley & Sons), New York-London, 1962.

[Lan02]    ———, *Algebra*, third ed., Graduate Texts in Mathematics, vol. 211, Springer-Verlag, New York, 2002. MR 1878556

[Laz04]    Robert Lazarsfeld, *Positivity in algebraic geometry. I-II*, Ergebnisse der Mathematik und ihrer Grenzgebiete. 3. Folge., vol. 48–49, Springer-Verlag, Berlin, 2004.

[LN59]     Serge Lang and André Néron, *Rational points of abelian varieties over function fields*, Amer. J. Math. **81** (1959), 95–118.

[LO21]     Max Lieblich and Martin Olsson, *Derived categories and birationality*, https://arxiv.org/abs/2001.05995v2, 2021, preprint.

[Lur17]    Jacob Lurie, *Higher algebra*, preprint dated September 18, 2017.

[Mat52]    Teruhisa Matsusaka, *On the algebraic construction of the Picard variety. II*, Jap. J. Math. **22** (1952), 51–62.

[Mil17]    James S. Milne, *Algebraic groups*, Cambridge Studies in Advanced Mathematics, vol. 170, Cambridge University Press, Cambridge, 2017, The theory of group schemes of finite type over a field.

[Mil08]     ———, *Abelian varieties (v2.00)*, 2008, Available at `www.jmilne.org/math/`, p. 172.

[MP12]      Davesh Maulik and Bjorn Poonen, *Néron-Severi groups under specialization*, Duke Math. J. **161** (2012), 2167–2206.

[Mum66]     David Mumford, *Lectures on curves on an algebraic surface*, With a section by G. M. Bergman. Annals of Mathematics Studies, No. 59, Princeton University Press, Princeton, N.J., 1966.

[Mum70]     ———, *Abelian varieties*, Tata Institute of Fundamental Research Studies in Mathematics, No. 5, Published for the Tata Institute of Fundamental Research, Bombay; Oxford University Press, London, 1970.

[Nér52a]    André Néron, *La théorie de la base pour les diviseurs sur les variétés algébriques*, Deuxième Colloque de Géométrie Algébrique, Liège, 1952, Georges Thone, Liège; Masson & Cie, Paris, 1952, pp. 119–126.

[Nér52b]    ———, *Problèmes arithmétiques et géométriques rattachés à la notion de rang d'une courbe algébrique dans un corps*, Bull. Soc. Math. France **80** (1952), 101–166.

[Neu54]     Bernhard H. Neumann, *Groups covered by finitely many cosets*, Publ. Math. Debrecen **3** (1954), 227–242.

[Per06]     Jorge Vitório Pereira, *Fibrations, divisors and transcendental leaves*, J. Algebraic Geom. **15** (2006), no. 1, 87–110, With an appendix by Laurent Meersseman. MR 2177196

[Pic1895]   Émile Picard, *Sur la théorie des groupes et des surfaces algébriques*, Rend. Circ. Math. Palermo **9** (1895), 244–255.

[Poo01]     Bjorn Poonen, *Points having the same residue field as their image under a morphism*, J. Algebra **243** (2001), no. 1, 224–227.

[Poo04]     ———, *Bertini theorems over finite fields*, Ann. of Math. (2) **160** (2004), no. 3, 1099–1127.

[Poo08]     ———, *Smooth hypersurface sections containing a given subscheme over a finite field*, Math. Res. Lett. **15** (2008), no. 2, 265–271.

[Pop94]     Florian Pop, *On Grothendieck's conjecture of birational anabelian geometry*, Ann. of Math. (2) **139** (1994), no. 1, 145–182. MR 1259367

[Rey1866]   Theodor Reye, *Die Geometrie der Lage, 2 vols.*, Hannover, Rümpler, 1866.

[Ros98]      Alexander L. Rosenberg, *The spectrum of abelian categories and reconstruction of schemes*, Rings, Hopf algebras, and Brauer groups (Antwerp/Brussels, 1996), Lecture Notes in Pure and Appl. Math., vol. 197, Dekker, New York, 1998, pp. 257–274. MR 1615928

[RS06]       Girivaru V. Ravindra and Vasudevan Srinivas, *The Grothendieck-Lefschetz theorem for normal projective varieties*, J. Algebraic Geom. **15** (2006), no. 3, 563–590.

[RS09]       _____, *The Noether-Lefschetz theorem for the divisor class group*, J. Algebra **322** (2009), no. 9, 3373–3391.

[Rus1903]    Bertrand Russell, *The Principles of Mathematics*, Cambridge University Press, 1903.

[Sch03]      Stefan Schröer, *The strong Franchetta conjecture in arbitrary characteristics*, Internat. J. Math. **14** (2003), no. 4, 371–396.

[Ser59]      Jean-Pierre Serre, *Morphismes universels et variété d'albanese*, Séminaire Claude Chevalley **4** (1958-1959), 1–22 (fr), talk:10.

[Ser89]      _____, *Lectures on the Mordell-Weil theorem*, Aspects of Mathematics, E15, Friedr. Vieweg & Sohn, Braunschweig, 1989, Translated from the French and edited by Martin Brown from notes by Michel Waldschmidt.

[Sev1906]    Francesco Severi, *Sulla totalità delle curve algebriche tracciate sopra una superficie algebrica*, Math. Ann. **62** (1906), no. 2, 194–225.

[Sha66]      Igor R. Shafarevich, *Lectures on minimal models and birational transformations of two dimensional schemes*, Notes by C. P. Ramanujam. Tata Institute of Fundamental Research Lectures on Mathematics and Physics, No. 37, Tata Institute of Fundamental Research, Bombay, 1966.

[Sha74]      _____, *Basic algebraic geometry*, Springer-Verlag, New York, 1974, Die Grundlehren der mathematischen Wissenschaften, Band 213.

[Sil83]      Joseph H. Silverman, *Heights and the specialization map for families of abelian varieties*, J. Reine Angew. Math. **342** (1983), 197–211.

[Sta22]      The Stacks Project Authors, *Stacks Project*, http://stacks.math.columbia.edu, 2022.

[Ter85]      Tomohide Terasoma, *Complete intersections with middle Picard number 1 defined over* $\mathbb{Q}$, Math. Z. **189** (1985), 289–296.

[Top16]      Adam Topaz, *Reconstructing function fields from rational quotients of mod-$\ell$ Galois groups*, Math. Ann. **366** (2016), no. 1-2, 337–385.

[Top17]      ———, *A Torelli theorem for higher-dimensional function fields*, 2017.

[Tot00]      Burt Totaro, *The topology of smooth divisors and the arithmetic of abelian varieties*, Michigan Math. J. **48** (2000), 611–624, Dedicated to William Fulton on the occasion of his 60th birthday. MR 1786508

[vdW37]      Bartel Leendert van der Waerden, *Zur algebraischen Geometrie. X; Über lineare Scharen von reduziblen Mannigfaltigkeiten*, Math. Ann. **113** (1937), no. 1, 705–712.

[Veb1906]    Oswald Veblen, *Finite projective geometries*, Trans. Amer. Math. Soc. **7** (1906), 241–259.

[Voe90]      Vladimir A. Voevodskiĭ, *Étale topologies of schemes over fields of finite type over* **Q**, Izv. Akad. Nauk SSSR Ser. Mat. **54** (1990), no. 6, 1155–1167.

[vS1857]     Karl Georg Christian von Staudt, *Beiträge zur Geometrie der Lage, Heft 2*, Kornschen Buchhandlung, Nürnberg, 1857.

[VY1908]     Oswald Veblen and John Wesley Young, *A Set of Assumptions for Projective Geometry*, Amer. J. Math. **30** (1908), no. 4, 347–380.

[Wan96]      Zhe-Xian Wan, *Geometry of matrices*, World Scientific Publishing Co., Inc., River Edge, NJ, 1996, In memory of Professor L. K. Hua (1910–1985). MR 1409610

[Wei62]      André Weil, *Foundations of algebraic geometry*, American Mathematical Society, Providence, R.I., 1962. MR 0144898

[Wei82]      Rainer Weissauer, *Der Hilbertsche Irreduzibilitätssatz*, J. Reine Angew. Math. **334** (1982), 203–220.

[Whi1906]    Alfred North Whitehead, *The axioms of projective geometry*, Cambridge University Press, 1906.

[WK81]       Roger Wiegand and William Krauter, *Projective surfaces over a finite field*, Proc. Amer. Math. Soc. **83** (1981), no. 2, 233–237.

[Zar41]      Oscar Zariski, *Pencils on an algebraic variety and a new proof of a theorem of Bertini*, Trans. Amer. Math. Soc. **50** (1941), 48–70.

[Zha98]      Shou-Wu Zhang, *Equidistribution of small points on abelian varieties*, Ann. of Math. (2) **147** (1998), no. 1, 159–165.

[Zil14]      Boris Zilber, *A curve and its abstract Jacobian*, Int. Math. Res. Not. IMRN **2014** (2014), no. 5, 1425–1439.

# Index of Notation

$(-)^{\mathrm{car}}$, Cartier locus, 208

$(-)^{\mathrm{lin}}$, $(-)^{\mathrm{prop}}$,$(-)^{\mathrm{tor}}$,$(-)^{\mathrm{unip}}$, maximal connected linear, resp. proper, resp. multiplicative, resp. unipotent subgroup, 209

$\equiv_{\mathrm{s}}$, numerical similarity, 98

$\sim$, linear equivalence, 139

$\sim_{\mathrm{s}}$, linear similarity, 98, 146

$\sim_{\mathrm{sa}}$, linear similarity of ample divisors, 150

$\mathbf{Alb}(X)$, Albanese variety, classical version, 204

$\mathbf{Alb}^{\mathrm{gr}}(X)$, Albanese variety, Grothendieck version, 204

$\mathbf{Alb}(X, \Sigma)$, Albanese variety with respect to $\Sigma$, 206

$\mathrm{BH}(k)$, Bertini-Hilbert dimension of $k$, 154

$\mathrm{Chow}_d^1(X/S)$, Chow variety, 95

$\mathrm{Cl}(X, \Sigma)$, Weil divisors Cartier along $\Sigma$, 202

$\mathrm{Cl}^\circ(X)$, divisors alg. eq. to 0, 202

$\mathbf{Cl}^\circ(X, \Sigma)$, identity component of $\mathrm{Cl}(X_{\overline{k}}, \Sigma_{\overline{k}})$, 203

$\mathbf{CL}(k)$, set of curves with ample line bundle over $k$, 169

$\mathrm{Cox}(X, M)$, Cox ring with respect to a monoid, 147

$\mathrm{Cox}(X, |\mathbf{Q}D|)$, Cox ring with respect to a divisor, 148

$\mathscr{C}_{X,A}$, category of constructible étale $A$-modules, 177

$\mathfrak{d}_{|D|}(C)$, intersection number of $C$ with a pencil, 109

$\mathscr{DP}$, category of divisorially proper varieties, 56

$\mathrm{Div}(X)$, divisors of $X$, 55

$\Gamma^{\subset B}(Y, \mathscr{L})$, sections with support in $B$, 143

$\Gamma^B(Y, \mathscr{L})$, sections with support $B$, 143

$\mathrm{genmin}(g)$, generic minimum, 111

$\mathrm{Gr}(1, \mathbf{P}(V)), \mathbb{Gr}(1, \mathbb{P}(V^\vee))$, 22

$H^0(C, \mathscr{L}, s_Z)$, sections restricting to a multiple of $s_Z$, 153

$\mathscr{H}_n^{\mathrm{def},B}$, set of lines that are definable or given by $B$, 84

$|H_i, \mathrm{LC}|$, linear system with local conditions, 191

$|\mathscr{L}, s_Z|$, subsystem restricting to a multiple of $s_Z$, 153

$|\mathscr{L}|^{\mathrm{set}}$, linear system as set, 23

$|\mathscr{L}|^{\mathrm{var}}$, linear system as projective variety, 23

$|\mathscr{L}|$, linear system as discrete projective space, 23

$\overline{\boldsymbol{\mu}}(\mathscr{P})$, proportion in $\mathscr{P}$ using sup, 81

$\boldsymbol{\mu}_B$, proportion of a set $B$, 81

$\boldsymbol{\mu}(\mathscr{P})$, proportion in $\mathscr{P}$, 81

$\mathrm{NS}(X)$, Néron-Severi group, 201

$\mathrm{NS}^{\mathrm{cl}}(X)$, Néron-Severi class group, 202

# Index of Terminology

abelian variety, 200
absolutely scip with finite defect,
  125
admissible collection, 35
Albanese map, 206
Albanese variety, 204
algebraic group, 209
algebraic pencil of divisors, 97
ample degree function, 109
ample t-pencil, 107
ample-ci, 195
ample-sci, 195
anti-Mordell-Weil field, 135

base locus of local conditions, 191
base locus of pencil, 97
base locus of t-pencil, 107
Bertini-Hilbert dimension, 154

Cartier locus, 208
Chow variety, 95
ci, complete intersection, 194
class rank, 202
compatible pencils, 111
complete $H$-intersection, 195
complete intersection, 194
complete intersection property, 184
composite pencil, 98
constant field, 56

**D**-good, 35
definable projective space, 26
definable subspace, 62
degree function, 109
detects linear similarity, 148
dimension of definable projective

space, 26
divisorial structure, 56
divisorially proper variety, 55
  category of, 56

essential open subscheme, 61
exact modulo torsion, 200

fiber of pencil, 97
field-locally thin subset, 145

general position, t-pencil, 110
generalized Jacobian, 210
generically scip, 118, 126
geometric member of a pencil, 98

$H$–isci, irreducible-sci, 195
$H$-ci, complete intersection, 195
$H$-sci, set-theoretic, 195
Hilbertian field, 136

inherited by general complete
  $|H_i|$-intersections, 190
irrational pencil, 97
irreducible-sci, isci, 195
isci, irreducible-sci, 195
isogeny, 200

Jacobian, 210

$\mathscr{L}$-linked, 140, 154
$\mathscr{L}$-linking is free, 156
$\mathscr{L}$-linking is minimally restrictive,
  159
$\mathscr{L}$-linking on $Z \cup W_2$ determines
  $\mathscr{L}$-linking on $Z \cup W_1$, 159
linear algebraic group, 209

Milton Keynes UK
Ingram Content Group UK Ltd.
UKHW051053300624
444875UK00008B/274